Cocoa Touch™ for iPhone® OS 3

Cocoa Touch™ for iPhone® OS 3

Jiva DeVoe

Wiley Publishing, Inc.

Cocoa Touch™ for iPhone® OS 3

Published by
Wiley Publishing, Inc.
10475 Crosspoint Boulevard
Indianapolis, IN 46256
www.wiley.com

Copyright © 2010 by Wiley Publishing, Inc., Indianapolis, Indiana

Published by Wiley Publishing, Inc., Indianapolis, Indiana

Published simultaneously in Canada

ISBN: 978-0-470-48107-3

Manufactured in the United States of America

10 9 8 7 6 5 4 3 2 1

No part of this publication may be reproduced, stored in a retrieval system or transmitted in any form or by any means, electronic, mechanical, photocopying, recording, scanning or otherwise, except as permitted under Sections 107 or 108 of the 1976 United States Copyright Act, without either the prior written permission of the Publisher, or authorization through payment of the appropriate per-copy fee to the Copyright Clearance Center, 222 Rosewood Drive, Danvers, MA 01923, (978) 750-8400, fax (978) 646-8700. Requests to the Publisher for permission should be addressed to the Permissions Department, John Wiley & Sons, Inc., 111 River Street, Hoboken, NJ 07030, 201-748-6011, fax 201-748-6008, or online at http://www.wiley.com/go/permissions.

LIMIT OF LIABILITY/DISCLAIMER OF WARRANTY: THE PUBLISHER AND THE AUTHOR MAKE NO REPRESENTATIONS OR WARRANTIES WITH RESPECT TO THE ACCURACY OR COMPLETENESS OF THE CONTENTS OF THIS WORK AND SPECIFICALLY DISCLAIM ALL WARRANTIES, INCLUDING WITHOUT LIMITATION WARRANTIES OF FITNESS FOR A PARTICULAR PURPOSE. NO WARRANTY MAY BE CREATED OR EXTENDED BY SALES OR PROMOTIONAL MATERIALS. THE ADVICE AND STRATEGIES CONTAINED HEREIN MAY NOT BE SUITABLE FOR EVERY SITUATION. THIS WORK IS SOLD WITH THE UNDERSTANDING THAT THE PUBLISHER IS NOT ENGAGED IN RENDERING LEGAL, ACCOUNTING, OR OTHER PROFESSIONAL SERVICES. IF PROFESSIONAL ASSISTANCE IS REQUIRED, THE SERVICES OF A COMPETENT PROFESSIONAL PERSON SHOULD BE SOUGHT. NEITHER THE PUBLISHER NOR THE AUTHOR SHALL BE LIABLE FOR DAMAGES ARISING HEREFROM. THE FACT THAT AN ORGANIZATION OR WEBSITE IS REFERRED TO IN THIS WORK AS A CITATION AND/OR A POTENTIAL SOURCE OF FURTHER INFORMATION DOES NOT MEAN THAT THE AUTHOR OR THE PUBLISHER ENDORSES THE INFORMATION THE ORGANIZATION OR WEBSITE MAY PROVIDE OR RECOMMENDATIONS IT MAY MAKE. FURTHER, READERS SHOULD BE AWARE THAT INTERNET WEBSITES LISTED IN THIS WORK MAY HAVE CHANGED OR DISAPPEARED BETWEEN WHEN THIS WORK WAS WRITTEN AND WHEN IT IS READ.

For general information on our other products and services or to obtain technical support, please contact our Customer Care Department within the U.S. at (877) 762-2974, outside the U.S. at (317) 572-3993 or fax (317) 572-4002.

Library of Congress Control Number: 2009937274

Trademarks: Wiley and the Wiley logo are registered trademarks of John Wiley & Sons, Inc. and/or its affiliates, in the United States and other countries, and may not be used without written permission. Cocoa Touch and iPhone are trademarks or registered trademarks of Apple, Inc. All other trademarks are the property of their respective owners. Wiley Publishing, Inc., is not associated with any product or vendor mentioned in this book. Cocoa Touch for iPhone OS 3 is an independent publication and has not been authorized, sponsored, or otherwise approved by Apple, Inc.

Wiley also publishes its books in a variety of electronic formats. Some content that appears in print may not be available in electronic books.

For my wife, Dawn, and my children, Robert, Alex, and Izzy. You are, and always will be, the most important people in the world to me.

About the Author

Jiva DeVoe has been writing software for nearly 25 years, starting with his Commodore VIC-20 using BASIC and assembly language, and gradually working his way through C, C++, Python, Ruby, Java, and finally, Objective-C. In 2001, he founded Random Ideas, LLC, a software company dedicated to building great applications for the Mac. When the iPhone SDK was announced, he was honored to be selected as one of the earliest developers to have access to the SDK, and using it, he developed several applications that were available in the iTunes App Store when it launched on July 11, 2008. Since that time, his iPhone applications have received several awards — including being chosen as among the Top 100 apps and games in the App Store — and been featured as Apple Staff Picks and in Apple advertisements. Today, Jiva continues to work full time for his company, developing great iPhone and Mac applications. He lives with his wife, three children, and two basset hounds in the desert of Arizona.

Credits

Acquisitions Editor
Aaron Black

Executive Editor
Jody Lefevere

Project Editor
Martin V. Minner

Technical Editor
Dallas Brown

Copy Editor
Marylouise Wiack

Editorial Director
Robyn Siesky

Editorial Manager
Cricket Krengel

Business Manager
Amy Knies

Senior Marketing Manager
Sandy Smith

Vice President and Executive Group Publisher
Richard Swadley

Vice President and Executive Publisher
Barry Pruett

Project Coordinator
Katie Crocker

Graphics and Production Specialists
Andrea Hornberger
Jennifer Mayberry
Mark Pinto

Quality Control Technician
Rebecca Denoncour

Proofreading
Christine Sabooni

Indexing
BIM Indexing & Proofreading

Media Development Project Manager
Laura Moss

Media Development Assistant Project Manager
Jenny Swisher

Contents

Preface xix

Acknowledgments xxi

Part I: Getting Started with the iPhone 1

Chapter 1: Getting to Know Your Tools 3
Introducing Xcode 3
Introducing Interface Builder 5
Summary 6

Chapter 2: Building a Basic iPhone Application 7
Creating a Project from a Template 7
Building a Simple GUI 11
Writing a Simple UIViewController 14
Launching on the iPhone 16
Summary 18

Chapter 3: Exploring the Simulator in Depth 19
Exploring the Simulator UI 19
    Simulating multi-touch 20
    Simulating Core Location 21
    Turning the simulator on its side 23
    Digging deeper into the simulator filesystem 23
Understanding the Differences between the Simulator and the Real Thing .. 25
    Avoiding APIs that won't work on a real device 25
    Understanding performance differences 26
Summary 27

Part II: Building GUIs 29

Chapter 4: Understanding the Basics of an iPhone User Interface 31
Looking at the iPhone Screen 32
Creating a Cinematic User Experience 33
Looking at the Built-in Applications 34
Understanding the Basics of iPhone Navigation 36
Summary 37

Chapter 5: Introducing UIViewController ... 39

Understanding the Basics of Model, View, Controller ... 40
Exploring UIViewController ... 41
Overriding Methods on UIViewController ... 41
    Handling view rotations ... 43
    Handling memory warnings ... 44
Working with UIViewController ... 44
    Creating the code ... 45
    Creating the UIViewController header ... 46
    Creating the UIViewController implementation ... 47
    Adding the controller to Interface Builder ... 51
    Understanding the UIApplicationDelegate ... 53
    Creating the model in the App Delegate ... 53
    Adding the view to the window ... 55
Summary ... 57

Chapter 6: Using UITableView ... 59

Understanding UITableView Data Sources and Delegates ... 59
    Understanding NSIndexPath and how it works with UITableViews ... 60
    Exploring UITableViewDataSource ... 60
    Exploring UITableViewDelegate ... 64
Adding a UITableView to an Application ... 66
    Making a UITableViewDataSource ... 67
    Seeing the basic application run ... 73
    Taking action when a row is touched ... 75
Configuring a UITableView to Be Grouped ... 78
    Adding the state to your model ... 78
    Updating your UITableViewDataSource ... 79
Doing Advanced UITableView Configuration ... 83
    Adding an alphabetical list to the side ... 83
    Adding search ... 84
Summary ... 84

Chapter 7: Working with UITableViewCells ... 85

Understanding the Parts of a UITableViewCell ... 85
Adding Images to a UITableViewCell ... 87
Performing Deeper Customization of UITableViewCells ... 90
Thinking about Performance in Custom Cells ... 94
Reusing UITableViewCells ... 96
    Understanding the reuse identifier ... 97
Summary ... 99

Chapter 8: Working within the UINavigationController Model ... 101

Understanding the UINavigationController Navigational Model ... 101
Creating a UINavigationController ... 102
Configuring a UINavigationController ... 105

Pushing and Popping UIViewControllers 106
Adding a UIToolbar 107
Allowing Editing of the Rows 107
    Moving MyView into a UINavigationController 107
    Making Your EditViewController 109
    Editing rows 111
    Adding rows 115
    Deleting rows 116
Summary 116

Chapter 9: Exploring the Purpose of the UIApplicationDelegate 117

Handling Startup and Shutdown 118
    Understanding launch methods 118
    Understanding the applicationWillTerminate method 119
Receiving Notifications while Your Application Is Running 120
    Handling resource alerts 121
    Handling phone calls and sleep 122
    Changing status 122
    Handling remote notifications 123
Summary 123

Chapter 10: Applying Custom UIViews 125

Understanding Cocoa Touch View Geometry 125
Adding Custom Views to a Project 126
Implementing the Custom View Code 127
    Working with Core Graphics 128
    Implementing drawRect 133
Summary 136

Chapter 11: Handling Touch Events 137

Handling Touch Events in a Custom UIView 137
    Implementing touchesBegan:withEvent: 138
    Working with the touches NSSet 139
    Implementing touchesMoved:withEvent: 140
    Implementing touchesEnded:withEvent: 140
    Handling multi-touch events 140
Updating Your Custom View with Touch Events 140
    Moving the circle with a touch 141
    Adding scaling 144
Summary 146

Chapter 12: Working with Other Controls 147

Finding What Other Controls Are Available in Cocoa Touch 147
Working with a UISlider 148
    Configuring a UISlider through Interface Builder 148
    Updating the status of your UISlider 151

Using UITabBar..151
    Configuring a UITabBar through Interface Builder 153
    Adding UITabBarItems to a UITabBar 153
    Doing advanced configuration of UITabBars 155
    Using UIPickerView 155
    Configuring UIPickerView through Interface Builder 155
    Creating the UIPickerViewDataSource 156
    Creating a UIPickerView delegate 157
Summary...157

Chapter 13: Handling Text Input. 159

Configuring the Keyboard through Interface Builder 159
    Configuring capitalization 161
    Enabling and disabling auto correction 162
    Setting the keyboard type 162
    Setting the behavior of the Return key 163
    Looking at other settings 163
Working with a Keyboard through Code..163
    Making the keyboard appear 163
    Understanding UITextInputTraits 164
Handling Events for the Keyboard 165
    Creating a UITextFieldDelegate 166
    Moving your view in response to the keyboard appearance 167
Summary...169

Chapter 14: Building Cinematic UIs with Core Animation. 171

Using the UIViews Animation Methods 171
Using Advanced Core Animation with CALayer..178
Animating UIImageViews...181
Summary...182

Chapter 15: Using OpenGL ES 183

Understanding the Capabilities of iPhone OpenGL ES..183
Creating a Very Basic OpenGL View 184
Summary...188

Chapter 16: Integrating Safari. 189

Opening URLs Using the iPhone Web Browser..189
Using UIWebView 190
    Adding a UIWebView to your application 191
    Loading a URL 191
    Implementing a UIWebViewDelegate 192
    Loading HTML content from the application bundle 193
Summary...193

Part III: Working with Data 195

Chapter 17: Storing User Defaults 197
Acquiring the NSUserDefaults Object 197
Reading and Writing Values to NSUserDefaults 198
Setting Your Default Defaults 201
Using the Settings App 203
    Adding a settings bundle 203
    Adding settings to your settings bundle 204
Summary 207

Chapter 18: Implementing a Database with Core Data 209
Understanding Core Data's Building Blocks 210
Adding Core Data Support to Favorite Cities 211
    Modifying the app delegate 212
    Working with the Managed Object Model 216
CRUD — Creating, Reading, Updating, Deleting 219
    Creating 220
    Reading 220
    Updating 221
    Deleting 222
    Bringing it together and updating your app 222
Understanding What Core Data Makes Easier 230
Understanding What Core Data Is Not Good For 230
Summary 231

Chapter 19: Connecting to the World with Networking 233
Accessing the Web 234
    Using URLs with foundational classes 234
    Using NSURLRequest and NSURLConnection 235
Sending E-mail from within Your App with the Message UI Framework 239
Using Core Foundation Sockets 243
    Exploring CFSocket 244
    Getting host info with CFHost 247
    Using CFStreams 249
Exploring Bonjour 252
    Browsing for Bonjour services 253
    Using NSNetServices 255
Summary 256

Chapter 20: Using the Push Notification Service 257
Understanding the Push Notification Service Architecture 258
    Understanding the push notification communications 259
    Understanding push notification security 264

Acquiring Push Notification Certificates..266
Developing Your Server-Side Push Notification Service 269
    Implementing a Ruby push notification supplier..269
    Pushing notifications..272
    Checking delivery using feedback...272
Integrating Push Notifications with Your iPhone Client..273
    Registering for notifications 273
    Receiving notifications in your application 274
Summary..275

Chapter 21: Using the Game Kit API 277

Providing Peer-to-Peer Connectivity 277
    Finding peers...278
    Working with sessions...281
Providing In-Game Voice 284
Summary..288

Chapter 22: Implementing Cut, Copy, and Paste 289

Copying and Pasting with Standard Controls...289
Understanding Pasteboard Types..290
Interacting with UIPasteboard 291
Implementing Cut, Copy, and Paste on a Custom View 292
    Implementing touchesEnded:withEvent: to display the menu...........................292
    Implementing the copy: method 293
    Implementing the paste: method 294
    Understanding the interactions 294
Summary..295

Chapter 23: Using the Maps API 297

Showing an Embedded Map with MKMapView..297
    Creating an MKMapView 297
    Specifying the map region...299
Annotating Maps...301
    Thinking about performance with annotations 303
Converting Coordinates 303
Summary..305

Part IV: Working with Media 307

Chapter 24: Exploring the Supported Media Types..................... 309

Supported Audio Formats..309
    Compressed audio 309
    Uncompressed audio 310
Supported Video Formats..310
Summary..310

Chapter 25: Playing Audio 311

Using the AV Foundation Framework 311
    Setting your audio configuration with AVAudioSession 311
    Using an AVAudioSessionDelegate 314
    Playing audio with AVAudioPlayer 315
    Using an AVAudioPlayerDelegate 317
Playing Audio with OpenAL 317
Summary 318

Chapter 26: Accessing the iPod Library 319

Working with the Media Player Framework 319
Accessing the Media Library 320
    Using the MPMediaPickerController 320
    Searching for media 323
Working with Player Controllers 327
Accessing Media Artwork 328
Summary 328

Chapter 27: Recording Audio 329

Setting up Your AVAudioSession 329
Allocating an AVAudioRecorder 330
Creating a Voice Recorder 332
Summary 337

Chapter 28: Playing Video in Your Application 339

Playing Video Files Contained in the App Bundle 339
Playing Video from the Internet 341
Summary 342

Part V: Working with the iPhone Hardware 343

Chapter 29: Discovering Information about the Device 345

Accessing the Battery State 345
Accessing the Proximity Sensor 346
Working with Device Metadata 346
Summary 347

Chapter 30: Getting Your Location Using Core Location 349

Finding Where You Are 349
    Allocating a CLLocationManager 351
    Setting yourself as the delegate 352
    Implementing the CLLocationManagerDelegate protocol 352
    Starting and stopping location updates 353
    Viewing your location on Google Maps 354

Narrowing the Accuracy of the Coordinates 356
Filtering Location Updates 356
Looking at the Final Code 357
Working with the iPhone 3GS Compass 359
Summary 360

Chapter 31: Working with the Accelerometer 361

Determining Which Way Is Up 361
    Building a bubble level 364
    Understanding the UIAcceleration object 367
Capturing Shake Events 367
    Building a "shake to break" detector 368
    Implementing drawRect 368
    Implementing motionBegan and motionEnded 370
    Cancellation of motion events 370
Summary 371

Chapter 32: Interfacing with Peripherals 373

Creating Accessories That Interface with iPhone 373
Finding Accessories That Are Connected Using EAAccessoryManager 374
Understanding the EAAccessory Class 375
Working with EASession 375
Talking to Your Device Using NSStreams 376
Summary 377

Part VI: Handling Distribution 379

Chapter 33: Code Signing Your Apps 381

Acquiring a Development Certificate 381
Understanding the Provisioning Process 382
    Understanding development, distribution, and ad hoc 382
    Installing provisioning profiles 383
    Exploring what happens when an app is signed 384
    Setting up your build to be signed 385
Doing Ad Hoc Builds 386
    Configuring the build in Xcode 386
    Distributing to users 388
Summary 388

Chapter 34: Expanding Your Application Using the In-App Purchase Support 389

Knowing the Types of Things You Can Sell 389
Working with Unlockable Content 390
Setting up Purchasable Content in iTunes Connect 391

Working with Store Kit	392
Verifying app purchase availability	393
Presenting your store	393
Making the purchase	395
Processing the payment	395
Verifying the transaction	397
Unlocking the content	397
Restoring purchased content	398
Understanding In-App Purchasing testing	399
Summary	399
Index	**401**

Preface

My goal in writing this book was to provide a comprehensive toolkit for both new and experienced iPhone developers. Its focus is intended to be primarily on the new technologies of iPhone OS 3, but it includes enough general iPhone development material that even a new developer to the platform will receive a great deal of benefit from reading it.

Writing a book about Cocoa Touch programming is an incredible challenge. It's very difficult to judge the technical capability of the typical reader of a book like this. Does the reader already know Objective-C? Has he or she already been developing for Mac OS X? These are the first questions that I had to ask myself when I began to work on this project.

In the end, I decided that with the development of iPhone OS 3, it was a unique opportunity where the new and experienced reader intersected and therefore, I thought it was an ideal time to write a book that would be useful to them both.

So I decided that this book would contain some introductory material, but that it would primarily focus on the new features of iPhone OS 3. In this way, it provides an excellent bridge for learning the technologies of the new operating system, as well as an introduction to general iPhone development.

As a reader, you are expected to already be somewhat familiar with Objective-C. Additionally, you should be somewhat familiar with either Mac OS X, or iPhone development, though you need not be an expert.

If you're completely new to the platform, I suggest picking up a book specifically on Objective-C to complement this book. There are several such books on the market, but I recommend one that approaches Objective-C from the point of view of learning the core language itself, rather than one that mixes an introduction to Objective-C with other topics. After you've worked through enough of that book to feel comfortable with the syntax of the language, you can begin this book at Chapter 1.

If you've already been developing for Mac OS X, but haven't done any iPhone development, you probably already know Objective-C and are familiar with many of the tools used in iPhone development. However, while iPhone development is similar to Mac OS X development, it's not exactly the same. So I suggest that you skip Chapter 1, and start reading Chapter 2, where you build a basic iPhone app from scratch.

Finally, if you're already an experienced iPhone developer, you already know all the basics involved in iPhone development; you're just here for the iPhone OS 3 material. I suggest you start at Chapter 5. All the chapters were written with the new and updated iPhone OS 3 API in mind, and so even if you are already familiar with UIViewController and friends, you will find new material there.

In these ways, this book provides an excellent extension to your library if you already have books on any of these three subjects.

With regard to conventions used within this book, I've tried to be reasonably consistent, and also tried to generally err on the side of Apple conventions when prudent. The only notable exception has been in my use of the term "method" to indicate functions on instances and classes. Apple generally prefers the term message. This is in part due to the influence of Smalltalk on Objective-C.

Also with regard to method calling conventions, Objective-C is known for being particularly verbose in its method names. As a result, some abbreviation conventions have been adopted when writing about particular methods. I have chosen to follow two standards. The first, I use when the usual use of the method in question is in implementing your own, or in overriding it in a subclass. This is most commonly the case when dealing with delegate methods. In these cases, because it's helpful to know the entire signature of the method, in order to write your own implementation, I have chosen to include the entire method signature. So, for example, these will be written as `-(BOOL)foo:(NSString *)bar withBaz:(NSString *) baz`. As you can see, in this example, we have a method called `foo`, which takes two parameters of type `NSString *`, one called `bar` and a second called `baz`, and which returns a `BOOL` value.

Alternatively, in the case where your typical use will be only to use the method in your own code, since Xcode generally automatically completes the types of the parameters for you, I have chosen to use the abbreviated form of writing the method signatures. In the case of the previous method, that means it's written `foo:withBaz:`. Notice that wherever a parameter is inserted, a colon holds the place of the parameter.

When referring to keyboard shortcuts, I opted to use the term Command key or the ⌘ symbol to indicate keyboard shortcuts using the key directly to the left of the space bar on most Apple keyboards. You may also know this as the "Apple" key, as until only a few years ago, it included a small Apple logo on it. Additionally, the key next to the Command key has been called the Option key and the key next to that, the Control key. These should all be consistent with Apple documentation conventions.

When discussing the use of menus in Xcode, I've used the technique of separating the nested menu items using arrow notation. So, to describe the New File sub item of the File menu, it will be written as File ➪ New File.

Finally, with regard to sample code, in chapters where I have instructed you to build specific full projects, I have generally tried to include full listings for the code. In cases where I have not, you can always download the projects, complete with artwork and other supporting files, from the book's Web site, located at `http://www.wiley.com/go/cocoatouchdevref`. There are also chapters where it didn't really make sense to create a full project to demonstrate a technology. In these cases, the code listings are snippets that you can use as a basis for your own code. Because these snippets don't comprise fully functional projects, there will not be example projects for them on the Web site.

I hope that you find this book as enjoyable an experience to read as I had writing it. To me, the mark of a good technical book is that it doesn't sit on my shelf. It holds a place of honor by my desk because I keep returning to it, time and again. I hope that this book holds such prestige in your library, and that it becomes a dog-eared, cover-torn, page-scribbled-on reference that remains useful to you for years to come.

Jiva DeVoe

`book@random-ideas.net`

Acknowledgments

Writing this book has been one of the most challenging and exciting projects I have done in my career, but I could not have done it without the aid and support of some specific individuals whom I would like to thank.

First, I'd like to thank my friend and technical editor, Dallas Brown, of HashBang Industries, who took up the challenge of correcting my mistakes and keeping me honest. Your time spent, and your thoughtful comments were excellent.

Along that same line, I'd also like to thank my friend, Brad Miller, of Cynical Peak Software, who also provided welcome criticism and an extra set of eyes on several chapters, and who always seems to be awake and online at the same times I am.

For pressing forward with the book, even with a tight schedule, I'd like to thank all the folks at John Wiley & Sons. I look forward to working with you on new projects in the future.

For teaching me to marvel at the wonders of technology and encouraging me to pursue my dreams in computers, I'm thankful to my father, Robert A. DeVoe.

Many thanks to my children, who have endured these several months of my working late without complaint. You have earned your trip to Disneyland! It is for you that I do everything I do.

Finally, and most importantly, I'd like to thank my wife, for her unerring support, not just in this project, but in all my work. Without her, this book could not have been finished. You lift me when my spirits are low and tired, and inspire me to keep reaching for new accomplishments and goals. I can't thank you enough.

Getting Started with the iPhone

In This Part

Chapter 1
Getting to Know Your Tools

Chapter 2
Building a Basic iPhone Application

Chapter 3
Exploring the Simulator in Depth

Getting to Know Your Tools

They say that when a craftsman finds a tool that he loves, over time it becomes an extension of him. He learns its idiosyncrasies inside and out, backwards and forwards, and this leads to a relationship that transcends simple use of the tool and instead becomes more involved. The tool becomes an extension of his hand, an extension that he can guide with an almost extrasensory vision.

You are fortunate then, that Apple provides you with an excellent set of free tools for developing software for the iPhone. They are tools that have evolved over the last 15 years of Objective-C development, first on NeXT computers, then on Mac OS X, and now for iPhone.

In this chapter, you will take a brief look at these tools and learn where you can find more information about them. They are incredibly powerful tools that seem to be unique in software development, both for their ability to provide enough power to enable incredibly complex software systems to be developed, and also because they seem to know just when to stay out of your way and simply provide a great text-editing environment for you to write code in.

Unfortunately, a comprehensive description of every last feature of these applications is beyond the scope of this book, and so I won't be delving into them in great detail. My main goal here is simply to introduce you to the tools so that you're familiar with them.

In This Chapter

Becoming familiar with Xcode and Interface Builder

Introducing Xcode

The cornerstone of iPhone software development is the Xcode integrated development environment, or IDE. Xcode originated on NeXT Step computers as Project Builder. Over the years, it has gone through many revisions to finally arrive at the version that is available to you today. It uses GCC as its underlying compiler technology and provides many sophisticated features found in modern IDEs today, such as code completion, re-factoring, and sophisticated code navigation. Interestingly, it also has one of the best cross-platform compiling capabilities of any modern IDE. With it, you can compile for Intel, PowerPC, iPhone OS, or even (with third-party tools) Microsoft Windows. You can do all of this simply by configuring targets in the IDE.

NOTE
To download Xcode, all you need to do is sign up for a free developer account on the Apple Web site. You can do this at `http://developer.apple.com/iphone`.

Figure 1.1 shows the main Xcode window. In it, you can see the left panel, which shows the file organization view. From here, you can drag and drop files into your projects or organize them by groups. It also provides the ability to organize your files by Smart groups, which are built using search queries and can be useful for looking for particular files in your projects.

Figure 1.1

The Xcode interface

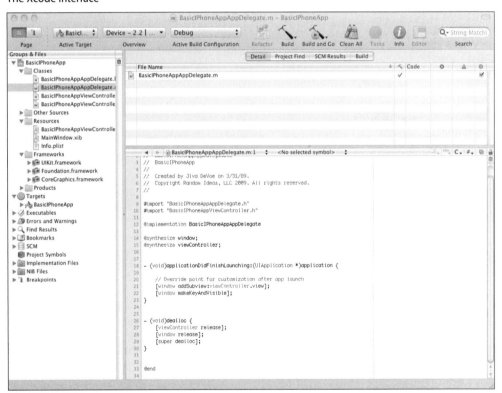

On the right side, you can see the main editing window. It is in this window that you will do the majority of your programming work. At the top of the text-editing window, you can see some drop-down menus that enable you to quickly jump to any recently opened files. The second drop-down menu from the left enables you to quickly jump to any method in the current file. Holding down the ⌘ key and using the right and left arrow keys enables you to quickly navigate backwards and forwards through the file history. Additionally, holding down the Option and ⌘ keys together and pressing the up arrow key enables you to quickly swap between the implementation and header files for the currently active compilation unit.

The Xcode editor is quite sophisticated and can be configured with a variety of shortcuts and hot-keys that make your editing much easier and faster. For example, using the Option key and the right and left arrow keys enables you to quickly jump from word to word in your code. Holding down the ⌘ key and using the left and right arrow keys enables you to quickly jump to the front or beginning of the current line.

It's a good idea to learn the intricacies of the Xcode editor inside and out, because it is such a fundamental tool to everything that you will be doing as an iPhone OS developer.

Introducing Interface Builder

The second major component of the Xcode programming environment is the graphical user interface (GUI) builder called Interface Builder, shown in Figure 1.2. It is this application that you will use to draw your GUI for your application and connect your buttons to actions in your code.

Many developers coming to iPhone development from other environments are sometimes confused by Interface Builder because it doesn't generate any code. However, this is an asset and not a liability. IDEs that simply generate code tend to be more difficult to work with over time, as the code that they generate becomes out of sync with the user interface. Interface Builder uses more of a metadata style approach. This means that you tell it that you want to instantiate an object of a given type, and when your nib is loaded, it goes and finds the class for that type, instantiates it, and attaches the outlets and actions that you have configured to the appropriate places. It does not serialize actual instances of your objects, nor does it generate code that is compiled.

NOTE
The files that Interface Builder saves are referred to as nib files. This stands for NeXT Interface Builder. The file extension of the nib files used with the iPhone is .xib. This is to differentiate them as containing XML versus the older, original NeXT format.

I will talk about Interface Builder in a bit more detail in Part II. For now, the important thing to know is that you can start Interface Builder either separately by simply launching it or by double-clicking any of the .xib files in your project.

Figure 1.2

Interface Builder

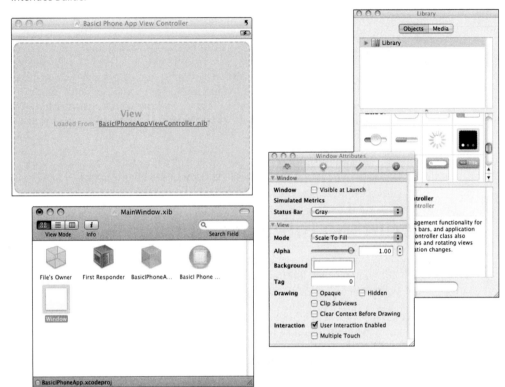

Summary

In this chapter, I introduced you to Xcode and Interface Builder. In the upcoming chapters, you will use these tools in much more depth, but I wanted to simply give you a brief overview of what they look like so that when you use them in the future you will be familiar with them. I encourage you to read through the documentation on them very carefully and try to become as intimately familiar with them as you can.

Building a Basic iPhone Application

Now that you're a bit more familiar with the tools that you're going to be using for developing on the iPhone, let's look at how to build a very basic iPhone application.

In this chapter, you're going to build a simple iPhone application that will have a text field to display some output and a button to show a small amount of interactivity. When you're finished with this application you will be familiar with the basics of launching an iPhone application from Xcode and running it in the simulator. Additionally, before you're finished, you will run the application on your iPhone to see it running on an actual device.

You begin by choosing one of the templates available from the New Project menu in Xcode.

Creating a Project from a Template

The first thing you'll do to create this basic iPhone application is open up Xcode and go to the New Project menu item, which you will find under the File menu. This presents you with the dialog shown in Figure 2.1. For the purposes of this demonstration, you will choose to create a View-Based Application.

Once you have created this project, Xcode already has files that correspond to the user interface, or UI, for your application, a controller for your view, and various other housekeeping files. It also has preconfigured targets for both the simulator and for a device. The files that are used to define the interface end with an extension of `.xib`. These are the files that you load into Interface Builder to define your user interface. The view controller handles the interaction with your buttons from your user interface. It has two files, a header file with an extension of `.h` and an implementation file with an extension of `.m`. Finally, it also includes a default

In This Chapter

Building your first iPhone Application

Using Interface Builder to build a basic GUI for an app

Implementing a very simple UIViewController

Running your basic app on your iPhone

implementation of an app delegate. This is the file that handles messages from the operating system to your application, such as notifying you that the user has requested that your application terminate, or that your application is running low on memory. Xcode, with all these files, is shown in Figure 2.2.

Figure 2.1

The New Project dialog

Figure 2.2

Xcode with the project files

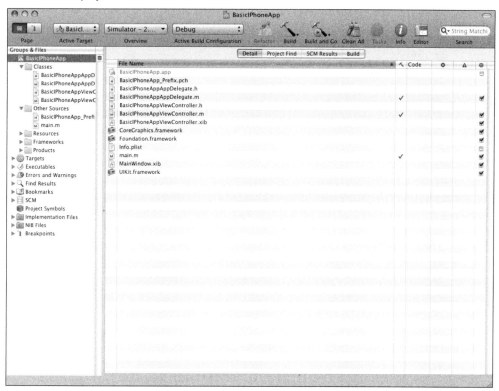

You will look at each of these files in more detail very shortly. For now, if you simply click Build and Go, the application launches and shows a very dull, blank screen on your iPhone simulator, as shown in Figure 2.3.

Figure 2.3
The app UI with nothing on it

Building a Simple GUI

I think you would agree that as applications go, this one is pretty dull (although, if your goal was to create a flashlight application, there you go, you're finished). However, you want to make something a little bit more interesting. So the next thing you're going to have to do is manipulate your user interface and add a couple of controls to your view. To do this, find the `Basic IPhoneAppViewController.xib` file in the file view in Xcode and double-click it. This starts Interface Builder and loads your view into it for manipulation, as shown in Figure 2.4.

Figure 2.4

Interface Builder with the app UI loaded

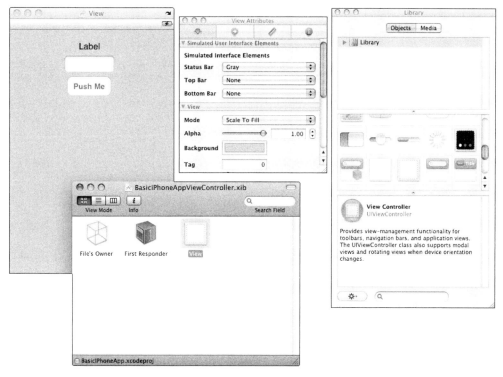

Interface Builder consists of three primary windows: the Library window, the Document window, and the Properties window. These allow you to manipulate the various objects in your interface. The fourth window that you see when you start Interface Builder is the blank window that represents your view. It is here that you drag and drop controls to create your user interface.

NOTE
The traditional example application is one that displays "Hello World." Yours will do slightly more than that, but not much.

Once you have your interface loaded into Interface Builder, browse through the object Library and find an instance of a Label. Drag and drop this instance onto your view. Next, find an instance of a text field, and drag and drop this instance beneath your label. Finally, find an instance of a Round Rect Button, and drag and drop this instance onto your view. When positioning each of these controls, try to position them in the upper third of the view. This enables the keyboard to pop up from the bottom when you touch the text field. Later, you will learn how to move your view when the keyboard comes into play, but for now you will just position it up and out of the way.

When finished, your user interface looks something like Figure 2.5.

Now, you're going to take the text that is typed into the text field, and when the button is pushed, you're going to copy that text into the label field. This is a bit more complicated than your basic "hello world," but it shows you all of the basic components needed to build a very simple graphical user interface, or GUI.

Right now, save your work in Interface Builder, and close the application. You're going to return to Xcode, where you will add some code to your view controller.

Figure 2.5
The user interface as designed

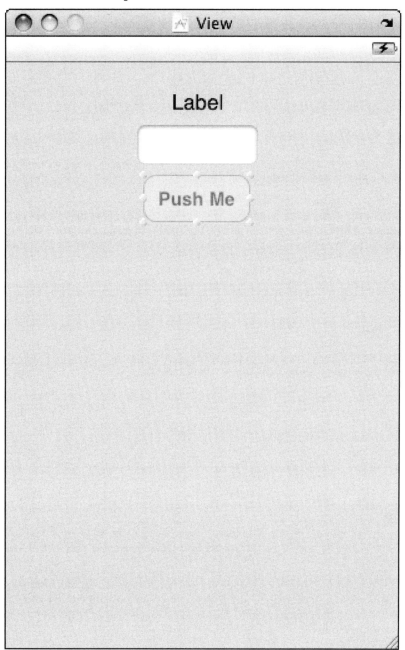

Writing a Simple UIViewController

In Chapter 5, I go into a lot more detail on what a `UIViewController` is and how you use it. For now, however, you're just going to fill in some methods and add some code to your header that will enable you to attach the controls that you put on your view to code, and allow you to take action when the button is pressed.

So at this point, you can go back into Xcode and find the files `BasicIPhoneAppView Controller.h` and `BasicIPhoneAppViewController.m`. These are the header and implementation files, respectively, for the code that will control your UI.

NOTE
When you use the templates, you get a fair amount of extra code that's provided for you, but commented out. Feel free to read through the comments, as they can be instructional.

If you single-click on `BasicIPhoneAppViewController.h`, then that file opens in the editing window of Xcode. What you need to do here is to add some Interface Builder outlets, or `IBOutlets`, for the controls that are on your view. To do that, make your code look like Listing 2.1.

Listing 2.1

View Controller Header

```
#import <UIKit/UIKit.h>
@interface BasicIPhoneAppViewController : UIViewController
{
    IBOutlet UILabel *myLabel;
    IBOutlet UITextField *myTextField;
}
-(IBAction)setLabelPushed:(id)sender;
@end
```

NOTE
`IBOutlet` is a hint to Interface Builder that enables it to know what variables you are expecting to connect to elements of the UI. Similarly, `IBAction` tells Interface Builder about methods you intend to attach actions to.

What you're going to do is attach your label and text field to the two variables represented here as `IBOutlets`. Then, you're going to attach the action from the button to the `setLabel:` method so that when the user presses the button, whatever has been typed into the text field will be put into the label.

To do all this, you're going to need to edit your implementation file to add the implementation for the `setLabel:` method. That code is shown in Listing 2.2.

Listing 2.2
View Controller Implementation

```
#import "BasicIPhoneAppViewController.h"
@implementation BasicIPhoneAppViewController
-(IBAction)setLabelPushed:(id)sender
{
    [myLabel setText:[myTextField text]];
}
- (void)didReceiveMemoryWarning
{
    [super didReceiveMemoryWarning];
}
(void)dealloc
{
    [myLabel release];
    [myTextField release];
    [super dealloc];
}
@end
```

When you first open this file, it is filled with a certain amount of boilerplate code. Some of this code is commented out, and some is not. For the purposes of simplicity, I have removed all of the commented-out boilerplate code and left in the code that was not commented out. The only thing that you actually have to modify here is the `setLabelPushed:` method, which you will have to write yourself from scratch. Just make it look exactly like this one, and you'll be fine.

Now the last thing that you have to do to tie all this together is go back into Interface Builder and attach your objects to your code. Double-click the `BasicIPhoneAppViewController.xib` file. This relaunches Interface Builder. Once you're back in Interface Builder, you single-click to select the File Owner object in the main Interface Builder window. This is actually attached to an implementation of your `BasicIPhoneAppViewController` class. So if you click on the identity tab on the info panel, you see that the class is set to `BasicIPhoneAppViewController` and you can see your `setLabelPushed:` method in the Actions section, and your label and text field variables in the Outlets section.

Interface Builder knows about the files in your Xcode project and automatically reads the header when you save it, parses it, and makes your changes available to you within its interface.

CAUTION

If your `IBOutlets` or `IBActions` aren't available in Interface Builder, then first make sure you saved your changes. If they're still not there, check for any typos. Interface Builder can't read your code if it's not syntactically correct.

You now need to hold down the Control key, click on the File Owner, and drag your mouse from the file owner to the label on your UI. Once you release the mouse button on the label, it brings up a Heads-Up-Display, or HUD-style window. In this window, you see an outlet for your label. Select it, and this connects the label instance variable to your label instance on the UI.

Next, do the same thing for the text field to attach it to your text field outlet.

Finally, you want to go in the opposite direction for your button. You hold down the Control key, and drag-and-drop from your button to the File Owner. This again brings up a HUD-style window, except this time, it shows your actions that you have defined. Because you've only defined one action, it shows that one action, which is `setLabelPushed:`. Select this action, and it attaches the button to that action.

At this point, you can exit Interface Builder, return to Xcode, and click the Build and Go button. This launches your application in the simulator.

You can now touch inside the text field, and it brings up a keyboard. If you then type something into the text field and then click the Push Me button, whatever you type into the text field is copied into the label.

There you go! Congratulations, you've written a basic iPhone application.

Launching on the iPhone

Now running on the iPhone simulator is great, but ultimately, you're going to need to test your application on actual hardware. Why? Well, it's important to realize that when the simulator is running your application, it's actually running it using the hardware capabilities of your computer. This means that performance-wise, your application can behave quite differently in the simulator on your computer versus on an actual iPhone. These differences in hardware can even make certain bugs only appear on an iPhone and not in the simulator.

Not only is running your app on an actual iPhone prudent for testing purposes, but there are also certain capabilities that are not available on the simulator that are available on the iPhone. These capabilities include the accelerometer, Core Location, and others. Conversely, there are application programming interfaces (APIs) that are available only on the simulator and not on an actual iPhone. In the latter situation, you should obviously avoid those APIs. I mention it here because it is a common mistake that new developers make when coming to the platform. They look at the documentation for Mac OS X on a particular API and assume that the API works the same on the iPhone. Because the simulator uses the desktop version of some libraries, when they test on the simulator, their code appears to work. The end result, however, is that the application works fine on the simulator, but when the developer goes to compile it for actual hardware, the compiler can't find the requested API, resulting in an unresolved symbols compilation error.

So the point here is that you need to run your application on the iPhone or iPod Touch that you're using for development. To do this, you need to obtain a development certificate from Apple, load the certificate into Xcode, and configure your build target to use it. Most of this process is closely related to the topic of distribution and uploading to the iTunes Store; therefore, I address this process in the chapter that handles that subject.

CROSS-REF
Chapter 33 contains information on how to get your distribution certificate and install it in Xcode.

If you do not already have a development certificate, then you should probably visit that chapter now and go through the process of getting one so that you can use it in the upcoming chapters. Once you have done this, return here and continue.

NOTE
It costs $99 per year to be a registered iPhone developer.

Have you got your certificate? Good, let's proceed.

Once you've installed your development certificate into Xcode, it's a simple process to sign your application with your certificate and install it on your iPhone. Just pull down the target drop-down menu from the top of the window and choose Device for your target.

At this point, if you click Clean and then Build and Go, it builds your executable, downloads it onto your iPhone, and runs it.

NOTE
If you're using the Debug build, then you can actually run the debugger and debug on your actual device.

Summary

In this chapter, you've been able to build a very simple iPhone application, and you also learned to run it in the simulator as well as on your actual testing hardware. This was by no means a complicated application, but it did show the basics of the entire development cycle, from designing your GUI to writing your controllers and launching the application.

In the next chapter, you're going to take a closer look at how the simulator works because it's important to be familiar with the intricacies of it when you're doing debugging. Later, you will spend a lot more time looking at how to build more complicated applications than this one.

Exploring the Simulator in Depth

You're going to take a little side trip here to explore in detail the inner workings of the iPhone simulator and how you can interact with it in your development.

The simulator is an incredibly powerful development asset because it enables you to do very rapid compile/test cycles. It's much faster and easier to run your build and then fire it up on the simulator than it is to wait for it to download onto an iPhone.

It's important to note, however, that the simulator is just that, a simulator, and not an emulator. The distinction is that it does not actually emulate the iPhone or iPod Touch hardware. It is simply simulating the iPhone user interface (UI) over the top of an iPhone application, which is actually being compiled for the processor of your computer. The end result of this is that the libraries that an application running inside the simulator links with are, in fact, the same libraries that are used for your desktop applications. This can lead to some interesting inconsistencies where certain functions are available in the simulator but not on the actual iPhone hardware. You will take a look at how to avoid problems with this, as well as numerous other differences between the simulator and the iPhone hardware in this chapter.

First, let's begin by looking at how you interact with the simulator UI.

Exploring the Simulator UI

The purpose of the simulator is to provide the closest possible simulation of working with the actual hardware that it can. This is a lofty goal, and very difficult to accomplish on a desktop computer. The iPhone has a variety of unique hardware features available that must be made available within the simulator.

In This Chapter

Working with the simulator UI

Rotating and manipulating the simulator

Exploring the simulator filesystem

Learning what works on the actual hardware, and what doesn't

The developers of the simulator chose to simply ignore some of these features, such as the accelerometer, altogether. This makes sense, because it would be extremely difficult to develop a way of simulating accelerometer interaction with a desktop computer. Even if you were developing on a laptop, imagine having to shake your laptop in order to interact with the accelerometer in the simulator. The iPhone has a solid-state drive, and so therefore it has nothing that can be damaged when it is shaken, tilted, or otherwise manipulated. Most computers do not have this luxury.

When it comes to other features, such as Core Location, the simulator allows testing using those features, but it simulates the operation of hardware in a test mode. This means your code will work, but it will have only test data fed to it.

Finally, there is the ubiquitous multi-touch, which is in a lot of ways the defining feature of iPhone and iPod touch interactivity. The simulator allows you to simulate multi-touch through a variety of hotkeys.

Let's visit each of these subjects in turn.

Simulating multi-touch

There probably isn't a more defining feature for iPhone applications than the multi-touch screen. This is such a fundamental interaction with iPhone applications that Apple has even patented it. As a result, it is not surprising that the simulator provides several methods for simulating multi-touch when working with your application.

To enable a multi-touch, hover your mouse over the interface in the simulator, and press and hold the Option key on your keyboard. This brings up two circles that indicate the locations where the multi-touch will be activated when you press the mouse button. To move these circles up and down and around the screen together, hold the Shift key while continuing to hold the Option key. This enables you to move the multi-touch interactor around the screen.

Once you have positioned the circles where you want them, press the mouse button, and you get a multi-touch event for the locations you chose.

CAUTION
Multi-touch events on the simulator are always simulated with both touches coming in at the beginning of the event in perfect synchronization with each other. In practice on actual hardware, the two fingers often do not actually touch the screen at precisely the same moment; as a result, you should be prepared to transition from a single touch event to a multi-touch event when you have detected that this has occurred.

CROSS-REF
Multi-touch events are discussed in more detail in Chapter 11.

Simulating Core Location

Core Location is the application programming interface, or API, that is used on the iPhone to determine your location using a combination of GPS, cellular triangulation, and Wi-Fi access point mapping.

CROSS-REF
I discuss Core Location in greater detail in Chapter 30.

Core Location is a somewhat complicated feature to test in the simulator. There aren't many (any?) desktop computers that have built-in GPS capabilities. As a result, when using Core Location in the simulator, it simulates only one location. That location, naturally, is Cupertino, California, the headquarters of Apple. Figure 3.1 shows the Core Location demo application running in the simulator, showing the location that Core Location returns when used on the simulator.

Unfortunately, this is all that the simulator is capable of doing. So if you need to test an application that requires more detailed interaction with Core Location, then you need to run it on actual hardware. Still, you can probably limit this in your application to just testing your location awareness code. Then, once you're certain that that's working, you can just run in the simulator and pretend that you work at Apple.

Figure 3.1

The Core Location demo app

Turning the simulator on its side

I originally said that the accelerometer was not supported on the simulator at all. This is not entirely true. It is possible to simulate turning your iPhone from portrait to landscape. This is available from the Hardware menu in the simulator under the "Rotate Left" and "Rotate Right" options. This causes the simulator to rotate left and right, respectively. Additionally, any events that would get fired by rotating the actual device are also fired when using these menu options. It's important to note, however, that this is not actually invoking the accelerometer API, and events associated with the accelerometer are not triggered by this interaction. There are, however, specialized events for transitioning from portrait to landscape, and vice versa, that will be triggered.

Figure 3.2 shows the simulator rotated to its left in landscape mode.

Figure 3.2
The simulator in landscape mode

Digging deeper into the simulator filesystem

When debugging your applications, it can be extremely useful to be able to navigate into the actual filesystem that's being used by the simulator to run your application. This is particularly useful, for example, when you want to see if files that you are writing are being written correctly or if your database schema is being created properly. Fortunately, the simulator's filesystem is easily accessible inside your home directory of your computer. The location of the iPhone filesystem is `/Users/<your username>/Library/Application Support/iPhone`

`Simulator/User`. Inside this folder, you'll find a set of system directories for the iPhone simulator. Most of these are unimportant to you. The one that is important, however, is the Applications folder, which contains the sandbox for your application when it is installed in the simulator.

NOTE
The application sandbox is the restricted filesystem that consists of your application folder, its documents folder, and several system folders. When your application runs, it is restricted to this filesystem in terms of reading and writing data.

Now if you navigate into this folder and take a look around, the first thing that you'll notice is that the folders inside the Applications folder are not named after the application names. They are, in fact, given a globally unique identifier, or GUID.

There really isn't any way to predict or associate the GUID with your application; instead, you need to look inside these folders to look for the .app folder corresponding to your application. So, for example, for your `BasicIPhoneApp`, you see a file inside one of these folders called `BasicIPhoneApp.app`. This is the application bundle, and the top-level (GUID) directory is the application sandbox for your application. Inside the application sandbox, there are several other folders, including Documents and Library, described in Table 3.1.

Table 3.1 Directories Inside the Simulator Filesystem

Directory Name	Description
Documents	The Documents folder is the writable location where you should be writing any data files that your application requires. So, for example, if you're working with a database, this is where you put your writable version. If you write a file out just as a raw text file, this is where it is written.
Library	The Library folder contains your application preferences. So when you write a preference using `NSUserDefaults`, this is where those preferences are written.
<Application Name>.app	The application bundle folder (in your case, `BasicIPhoneApp.app`) is the actual application bundle of your application binary and any resource files that you included in your bundle.

When your application is actually running, you are able to write to the Documents directory, but you are not able to write to anything inside of your application bundle. This becomes important when you deal with bundling and writing to a database.

CROSS-REF
I talk more about databases in Chapter 18.

Understanding the Differences between the Simulator and the Real Thing

I've mentioned several times now that the simulator is not a perfect replica of iPhone hardware and that the hardware does not necessarily support everything that the simulator does. As a developer of iPhone applications, it's important to be able to discern between what is available on the simulator and what is available on the hardware, and to know the differences in terms of performance between the two.

Avoiding APIs that won't work on a real device

The Cocoa and Cocoa Touch development environments can be defined in terms of layers. At the topmost layer, the layer that comes into direct interaction with the user, is the UIKit framework on Cocoa Touch, and the AppKit framework on Cocoa. These contain components such as buttons, views, list controls, and so forth. Originally, all of the Objective-C classes that descended from NSObject and that were provided by Apple were all prefixed with "NS". This stood for NeXTStep, and was a legacy holdover from the original platform used on the NeXT computers. Gradually, as Apple has been expanding Cocoa, it has been adding more and more classes that do not follow this convention. Cocoa Touch is one of these cases.

Inasmuch as they serve the same purpose and contain similar graphical elements (both of them have buttons, for example), many of the class names in `AppKit` and `UIKit` are similar if not exactly the same. Because of this, to differentiate the `UIKit` classes from the `AppKit` classes, Apple chose to preface the classes in `UIKit` with "UI" prefixes rather than the "NS" prefix used in `AppKit`. As a result, differentiating between `AppKit` documentation and classes and `UIKit` documentation and classes is relatively simple. Additionally, when compiling for the simulator, the compiler links with a special version of `UIKit` made specifically for the simulator; it is not linked with `AppKit`. So, when you're working with `UIKit` classes, you can be reasonably certain that anything that you call on the simulator will be available on the hardware as well. I'm going through a lot of effort to explain this because when you look at Foundation, you'll see that it's not quite so simple.

Foundation is the next layer of API beneath `AppKit` and `UIKit`. It includes `NSString`, `NSArray`, `NSDictionary`, and many other low-level classes.

When it comes to these classes, generally speaking, most things are supported on both Cocoa and Cocoa Touch. However, there are some notable exceptions, in particular with regard to drawing things. On Cocoa, for example, `NSString` has a category that gives it the ability to draw `NSString`s within a view. Cocoa Touch does not include this capability.

The best way to avoid using Foundation APIs that are not included in Cocoa Touch is to be sure when you're browsing the documentation on the API, that you are looking at the iPhone documentation as opposed to Mac OS X documentation.

Understanding performance differences

I have mentioned it before, but an important concept to understand when interacting with the simulator is that the simulator is actually running on the processor and hardware of your computer, and that it does not emulate the slower processor and graphics capabilities of the actual iPhone hardware. The end result is that it has very different performance capabilities in comparison to actual hardware. What this means is that you really can't trust running in the simulator as a valid gauge of how your app will actually perform when running on an actual device.

While Apple has always been rather tight-lipped about its hardware specifications, there are certain things you probably know about the hardware. Specifically, you probably know that the processor on the generation 1 and 2 iPhones is a 620 MHz ARM processor underclocked to 412 MHz. You also probably know that it has a PowerVR MBX Lite 3-D graphics processor. You should know that it has 128MB of memory, and a solid-state flash hard disk.

Table 3.2 shows the specifications of the iPhone as compared to a typical desktop computer in more detail.

Table 3.2 Comparison of iPhone Hardware to Desktop Machines

Specification	iPhone (iPhone 3GS)	Desktop
Processor	620 MHz ARM (833 MHz ARM)	3.0 GHz Xeon
Memory	128MB (256MB)	2.0GB +
Disk Space	8, 16, 32GB	300GB +
Screen Resolution	480x320	1920x1200

When testing your applications, the impact of this difference in hardware is pervasive.

NOTE
Though the iPhone 3GS adds significant new capabilities in terms of processor speed and memory, it's important to remember that the 3GS still represents a small percentage of overall iPhone OS users. If you truly want widespread adoption, you should try to target the lesser hardware specifications whenever possible.

Using only 128MB of memory means that you are very constrained in memory when working on the device. The 128MB that is included is shared not just with your app, but also with the operating system, and some of the built-in applications that run in the background even when they are shut down. This means that you can really only count on having about 15 to 22MB of memory available for your application. After this, you start to see memory warnings. The simulator never generates a warning unless you ask it to, and furthermore, apps running in the simulator utilize your system RAM, which includes swap and usually multiple gigabytes of available storage.

NOTE On the iPhone 3GS, the available memory for an application can be as high as 150MB as opposed to the 15 or 22 on the older phones.

The slower processor and graphics processor on the iPhone mean that algorithms and graphics operations that run very fast on the simulator can often be much slower when running on actual hardware. For example, operations that involve transparency are known to be particularly expensive on the iPhone hardware and should be avoided. These typically are not a problem when running on the simulator. Additionally, some operations that may seem trivial, for example, object allocation, can be very expensive when done on actual hardware. The simulator does not suffer from these issues.

What all this means is that there is simply no substitute for testing your app on actual hardware. Test, and test often. Don't wait until the last minute and test it as an afterthought. Do it daily.

Summary

In this chapter, you've looked at the simulator in all its glory. You've dug into its filesystem and looked at how to interact with it. You've seen what capabilities are supported on the simulator and what are not. You've looked at the performance differences between the simulator and actual iPhone hardware. During your career as an iPhone developer, you're going to find yourself returning to the simulator over and over again; this background information should make you feel comfortable doing further exploration and help you understand where to look when things go wrong in your application and you need to track the problem down.

Building GUIs

 In This Part

Chapter 4
Understanding the Basics of an iPhone User Interface

Chapter 5
Introducing the UIView Controller

Chapter 6
Using UITableView

Chapter 7
Working with UITableView Cells

Chapter 8
Working within the UINavigationController Model

Chapter 9
Understanding the UIApplicationDelegate

Chapter 10
Applying Custom UIViews

Chapter 11
Handling Touch Events

Chapter 12
Working with Other Controls

Chapter 13
Handling Text Input

Chapter 14
Building Cinematic UIs with Core Animation

Chapter 15
Using OpenGL ES

Chapter 16
Integrating Safari

Understanding the Basics of an iPhone User Interface

The iPhone is known for its beautiful user interface and ease of use. Studies have shown that while the users of other smart phones use perhaps one or two features of their smart phones on a daily basis, iPhone users use over ten. Making your application be one of those ten features requires a great user interface that delights and inspires the user.

As a mobile device app developer, you also have to assume that for the majority of cases, when a user wants to work with your application, the window of interaction is usually no more than about 45 seconds. That is to say, the user wants to pull out the iPhone, open your application, navigate to the information that they want, get the information or perform some action, and then exit the application and put the iPhone back to sleep — all of this in about 45 seconds or less. This makes designing a good user interface a tremendous challenge.

You'll begin your exploration of building iPhone UIs by looking at the UIs of the applications that ship with the iPhone out-of-the-box, as well as the iPhone home screen. Apple has some of the best user interface designers in the world, and I think that there is a great deal to be learned by studying what they have done in their applications and applying the same concepts to yours. Additionally, it is important as an iPhone developer to be sure that you are adhering to the Apple iPhone human user interface design guidelines, as Apple has been known to reject applications that ignore these guidelines.

Obviously, if you're designing an application like a game that relies on a unique cinematic UI in order to better immerse the user in the game experience, then you are probably justified in deviating from these standards. However, most basic applications follow these models, and it is useful to know them well.

First, I'll discuss the iPhone home screen.

In This Chapter

Exploring the UIs of various built-in applications

Understanding the basic components that make up an iPhone UI

Knowing the standards required for building an app that fits into the iPhone model

Looking at the iPhone Screen

Figure 4.1
The main iPhone screen

The iPhone home screen; here you begin your exploration of how iPhone user interaction works.

The first thing that you should recognize when looking at the iPhone screen is that, generally speaking, it is divided up into three, or sometimes four, sections. Starting from the top, the first section is the status bar. This provides information about the iPhone, including the wireless carrier indicator, the clock, the Wi-Fi indicator, whether or not the iPod is playing, and so forth.

NOTE
The status bar can be hidden in your application by setting the `UIStatusBarHidden` property in your application's `Info.plist` file.

Between the status bar and the bottom "favorite applications" dock is the main interaction area of the iPhone. This is where the majority of the application interaction occurs, and is also where the majority of information is displayed. On the home screen, this is where the application icons are located.

Some applications, including some of the ones that you will work with through this book, also display a navigation controller at the top of this application interaction area.

CROSS-REF

I will talk about the navigation controller a lot more in Chapter 8, because this is one of the fundamental iPhone user interface metaphors that you absolutely have to master in order to develop a good iPhone application.

Finally, at the bottom of the screen is the "favorite applications" dock. On the iPhone home screen, this is where you can put your most-used applications. They are available whenever you're on the home screen, even if you swipe to move to another set of applications. This is an important paradigm that I will revisit shortly when you are looking at an actual application, because this interaction is used again in the form of the Tab Control.

Creating a Cinematic User Experience

One of the concepts that Apple engineers like to talk about when they're talking about beautiful user interfaces is the concept a "cinematic user experience." A cinematic user experience essentially is a user interface that looks like something from a Hollywood movie. It looks futuristic and smooth, and it uses animation to enhance the feeling of working with physical objects.

Apple user interface designers have specifically given you a set of tools that enable you to build these types of cinematic user interfaces. Tools like Core Animation give you the power to build user interfaces that involve elements that slide in from off-screen rather than just appearing, and elements that scroll with an almost physical weight to them.

Most of these capabilities are simply built into the UI components that you use in designing your interfaces. However, in cases where you might be designing a UI yourself (for a game, for example), you should strive to maintain this same cinematic quality.

This means that you should use smooth transitions from one view to the next, rather than jarring, sudden ones. You should use only high-quality graphics for your buttons and backgrounds, and when possible, leverage the unique capabilities of the iPhone hardware such as multi-touch, the accelerometer, and Core Location. By doing this, you will build an application that not only fits well within the iPhone UI framework but also makes the user excited to use your application.

Looking at the Built-in Applications

Figure 4.2
The iPod application UI

Opening the iPod application, you'll see that the basic zones that I just described are maintained inside this application, as shown in Figure 4.2. Specifically, you still have the status bar at the top. Next you have the navigation controller. You then have the list of songs and artists in the main interaction area. Finally, at the bottom of the screen, you have a tab controller that enables you to switch between the major functions of the iPod, and that generally stays active through most interactions. As you can see, this is somewhat similar to how the main iPhone user interface works.

The iPod application is an excellent example of an application that has to manage and navigate trees and lists of hierarchical data. There are other types of user interfaces that you can look at here, as well, such as the Weather application, shown in Figure 4.3.

Figure 4.3

The Weather application

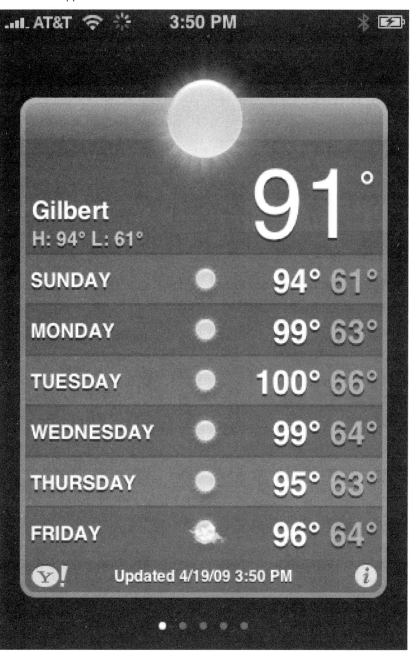

The Weather application doesn't have a lot of lists of data to show you, but instead is meant as a brief summary of information that you can quickly look at, get the information that you want, and then exit the application. In this case, the main screen contains all the data that the user is interested in. There is no need for any sort of complex navigation. To change the settings for the application, it provides a small info button in the lower-right corner that flips the user interface using Core Animation to allow the user to configure the cities that they want to see the weather in.

NOTE
The info icon is a standard graphic provided in Interface Builder. You can use it in your own applications.

This is a good model for applications where you don't have a lot of data that needs to be managed. Imagine, when considering this kind of user interface, that your application might be used for as little as ten seconds at a time.

Understanding the Basics of iPhone Navigation

In addition to understanding the big-picture issues surrounding user interface design on the iPhone, it's also important to take a good look at some of the smaller, often overlooked interactions.

For example, when designing custom buttons, it's important that you provide some immediate feedback to the user when a button is pressed, such as highlighting the button or providing an alternative image that displays. You will look at this in more detail in a future chapter, but for now, know that you can do this very easily through Interface Builder.

NOTE
Interface Builder allows you to highlight a button when it is selected, or use a separate image that you can provide to be displayed when the button is selected.

Another example of the finer details of interaction is to look at the way that the iPhone OS transitions views to navigate trees of data. Specifically, note that as you choose an item from a list, that item highlights, and then a detail view moves into the screen from the right to the left, revealing information about the selected item. Similarly, when returning from that detail view, the current view slides back out to the right, and the original view slides in from the left. This is an important interaction and one that you should mirror in your applications. It follows the way we read, left to right, and feels natural for us. It would be unnerving to a user to have it work some other way.

Finally, note that the iPhone has standardized the methods by which you edit data. That is to say, typically a view is in display mode by default, and when the user touches an edit button (typically in the upper-right corner), it changes the view to the editable version of that view. When the user is done editing the data on that screen, he touches the Done button and it transitions back to the display view.

NOTE
Watch carefully during these transitions and you'll see that they are usually animated. Judicious use of animation is one of the things that make the iPhone UI unique and cinematic.

None of these rules are set in stone, but it's useful when designing your application to ask yourself regularly, "What would Apple do?" and whatever the answer is, do that. Studies have even shown that when applications don't follow these conventions, users quickly become annoyed. They expect third-party apps to act like the built-in apps, so it can be worth your while to put forth the effort to make your apps fit in.

Summary

Perhaps the greatest compliment that I have seen given to an application is one where the user said, "This application makes me want to…(whatever the application does)…just so that I can use this application to do it." That should be what you strive to do in your apps.

In this chapter, we've taken a look at some of the elements that make iPhone user interaction unique and intuitive. The iPhone is a tremendous platform; you should strive to make your application worthy to run alongside the other great applications available for it, and worthy of your customers' time and money.

Introducing UIViewController

When developing the Cocoa Touch frameworks, Apple took the opportunity to standardize and codify its toolset for working with views and for navigating hierarchical lists of data. Many developers are already familiar with the Model, View, Controller design pattern that is in common use in Cocoa and Cocoa Touch, as well as in other programming languages such as Java. Cocoa has long supported this design pattern, particularly with regard to its views and model classes, but only recently has it really begun to solidify its concept of how a controller fits into its framework. Don't misunderstand; good developers working with Cocoa have always utilized controllers in their code, but with the introduction of Cocoa Touch, Apple has made a set of classes specifically designed to be inherited from when developing view controllers. These classes are `UIViewController` and `UITableViewController`.

In addition to providing these classes, which, by default, give you a certain amount of functionality straight out of the box, Apple has also provided an entire navigation framework that works with these controllers to make creating complex user interfaces, or UIs, that manipulate trees of hierarchical data incredibly easy to develop. The navigational classes associated with this framework are centered on the `UINavigationController` and its associated classes.

In This Chapter

Understanding how Cocoa Touch fits in the Model, View, Controller design pattern

Learning the basics of UIViewController

Building a basic iPhone app that uses UIViewController

CROSS-REF
I discuss `UINavigationController` in Chapter 8.

In this chapter, you'll take a look at how to write a `UIViewController` and how to use it to implement a Model, View, Controller design pattern using your view controller in your application. View controllers are a fundamental building block in Cocoa Touch and an important concept that you will need to know for future chapters.

Understanding the Basics of Model, View, Controller

Regarding `UIViewControllers`, the first thing you need to do is review the Model, View, Controller design pattern and see where the classes that you will be working with fit into it.

Figure 5.1 shows how the Model, View, Controller design pattern is typically described. As you can see, typically there are three major components to an application designed with this pattern in mind. On the far left, you see the model part of the diagram. This typically consists of data storage and manipulation classes such as arrays, dictionaries, or database/file storage classes. These are classes whose purpose is to handle the actual housekeeping of the storage of the data. This means that the model part is what actually handles the reading and writing of the data to the disk or the caching of that data in memory. It also contains methods for operating on that data to calculate raw values or other low-level manipulation of the data.

Figure 5.1

The Model, View, Controller design pattern

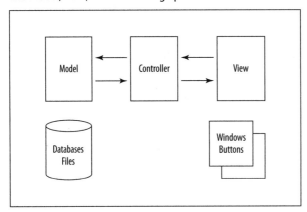

On the far right, you see the view part of the diagram. These are the objects and classes that interact directly with the user, presenting the data from the model in a form that the user can see and read. These classes contain methods whose purpose has to do specifically with the display of the data.

When implemented correctly, the view should have no data manipulation logic built into it and the model should have no display logic built into it. This separation of presentation from data is one of the key elements that make the Model, View, Controller design pattern powerful. It enables you to replace or change either the model or the view independently of each other. This flexibility is what it derives its power from.

So I've talked about the model and the view, but I have not yet talked about the controller. The controller is one of the often-misunderstood parts of the Model, View, Controller design pattern, but its functionality is actually very simple. The controller is essentially the gatekeeper between the model and the view. Its job is to provide the separation between the two of them and to provide the methods by which the model and the view can manipulate each other without being tightly coupled to each other. What this means is that the controller has references to both the model and the view, and when displaying the view the controller reads the data from the model and puts it into the view. Similarly, when the data is changed through the view, the controller is the one that reacts to those changes and sets the data in the model. Again, the goal here is to decouple the view from the model. The controller acts as that decoupling mechanism.

Exploring UIViewController

Before I get too far into the example for this chapter, it's important that you understand the role of the `UIViewController` in depth. The core reason for the existence of the `UIViewController` is to manage a view.

In Cocoa Touch, Apple has provided you with a base class from which you inherit your view controllers. That base class is the `UIViewController`. Let's take a look at how you write a basic `UIViewController` and how you connect it to your view.

Overriding Methods on UIViewController

`UIViewController` provides a variety of methods that can be overridden in subclasses. Some of these methods are shown in Table 5.1.

Table 5.1 Typically Overridden Methods from UIViewController

Method Signature	Description
`-(void)loadView`	Called to create the view that the controller manages. You must override this method if you intend to create your view yourself programmatically. Do not call this method manually yourself.
`-(void)viewDidLoad`	Called after the view has been loaded from the nib. This is used for initialization activities that require the objects in the nib.
`-(void)viewDidUnload`	Called when the view is unloaded from memory, for example, as a result of a low memory warning or when the view is being unloaded so it can be deallocated. You should use this for releasing resources allocated in `viewDidLoad`.

continued

Table 5.1 Continued

Method Signature	Description
`-(void)viewWillAppear:(BOOL)animated`	Called just before the view appears on-screen.
`-(void)viewWillDisappear:(BOOL)animated`	Called just before the view disappears from the screen.
`-(void)viewDidAppear:(BOOL)animated`	Called after the view appears on-screen.
`-(void)viewDidDisappear:(BOOL)animated`	Called after the view has disappeared from the screen.
`-(BOOL)shouldAutorotateToInterfaceOrientation:(UIInterfaceOrientation)interfaceOrientation`	Called when the device is being rotated to determine if you want the interface to allow rotation to the given orientation. If you support rotation to the given orientation in your view, you should return YES. The default implementation always returns NO for landscape modes.
`-(void)willRotateToInterfaceOrientation:(UIInterfaceOrientation)toInterfaceOrientation duration:(NSTimeInterval)duration`	Called just prior to the interface being rotated.
`-(void)willAnimateRotationToInterfaceOrientation:(UIInterfaceOrientation)interfaceOrientation duration:(NSTimeInterval)duration`	Called prior to the interface being rotated from within the animation block that will be used to animate the view. This means it's a perfect place for you to queue up additional animations that you want to occur during the animation.
`-(void)didRotateFromInterfaceOrientation:(UIInterfaceOrientation)fromInterfaceOrientation`	Called directly after the interface has rotated.
`-(void)willAnimateFirstHalfOfRotationToInterfaceOrientation:(UIInterfaceOrientation)toInterfaceOrientation duration:(NSTimeInterval)duration`	Called from within the animation block being used to animate the view rotation. Use it to add your own animations to the transition.
`-(void)didAnimateFirstHalfOfRotationToInterfaceOrientation:(UIInterfaceOrientation)toInterfaceOrientation`	Called after the first half of the rotation animation has completed.
`-(void)willAnimateSecondHalfOfRotationFromInterfaceOrientation:(UIInterfaceOrientation)fromInterfaceOrientation duration:(NSTimeInterval)duration`	Called from within the animation block being used to animate the view rotation. Use it to add your own animations to the transition.
`-(void)didReceiveMemoryWarning`	Called when your device is getting low on memory. Use it to purge cached data.

NOTE
For most of these methods, you should also be sure to call the superclass implementation of whatever method you're overriding to ensure that the default implementation of the call is also able to perform its functions.

CROSS-REF
To see more of the methods available from `UIViewController`, see the Cocoa Touch documentation in Xcode.

CAUTION
The view appearance methods (`viewWillAppear`, `viewWillDisappear`, and friends) are not called if you add the view to the hierarchy directly, so be sure that you call these methods manually when you do so.

Because the primary purpose of a `UIViewController` is managing a view, most of these methods have to do specifically with that purpose. The most common methods that you will need to override in your applications are `viewDidLoad`, `viewDidUnload`, `viewWill Appear`, `viewDidAppear`, `viewWillDisappear`, and `viewDidDisappear`. These are methods that you typically use to determine how your view reacts to each of these events. You would use these to perform tasks such as initializing variables, loading data from disk or (in the cases of unloading and disappearing) saving data to disk or setting values on the model.

Handling view rotations

The `UIViewController` can also provide basic handling of view rotations. To declare that you support orientations other than portrait, you must override the `shouldAutoRotateTo InterfaceOrientation` method to return YES. In this method, you should look at the orientation that is being requested and return YES for the orientations that your subclass supports. If your view needs more complex support for rotation, the additional methods `willRotate ToInterfaceOrientation` and `didRotateFromInterfaceOrientation` are provided as hooks for you to do more complex operations. These are called at the beginning and end, respectively, of the orientation event. So when the device is being rotated from portrait to landscape, you receive a `willRotateToInterfaceOrientation` call at the beginning of that process, telling you that the device is about to rotate to landscape mode. At the end of the rotation, you receive a `didRotateFromInterfaceOrientation` call, telling you that the device just rotated from portrait mode.

Typically, these methods would be overridden to disable processing during rotation, which might adversely impact the rotation animation itself. If you can imagine a view that needs to do some calculation to display, you might want to disable that processing during the actual animating from one orientation to the other. That's where you would use these kinds of methods.

Finally, if you want to provide special custom animations for the rotation transition, then you can override the method `willAnimateRotationToInterfaceOrientation:duration:`. This method is called within the animation block that is used to configure the animation for the rotation. This means that any changes to animatable properties that you do in this method will be added to that animation block and will occur at the same time that the user interface animation for the rotation is occurring. Typically, you would use this method to animate the motion of your views to new locations that are more appropriate for the new interface orientation.

CROSS-REF
I discuss more about Core Animation and animating views in Chapter 14.

NOTE
The method `willAnimateRotationToInterfaceOrientation:duration:` is mutually exclusive to the methods `willAnimateFirstHalfOfRotationToInterfaceOrientation:duration:` and `willAnimateSecondHalfOfRotationFromInterfaceOrientation:duration:`. If you override one, you should not override the others. The latter methods were used in iPhone OS 2.x. The former is considered the newer, "better" way to handle animations during rotations.

Handling memory warnings

As you know, memory on the iPhone is a finite resource. In a later chapter, I will visit memory management in greater detail, but it deserves a mention here that `UIViewController` provides a method that is called when your application receives a low memory warning. That method is `didReceiveMemoryWarning`. By default, the implementation of this method causes the view controller to unload and deallocate its view if the view is not currently a subview of another view. When it does this, it calls the `viewDidUnload` method to give you an opportunity to save any data that's appropriate or to release any objects you allocated in viewDidLoad. In addition to being mindful of this behavior, you should override the `didReceiveMemoryWarning` method and do your own purging of unneeded temporary data to free up memory. If you do this, it is particularly important that you also call the `[super didReceiveMemoryWarning]` method in your implementation. Good examples of data to free in `didReceiveMemoryWarning` might be cached data you can easily recreate or lazily loaded objects that will be recreated the next time they are accessed.

Working with UIViewController

Let's begin by creating a new project in Xcode. Name this project ViewControllerDemo. Create this example project using the Window-Based Application. This creates a very basic iPhone application with nothing more than a window. You will then use this project to build your application completely from scratch. By doing this, you will come to understand what the templates are giving you.

CAUTION
When creating the ViewControllerDemo project, be sure the "Use Core Data for Storage" check box is not selected in the project creation window.

NOTE
Xcode has templates for navigation-based applications and view-based applications, and there's absolutely nothing wrong with using these templates when you're creating your own applications. However, for the purposes of this example, I want you to understand exactly what's going on behind the scenes.

Creating the code

Once you've created the project, it will contain one nib file, the `MainWindow.xib`, which will be under the resources group, and your app delegate implementation, which consists of `ViewControllerDemoAppDelegate.h` and `ViewControllerDemoAppDelegate.m`, located under the classes group.

The first thing that you're going to do is to create a new `UIViewController` and add it to your project. You're going to do this by right-clicking on the classes group and choosing Add ⇨ New File. From here, you're going to choose the `UIViewController` subclass template. For the purposes of this example, call the file `MyViewController`. Be sure, when you do this, that the "With XIB for user interface" check box is not selected. This creates the `MyViewController` header and implementation files for you.

NOTE
I'm having you create the UIViewController here without the "With XIB for user interface" checked, because, again, I want you to understand the underlying things going on. In the future, you can certainly use that check box to have it automatically create the XIB for you. In this example, I have you create it manually.

In this example, you're going to have a very basic `NSMutableArray` to serve as your model. In this array, you're going to store the names of cities. You'll have two buttons. The first one will be a Next button that will advance you through the array's elements one by one. As you go through the array, you will set the text property of the `UILabel` to the names of the cities. When you come to cities that you like, you will be able to click the second button, Favorite, to show that you like that city. This will cause a "Favorite: YES" message to appear next to the label when that city is shown. Figure 5.2 shows how the application will look when it's finished.

Figure 5.2

The final application

Creating the UIViewController header

The main work of the application is going to occur in your UIViewController. To implement it, the first thing that you need to do is set up your `UIViewController` header file to include `IBOutlets` for each of the elements of the UI that I just described. Listing 5.1 shows exactly how you do this. You simply add the `IBOutlets` into the implementation section.

Listing 5.1

MyViewController Header

```
#import <UIKit/UIKit.h>
@interface MyViewController : UIViewController
{
    UILabel *currentCity;
    UIButton *nextButton;
    UIButton *favoriteButton;

    NSMutableArray *theData;
    int currentItem;
}
@property (nonatomic, retain) NSMutableArray * theData;
@property (nonatomic, retain) IBOutlet UILabel * currentCity;
@property (nonatomic, retain) IBOutlet UIButton * nextButton;
@property (nonatomic, retain) IBOutlet UIButton * favoriteButton;
-(IBAction)nextButtonTouched:(id)sender;
-(IBAction)favoriteButtonTouched:(id)sender;
@end
```

NOTE
You're making `theData` a property here so that you can set it later from your app delegate.

CAUTION
The property declarations here define the object ownership rules you wish to use for your IBOutlets, in this case, you want to retain them. On Mac OS X, IBOutlets without property declarations are not retained by default. This is the opposite of the behavior on iPhone. That means that on iPhone OS, if you do not declare a property, it is retained and thus must be released in your dealloc method. Due to this inconsistency, rather than just remembering which platform has which behavior, you can instead be in the habit of always declaring properties for all your IBOutlets. By doing this, you always have your ownership rule documented as part of your header file.

Creating the UIViewController implementation

The implementation file is a bit more involved. The first thing that you have to do is initialize your city label to show the first element of the list when the view is initially loaded. You do this in the `viewDidLoad` method. This method gets called when your view is loaded before displaying it. You do it here because you know that at this point in the code your outlets have been connected to the implementations that are in the nib files. Listing 5.2 shows what your implementation of the `viewDidLoad` method looks like.

CAUTION

Properties are cool. You should use them liberally in your code. However, they require declaration both in your header and your implementation file. The declaration in the implementation file typically will be of the form @synthesize <variablename> or @dynamic <variablename>. There are other options available to use with them, but for the purposes of this entire book, you can assume that *all* properties will use the @synthesize declaration. This means that, even if I don't mention it, if you see an @property in a header file, you should also have an @synthesize in the corresponding implementation file. In the case of this UIViewController, that means you should have @synthesize `theData` and so on in the implementation file.

Listing 5.2

viewDidLoad Implementation

```
-(void)viewDidLoad;
{
   currentItem = 0;
    NSDictionary *cityData = [theData objectAtIndex:currentItem];
    NSString *theInfo = [NSString stringWithFormat:@"%@ Favorite: %@",
                         [cityData objectForKey:@"NAME"],
                         [cityData objectForKey:@"FAVORITE"]];
    [currentCity setText:theInfo];
}
```

After setting things up in the `viewDidLoad` method, you need to write implementations for your event handlers for the Next button and for the Favorite button. Listing 5.3 shows what this code looks like.

Listing 5.3

Event Handler for the Next Button

```
-(IBAction)nextButtonTouched:(id)sender;
{
    currentItem++;
    if(currentItem >= [theData count])
        currentItem = 0;
    NSDictionary *cityData = [theData objectAtIndex:currentItem];
    NSString *theInfo = [NSString stringWithFormat:@"%@ Favorite: %@",
                         [cityData objectForKey:@"NAME"],
                         [cityData objectForKey:@"FAVORITE"]];
    [currentCity setText:theInfo];
}
```

In this code, when the user presses the Next button, you increment your `currentItem` index, which you then use to access the current city from your model class, the `theData` array. You take special care here to make sure that you don't increment your index to be higher than the actual number of items in the array. After getting the current item out of the array, you then set your text property of the `currentCity` label. Even though you haven't actually created the `UILabel` in code here, you still have an actual instance that is attached to that variable. This is because, as you will see in a moment, you will attach your `IBOutlet` for the `currentCity` variable to the actual instance of your `UILabel` in Interface Builder.

The key point of going through this explanation here is to make it clear that the controller is what is doing the manipulation of the view. Specifically, you're taking the data from the model and setting it in the view.

The event handler for the Favorite button `favoriteButtonTouched` is shown in Listing 5.4 and gives you a good look at how you take data from the UI and set it on the model. Specifically, what's happening here is that when the Favorite button is touched, this event fires; you then look at your current index, get the object from the array that is referred to by that index, set the favorite property on that object, and then make the favorite star visible on the UI.

Listing 5.4
Event Handler for the Favorite Button

```
-(IBAction)favoriteButtonTouched:(id)sender;
{
    NSMutableDictionary *cityData = [theData objectAtIndex:currentItem];
    if([[cityData objectForKey:@"FAVORITE"] isEqualToString:@"NO"])
        [cityData setObject:@"YES" forKey:@"FAVORITE"];
    else
        [cityData setObject:@"NO" forKey:@"FAVORITE"];

    NSString *theInfo = [NSString stringWithFormat:@"%@ Favorite: %@",
                        [cityData objectForKey:@"NAME"],
                        [cityData objectForKey:@"FAVORITE"]];
    [currentCity setText:theInfo];
}
```

So, bringing this all together, the entire implementation of MyViewController is shown in Listing 5.5. Remember, I didn't specifically say to do it, but you should have an @synthesize declaration for every @property, and you should remember to properly release all your instance variables in your `dealloc` method.

Listing 5.5

The Final Implementation of the View Controller

```objc
#import "MyViewController.h"
@implementation MyViewController
@synthesize theData;
@synthesize currentCity;
@synthesize nextButton;
@synthesize favoriteButton;
-(void)viewDidLoad;
{
    currentItem = 0;
    NSDictionary *cityData = [theData objectAtIndex:currentItem];
    NSString *theInfo = [NSString stringWithFormat:@"%@ Favorite: %@",
                        [cityData objectForKey:@"NAME"],
                        [cityData objectForKey:@"FAVORITE"]];
    [currentCity setText:theInfo];
}
-(IBAction)nextButtonTouched:(id)sender;
{
    currentItem++;
    if(currentItem >= [theData count])
        currentItem = 0;

    NSDictionary *cityData = [theData objectAtIndex:currentItem];
    NSString *theInfo = [NSString stringWithFormat:@"%@ Favorite: %@",
                        [cityData objectForKey:@"NAME"],
                        [cityData objectForKey:@"FAVORITE"]];
    [currentCity setText:theInfo];
}
-(IBAction)favoriteButtonTouched:(id)sender;
{
    NSMutableDictionary *cityData = [theData objectAtIndex:currentItem];
    if([[cityData objectForKey:@"FAVORITE"] isEqualToString:@"NO"])
        [cityData setObject:@"YES" forKey:@"FAVORITE"];
    else
        [cityData setObject:@"NO" forKey:@"FAVORITE"];

    NSString *theInfo = [NSString stringWithFormat:@"%@ Favorite: %@",
                        [cityData objectForKey:@"NAME"],
                        [cityData objectForKey:@"FAVORITE"]];
    [currentCity setText:theInfo];
}
- (void)dealloc
{
```

```
    [self setCurrentCity:nil];
    [self setNextButton:nil];
    [self setFavoriteButton:nil];
    [self setTheData:nil];
    [super dealloc];
}
@end
```

Adding the controller to Interface Builder

I mentioned a moment ago that I would talk about how you pull your code into Interface Builder and attach your `IBOutlets` to your actual objects on the UI, so let's take a look at that now.

First, you need to create a new nib to contain your view. To do this, right-click on the Resources folder in your project and choose Add ⇨ New File. In the New File dialog, choose User Interfaces as the category, and then choose View XIB. Call this file `MyView.xib`. After you click Finish, the file is created and added to the resources group. Double-click it to launch Interface Builder.

Before, when you used Interface Builder, you used it in conjunction with the template-generated nib files that you had in your template-generated project. In this case, you're going to do it all manually so that you understand all the details of what's going on behind the scenes. To do this, the first thing that you need to do is to set the file owner object to be an instance of your `MyViewController` class. You do this by selecting the File Owner object, and then in the properties window, selecting the last tab, the Identity tab, and in the Class drop-down menu, choosing the `MyViewController` class. What this does is enable you to see the `IBOutlets` that you created previously in your view controller and allow you to connect those `IBOutlets` to the objects that you are about to put on the user interface.

Once you've done this, drag-and-drop a Label and two Round Rect Buttons from the Library window onto your view, so that it looks like Figure 5.3.

Just like you did in your basic iPhone application, you Control-click and drag from the File Owner to each of the controls on the UI to connect them from your `IBOutlets` to the actual objects. Then you Control-click and drag from your buttons to the File Owner and choose the `IBActions` that you set up for those buttons. Control-clicking and dragging from buttons to IBActions connects the default event "Touch Up Inside" to the IBAction.

`UIViewController` provides you with a property that you can use to set your view to. This property is called, appropriately enough, view, and you should Control-click and drag from the File Owner to your main view to connect this as well.

Figure 5.3

Designing the interface

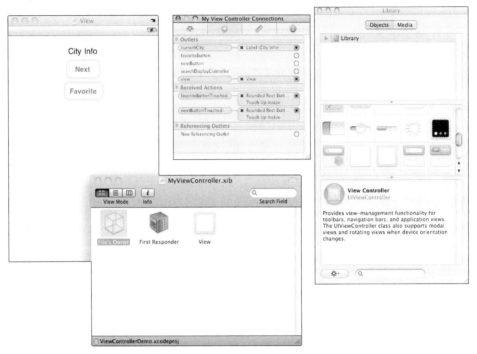

Finally, you need to do a little bit of work in the MainWindow nib file, so save the MyView nib file and close Interface Builder. Then, edit the ViewControllerDemoAppDelegate.h file to look like Listing 5.6.

Listing 5.6

ViewControllerDemoAppDelegate Header

```
#import <UIKit/UIKit.h>
#import "MyViewController.h"
@interface ViewControllerDemoAppDelegate : NSObject <UIApplicationDelegate>
{
    UIWindow *window;
    MyViewController *viewController;
}
@property (nonatomic, retain) IBOutlet UIWindow *window;
@property (nonatomic, retain) IBOutlet MyViewController * viewController;
@end
```

You have two `IBOutlets`, one that is connected to your base window of your application and the other that is connected to your view controller. When the application launches, you take the view from the view controller and add it to your window as a subview. To do this, flip over to the `ViewControllerDemoAppDelegate.m` file; I will then show you how to tie your model to your view controller.

Understanding the UIApplicationDelegate

It's important here to take a short side trip and talk about the `UIApplicationDelegate`. Every iPhone application has a `UIApplicationDelegate`. Like a main function, the `UIApplicationDelegate` is where the application is first notified that it is running, as well as being notified that the user would like to exit the application, that memory is low, and so on. It provides numerous useful methods for you to override, all of them optional. For the purposes of your application, you will be working with the `applicationDidFinishLoading` method, which is called when the application finishes loading. You will look at some of these other methods in more depth later.

Creating the model in the App Delegate

This application is going to be very simple. You're not going to do anything special with saving the data out to disk or anything like that. For the purposes of this demonstration, all you're going to do is to create an array that contains dictionaries inside of it for each of your cities. The dictionaries will have two objects in them, the name of the city and whether or not it's a favorite. In this way, your model is effectively an array of property lists.

CAUTION
Arrays of dictionaries make good book models but not usually good actual models. Use your judgment in making models that fit your applications well.

To create this model, you add the method shown in Listing 5.7 to your implementation file for the app delegate.

Listing 5.7

The makeModel Method

```
-(NSMutableArray *)makeModel;
{
    NSArray *cityNames = [NSArray arrayWithObjects:@"New York", @"Chicago", @"San
Francisco", @"Los Angeles", @"Phoenix", nil];

    NSMutableArray *ret = [NSMutableArray array];
    for(NSString *cityName in cityNames)
    {
        NSMutableDictionary *cityObject = [NSMutableDictionary dictionary];
```

continued

Listing 5.7   *(continued)*

```
        [cityObject setObject:cityName forKey:@"NAME"];
        [cityObject setObject:@"NO" forKey:@"FAVORITE"];
        [ret addObject:cityObject];
    }

    return ret;
}
```

It's important to add this method above the `applicationDidFinishLaunching` method so that you don't get a warning that the method is not defined when you call it.

NOTE
Placing this method above the applicationDidFinishLaunching method enables you to use it without a warning. Another way to do this is to declare this method in your header file. For me, I prefer to expose only methods that I would expect others to need to know about into my interface. In this way, the interface is not cluttered with methods that should not be called externally.

Once you've added this method, you edit the `applicationDidFinishLaunching` method to look like Listing 5.8.

Listing 5.8

The Implementation of applicationDidFinishLaunching

```
- (void)applicationDidFinishLaunching:(UIApplication *)application
{
    [viewController setTheData:[self makeModel]];
    [window addSubview:[viewController view]];
    [window makeKeyAndVisible];
}
```

What you're going to do here is to create your model by calling your `makeModel` method, and you're going to pass that model to your view controller. This is done with the call to `[viewController setTheData:cities]`.

When you edit the main window nib file, you are going to create a proxy to your view controller. This proxy will be attached to your view controller `IBOutlet` here in your application delegate. What this means is that when this method gets called, your member variable of `viewController` is attached to your view controller in the other nib file, which is attached to your view. Your `mainWindow` member variable is attached to your main window. With these, you take the view from the view controller and add it to the main window so that when the main window displays, the view shows up. This is done with your call to [mainWindow addSubview:[viewController view]]. It may sound like a lot of work, but it's not. It's really just one line of code.

Adding the view to the window

Last but not least, you need to open the main window nib file and work your Interface Builder magic in there to create the proxy for the view controller that you're going to connect to your view controller IBOutlet in your app delegate.

To do this, you double-click the MainWindow.nib file in Xcode, and then go to the Library window again and drag-and-drop a View Controller object from the library into your Document window. This creates a UIViewController instance in this nib file. Select this UIViewController instance in the Document window and then go to the Properties window, and from the first tab (the Attributes tab), set the NIB Name property to be "MyView." Figure 5.4 shows what this looks like. Then go to the last tab (the Identity tab) and set the Class property to be MyViewController.

Figure 5.4

Adding the view controller to the MainWindow nib file

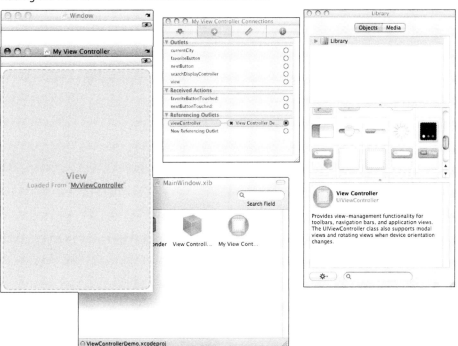

Doing this creates an object that you can use inside of your app delegate to access your view controller. You still have to tie it to your IBOutlet, though. To do this, again, Control-click and drag, but this time do it from the application delegate object to the view controller. This enables you to connect the view controller outlet to the view controller.

Listing 5.9 shows the final implementation of the app delegate.

Listing 5.9

The Final App Delegate Implementation

```objc
#import "ViewControllerDemoAppDelegate.h"
@implementation ViewControllerDemoAppDelegate
@synthesize window;
@synthesize viewController;
-(NSMutableArray *)makeModel;
{
    NSArray *cityNames = [NSArray arrayWithObjects:@"New York", @"Chicago",
                          @"San Francisco", @"Los Angeles", @"Phoenix", nil];

    NSMutableArray *ret = [NSMutableArray array];
    for(NSString *cityName in cityNames)
    {
        NSMutableDictionary *cityObject = [NSMutableDictionary dictionary];
        [cityObject setObject:cityName forKey:@"NAME"];
        [cityObject setObject:@"NO" forKey:@"FAVORITE"];
        [ret addObject:cityObject];
    }

    return ret;
}
- (void)applicationDidFinishLaunching:(UIApplication *)application
{
    [viewController setTheData:[self makeModel]];
    [window addSubview:[viewController view]];
    [window makeKeyAndVisible];
}

- (void)dealloc
{
    [self setWindow:nil];
    [self setViewController:nil];
    [super dealloc];
}
@end
```

Once you've done all of this, if you've done it correctly, you can build and run your application, see your view load, and click your buttons and have the application behave as you expect it to.

NOTE
Remember, you can download all the code for this chapter from the book's Web site.

Summary

In this chapter, I introduced you to the fundamental application building block that is the `UIViewController`. Virtually every iPhone application has at least one `UIViewController` or a class that is derived from `UIViewController`. It's important that you understand this essential class and how it fits in with the Model, View, Controller design pattern in order to build iPhone applications that are flexible and easy to maintain. In addition to this, you also learned the basics of handling orientation changes and memory warnings. I will revisit many of these subjects in upcoming chapters, but this has been a good way to get introduced to these subjects.

Using UITableView

The application example in the previous chapter got the job done, but it was certainly not an example of an elegant user interface. It was useful to demonstrate some of the intricacies of `UIViewController`, but if you consider the type of data that you were working with (a list of cities), then it's easy to recognize that a different kind of user interface might be more effective at navigating this data. Fortunately for you, there is another type of view that is uniquely suited to lists of data. That view is a `UITableView`, and it is this view that you will focus on now.

Table views are not new in the realm of GUI toolkits. On the iPhone, however, because of its small screen size, there are a variety of constraints that make it impossible to implement a traditional table view with rows and columns. Instead, Apple provides the `UITableView`, which can be used to display a variety of list-oriented data. One way to think of the `UITableView` is in terms of representing a single-column subset of your data, but it can also be used to represent hierarchical data. In this chapter, you'll see how to display your favorite cities as a list in a `UITableView`.

Understanding UITableView Data Sources and Delegates

The first thing that you need to do before you add a `UITableView` to your application is to create a data source for your model so that you can feed the data from your model to the table view. To do this, you need to understand the roles of two key protocols: the `UITableViewDataSource` protocol and the `UITableViewDelegate` protocol. They are used to provide data and customization, respectively, to your `UITableView` instances.

The `UITableViewDataSource` is used to provide data from your model class to the `UITableView`. It has methods that you are expected to implement that are called by the `UITableView` when it needs to load its data. These methods allow you to feed your data to the `UITableView` without coupling the `UITableView` directly to your model.

 In This Chapter

Exploring the methods of a UITableView

Learning how a UITableView accesses data from your model without being tightly coupled to it

Implementing a delegate and data source for a UITableView

Expanding the sample application to use a UITableView

The `UITableViewDelegate` protocol is used to provide customization for how the `UITableView` behaves. For example, there are methods on the `UITableViewDelegate` protocol that allow you to customize whether or not particular rows can be edited. In this way, imagine that the `UITableView`, whenever it needs to know how you want it to behave, asks the `UITableViewDelegate` what to do.

Let's take a look at what each of these protocols look like.

Understanding NSIndexPath and how it works with UITableViews

Before I talk about the data source and the delegate, I need to talk about `NSIndexPath`. The reason that this is important is because most of the methods that are in these protocols take an `NSIndexPath` as an argument, and so you need to know how to work with them.

In its simplest form, an `NSIndexPath` represents the path required to access a specific node in a tree of nested array structures. By default, an `NSIndexPath` can be used to represent a path of virtually limitless depth, but for when working with `UITableView` you are really only interested in using it to represent the location (row) of an item within a section. Because they are used so extensively in `UITableView`, there is a set of extensions that have been added to `NSIndexPath` specifically for use with `UITableView`. These are the section method and the row method, and they represent the section index and the row index, respectively. Therefore, to determine what item a particular `NSIndexPath` represents, you can ask it for its section, and then you can ask it for its row. You can then use this information to look within your table view.

Figure 6.1 shows an example of how an `NSIndexPath` can be used to represent a particular row within a particular section.

Now that you understand a bit about `NSIndexPath`, let's take a look at the `UITableViewDataSource` protocol.

Exploring UITableViewDataSource

The `UITableViewDataSource` is probably the most important component of working with a `UITableView`. Because of this, it's very important for you to fully understand exactly how it works and how you need to use it in your code.

The key element to understand is that when the `UITableView` needs to load its data, it asks the `UITableViewDataSource` for that data. This includes the following: how many groups you want in your table; what the headers for those groups are; what data to show in a particular cell; and what kind of custom cell to use for a particular row. As you can see, this is really the core of your data and thus also your table.

Figure 6.1

Using an NSIndexPath to locate a row

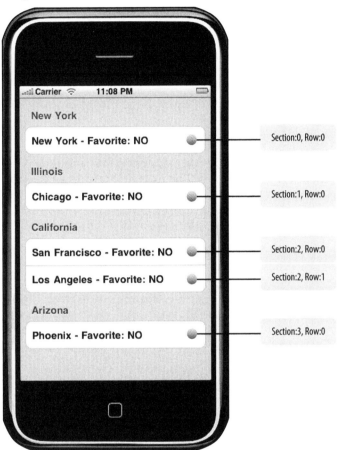

Understanding the core methods in the UITableViewDataSource

The first thing to recognize is that there is a set of methods that are virtually required for minimal functionality of your data source. Those methods are listed in Table 6.1.

Table 6.1 Required Methods on a UITableViewDataSource

Method	Description
`- (NSInteger) tableView: (UITableView *) tableView numberOfRowsInSection: (NSInteger) section`	Returns the number of rows for the given section
`- (UITableViewCell *) tableView: (UITableView *) tableView cellForRowAtIndexPath: (NSIndexPath *) indexPath`	Returns the cell to display your data
`- (NSInteger) numberOfSectionsInTableView: (UITableView *) tableView`	Returns the number of sections in your table view

When asking the data source for your data, the first method that the `UITableView` calls is the `numberOfSectionsInTableView:` Method. This method should return one section in the case where there is only one section in your table view.

NOTE
All table views have at least one section. In the case where you have multiple sections in your table view, you would return more than one.

Related to this method are the `tableView:titleForHeaderInSection:` and `tableView:titleForFooterInSection:` methods, which allow you to customize the header and footer, respectively, for each of the sections. Figure 6.2 shows where these headers and footers appear in a typical `UITableView`.

After the `UITableView` has asked for the information about the sections, it then asks for a count of the rows in each of the sections. The method that you implement to give the table view this information is the `tableView:numberOfRowsInSection:` method. This method should return the number of rows in your model for the given section.

Finally, perhaps the most important of these methods is the `tableView:cellForRowAtIndexPath:` method, which provides the actual data from your model to the table view. It does this by creating a `UITableViewCell` and then placing the data into that cell, which is then returned.

Figure 6.2

A typical `UITableView`, in this case, from the iPod Application

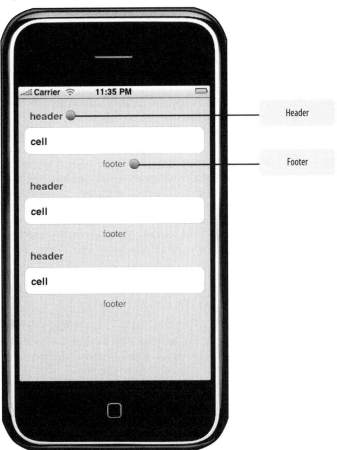

Understanding behavioral methods in UITableViewDataSource

In addition to the data-oriented methods in the data source, there are several methods in `UITableViewDataSource` whose purpose is to allow you to take action on your data in response to particular sorts of editing conditions.

The methods `tableView:commitEditingStyle:forRowAtIndexPath:`, `tableView:canEditRowAtIndexPath:`, and `tableView:didUpdateTextFieldForRowAtIndexPath:withValue:` are used for handling editing and deleting of rows, while the methods `tableView:canMoveRowAtIndexPath:` and `tableView:moveRowAtIndexPath:toIndexPath:` are used for handling the reordering of rows.

When it comes to handling editing and deleting, it's important to recognize that the `UITableViewController` is able to be in either an editing or not-editing state. This is controlled via the `setEditing:animated:` method, which can be set programmatically by you or set automatically by a parent `UINavigationControllers` edit button. I'll talk more about this later, but for now just recognize that when the `UITableViewController` is placed into editing mode, it asks the `UITableViewDataSource` whether each of the currently visible rows are editable or not via the `tableView:canEditRowAtIndexPath:` method. If this method returns YES, then it asks the `UITableViewDelegate` what type of editing is appropriate for that particular row and displays the appropriate editing controls for it. Once the editing is complete, the `UITableView` calls the `tableView:commitEditingStyle:forRowAtIndexPath:` method, which allows you to update the model as well as update the `UITableView` using the methods `insertRowsAtIndexPaths:withRowAnimation:` or `deleteRowsAtIndexPaths:withRowAnimation:` as needed.

NOTE
By default, all table views implement the "Swipe to delete" feature unless you explicitly disable editing of your rows by returning NO from `tableView:canEditRowAtIndexPath`.

The last type of editing that is supported by `UITableView` by default is the ability to reorder the rows. Again, the `UITableView` controller has to be in editing mode in order to enable reordering of rows, but once it is, if the method `tableView:canMoveRowAtIndexPath:` returns YES, then the user can move the rows up and down in the table view. When the user decides where she wants to place the cell and lifts her finger, the method `tableView:moveRowAtIndexPath:toIndexPath:` is called with the index path of where the row originally was and the index path of where the user wants to put it as arguments. If you want to disallow moving this row to the new location, the method `tableView:targetIndexPathForMoveFromRowAtIndexPath:toProposedIndexPath:` is called as the user is moving the row up and down in the table, and you can choose to return NO from this method to prevent the user from dropping the row in locations where you don't want it. I'll talk about this more when I talk about editing your data.

Exploring UITableViewDelegate

The `UITableViewDelegate` is used to customize the behavior of your `UITableView`. It controls things like how to react when a cell is touched, or whether a particular cell can or cannot be edited. Like the `UITableViewDataSource`, it has a variety of methods that you are able to implement, depending upon your needs. Most of the methods are optional. Of these methods, the ones that you'll most frequently utilize manage selections and allow for editing of table rows.

Customizing selection behavior

To allow or disallow a user from selecting an item in the UITableView, you implement the method tableView:willSelectRowAtIndexPath:. This method is passed the index Path of the row that the user would like to choose. If you want to allow the user to select the row, then you return that index path. Alternatively, if you want to redirect their selection to another row, you can return a different index path that points to the row you want them to select. Finally, if you want to disallow the selection altogether, simply return nil.

Once the user has made their selection, the method tableView:didSelectRowAtIndex Path: is called. It's from this method that you should allocate the view controller for the new view that the user will be navigating to from the selection, and then push it on to the navigation controller.

NOTE
When you have finished creating the new view, and have pushed it on to the navigation controller, you need to explicitly deselect the selected row in the table using the method deselectRowAtIndexPath:animated: on the UITableView.

Customizing editing behavior

When it comes to handling editing of rows in the UITableView, the methods provided by the UITableViewDelegate center primarily on handling the behavior surrounding deletion and reordering of the rows. Perhaps the most important of these methods is tableView: editingStyleForRowAtIndexPath:, which allows for the customization of the editing style for a particular cell. When overriding this method, you are expected to return an appropriate UITableViewCellEditingStyle value. The possible values for this are listed in Table 6.2.

Table 6.2 Possible Return Values for Editing Style

Setting	Description
UITableViewCellEditingStyleNone	No editing control is displayed.
UITableViewCellEditingStyleDelete	A minus sign within a red circle is displayed next to the row.
UITableViewCellEditingStyleInsert	A plus sign within a green circle is displayed next to the row.

NOTE
Displaying no edit control doesn't mean the cell disallows editing, only that no control is displayed.

One of the unique gestures utilized on the iPhone user interface is that of the "swipe to delete" gesture. This is enabled on the UITableView to allow deletion of rows and is initiated by the user when she swipes her finger horizontally across a row. When this occurs, the method tableView:willBeginEditingRowAtIndexPath: is called, allowing you to make adjustments to your user interface to account for the delete button or indentation of the rows. When editing has finished, the user interface can subsequently be readjusted back again by implementing the method tableView:didEndEditingRowAtIndexPath: and making your adjustments there.

Finally, if you need to customize the title of the delete button when utilizing this gesture, the method tableView:titleForDeleteConfirmationButtonForRowAtIndexPath: can be implemented to return an appropriate title.

NOTE
If you choose to customize the title for the delete confirmation button, be sure to localize it, too!

Handling reordering of rows

The last customizable interaction that I want to discuss is that of moving rows from one location to another. The method tableView:targetIndexPathForMoveFromRowAtIndexPath:toProposedIndexPath: can be implemented to perform your actual movement of the row, both in the model and in the table view.

Adding a UITableView to an Application

Now that you've reviewed the basics of the UITableViewDelegate and UITableViewDataSource and seen some of the most important methods that you'll need to implement on them, let's take a look at how you might add a UITableView to the favorite cities example application. For the first attempt at this, you'll just display your data and allow favorites to be set. Later, when you add a UINavigationController to your application, you'll also add an edit button and enable editing the rows and moving them around.

Just like with your UIView-based version of the application, you're going to need to create a view controller for your new UITableView. In this case, however, instead of inheriting from

`UIViewController`, you will inherit from `UITableViewController`, which is a subclass of `UIViewController`. Because it's a subclass, it can be used in the same way that you used the `UIViewController` before. The only difference is that instead of controlling a plain old `UIView`, it will control your `UITableView`.

Under many circumstances, it's typical for developers to also make the `UITableView Controller` act as the `UITableViewDelegate` or the `UITableViewDataSource`, or even both. This is convenient and can also result in cleaner code. In this particular case, however, I'm not going to do that because I want to show each of these different roles individually, and I want it to be very clear when you are working on each of the different components. So in this case, I'm going to have separate objects for both the `UITableViewDelegate` and the `UITableViewDataSource`. These two objects will both be owned by the `UITable ViewController`.

Making a UITableViewDataSource

The first thing you need to do is to create the `UITableViewDataSource`, which will wrap around your city list model. To do this, you'll create an object that has a reference to your model array, and then in your implementation you will use the model array to implement each of the important methods that you saw in the previous section for serving data to your `UITableView`.

So add a new Objective-C class to your project and call it MyViewDataSource. The interface for this class is shown in Listing 6.1.

Listing 6.1

The Data Source Interface

```
#import <Foundation/Foundation.h>
@interface MyViewDataSource : NSObject <UITableViewDataSource>
{
    NSMutableArray *favoriteCities;
}
@property (retain, nonatomic) NSMutableArray * favoriteCities;
-(id)initWithFavoriteCities:(NSMutableArray *)inFavoriteCities;
- (NSInteger)numberOfSectionsInTableView:(UITableView *)tableView;
- (NSInteger)tableView:(UITableView *)tableView numberOfRowsInSection:(NSInteger)
    section;
- (UITableViewCell *)tableView:(UITableView *)tableView cellForRowAtIndexPath:(NS
    IndexPath *)indexPath;
@end
```

NOTE

I'm having you change the member variable `theData` to `favoriteCities` because that's more descriptive of what we're working with here. Later, you'll need to update your application delegate to account for this. I'll show you how.

Note that you are declaring that you are implementing the `UITableViewDataSource` protocol, and declaring the methods that you are implementing.

Writing the UITableViewDataSource implementation

Jumping over to the implementation file for your `UITableViewDataSource`, you will go through each of the important methods that you previously discussed to put data into your `UITableView`.

The first method to implement is the numberOfSectionsInTableView method, which we will implement to return one. Remember again that all `UITableView`s have at least one section. Later I will show how you implement the `UITableView` with multiple sections, but for now, you will work with one. Listing 6.2 shows the implementation of this method for your application.

Listing 6.2

Code to Return the Number of Sections

```
- (NSInteger)numberOfSectionsInTableView:(UITableView *)tableView
{
    return 1;
}
```

For this version of your application, because you only have one section, you don't need to do anything with headers or footers for that section. So the next thing on the list that you need to implement is the method that returns the number of rows in your section, `tableView:numberOfRowsInSection:`. In this case, you're just going to return the count from the array. You don't even have to necessarily look at the section number, because you know that your `UITableView` only has one section. The implementation of this method is shown in Listing 6.3.

Listing 6.3

Implementation of How Many Rows Are in Your Section

```
- (NSInteger)tableView:(UITableView *)tableView
    numberOfRowsInSection:(NSInteger)section
{
return [favoriteCities count];
}
```

This now brings you to the most important method, the one that actually returns the cells for your `UITableView`. That method is the `tableView:cellForRowAtIndexPath:` method. To implement this method, you take the index path and use it to look up the appropriate row in your array. You then use the property list from that row to create a `UITableViewCell` with the appropriate text and return it. Implementation for this method is shown in Listing 6.4.

Listing 6.4

Implementation of the tableView:cellForRowAtIndexPath Method

```
- (UITableViewCell *)tableView:(UITableView *)tableView
        cellForRowAtIndexPath:(NSIndexPath *)indexPath
{
    static NSString *CellIdentifier = @"Cell";

    UITableViewCell *cell = [tableView
                    dequeueReusableCellWithIdentifier:CellIdentifier];
    if (cell == nil)
    {
        cell = [[[UITableViewCell alloc]
                    initWithStyle:UITableViewCellStyleDefault
                    reuseIdentifier:CellIdentifier] autorelease];
    }
    NSDictionary *city = [favoriteCities objectAtIndex:[indexPath row]];

    NSString *cityName = [city objectForKey:@"NAME"];
    NSString *favorite = [city objectForKey:@"FAVORITE"];
    NSString *cellText = [NSString
            stringWithFormat:@"%@ - Favorite: %@", cityName, favorite];
    [[cell textLabel] setText:cellText];
    return cell;
}
```

An important aspect of this method that you need to recognize here is that you are checking to see if you can dequeue an instance of a `UITableViewCell` of the appropriate type before creating a new one. Creating new `UITableViewCells` is actually a relatively expensive process and takes up a great deal of memory, and so it is recommended that you reuse the cells and simply place new data into them as you scroll through your list. Apple has provided the `UITableViewCell` caching mechanism, shown here for this purpose, and this is how you use it.

CAUTION
The cell identifier can be any string here, as long as it is unique to each type of cell you have in your table.

For this first version, you are simply going to say "Favorite:" next to the cities that are marked as favorites. In an upcoming version, you'll do something a little bit fancier by placing a star next to it.

Finally, you need to implement an initializer for your new data source that will take the model object so that it can be passed in when it's initialized. This will be done in the method `initWithFavoriteCities:`, which is shown in Listing 6.5.

Listing 6.5
The Data Source's Initializer

```
-(id)initWithFavoriteCities:(NSMutableArray *)inFavoriteCities;
{
    if(self = [super init])
    {
        [self setFavoriteCities:inFavoriteCities];
    }
    return self;
}
```

Configuring the UITableView in Interface Builder

At this point, you've implemented enough of your data source that if you go ahead and create the `UITableView` in Interface Builder and replace your `UIView` with it, you should be able to actually see some data in your table. So let's go ahead and do this.

To do this, you are going to open your `MyView` nib file in Interface Builder by double-clicking the file inside of Xcode. Once the file loads, you're going to click on the view in the document window and click the delete key to delete it. Then you're going to go to the browser window again, and you're going to find a `UITableView` instance and drag-and-drop that over to the document window.

NOTE
If you had made your `UITableViewController` also be your `UITableViewDataSource` or `UITableViewDelegate`, you could connect the outlet from the `UITableView` to your view controller. Because you are actually creating additional objects for this purpose, you will set them up programmatically in your `viewDidLoad` method.

CAUTION
Be sure to connect the view outlet from the view controller to the `UITableView`.

Once you've set all of this up, save the nib and close Interface Builder.

Now you're going to take a look at your view controller. The first thing that you need to do is change the parent class of your controller from `UIViewController` to `UITableViewController`. This is done in the interface file and is shown in Listing 6.6.

Listing 6.6

The MyViewController Interface

```
#import <UIKit/UIKit.h>
#import "MyViewDataSource.h"
@interface MyViewController : UITableViewController
{
    NSMutableArray *favoriteCities;
    MyViewDataSource *dataSource;
}
@property (retain, nonatomic) NSMutableArray * favoriteCities;
@property (retain, nonatomic) MyViewDataSource * dataSource;
@end
```

NOTE
Don't forget to add the @synthesize for the properties we've defined here.

Next, flip over to the implementation file and find the `viewDidLoad` method. Previously, you were taking the model and simply setting the data from the first element of the model in your view. Now, you will create an instance of your data source object, passing your model to the data source, and then you will set the data source on your `UITableView`. All of this is shown in Listing 6.7.

CAUTION

Remember that to instantiate your MyViewDataSource object, you have to import its header in your MyViewController implementation first. Do this at the top of the file before creating the viewDidLoad method.

Listing 6.7

The viewDidLoad Method

```
-(void)viewDidLoad
{
    [[self navigationItem] setRightBarButtonItem:[self editButtonItem]];
    [self setDataSource:[[MyViewDataSource alloc]
                        initWithFavoriteCities:favoriteCities]];
    [[super tableView] setDataSource:dataSource];
    [super viewDidLoad];
}
```

So, the implementation for your UITableViewController should now look like Listing 6.8. In this code, I've also added the @synthesize declarations for the properties, and I'm making sure to release my member variables appropriately. This is just the first stage of the implementation, but it should be enough to see the application run.

Listing 6.8

UITableViewController Implementation

```
#import "MyViewController.h"
#import "MyViewDelegate.h"
#import "MyViewDataSource.h"
@implementation MyViewController
@synthesize favoriteCities;
@synthesize dataSource;
- (void)dealloc
```

```
{
    [self setFavoriteCities:nil];
    [super dealloc];
}
-(void)viewDidLoad
{
    [[self navigationItem] setRightBarButtonItem:[self editButtonItem]];
    [self setDataSource:[[MyViewDataSource alloc]
                        initWithFavoriteCities:favoriteCities]];
    [[super tableView] setDataSource:dataSource];
    [super viewDidLoad];
}
- (void)viewDidUnload
{
    [self setDataSource:nil];
}
@end
```

Seeing the basic application run

Because in this new view controller you changed the name of "theData" to "favoriteCities," you need to update your application delegate accordingly. You need to change the code in the method application DidFinishLaunching: to call the appropriate accessor for the favorite cities list. Listing 6.9 shows the updates for this method.

Listing 6.9

Updated Application Delegate

```
- (void)applicationDidFinishLaunching:(UIApplication *)application
{
    favoriteCities = [[self makeModel] retain];
    [viewController setFavoriteCities:favoriteCities];
    [window addSubview:[viewController view]];
    [window makeKeyAndVisible];
}
```

Once you've done all this, you can compile the application and run it, and then see the data displayed in the list. It looks something like Figure 6.3.

So this is great; you can see the list of your cities and scroll the list up and down. But you need to be able to set your favorites. This is exactly what you're going to do next.

Figure 6.3
The application, showing your cities in the list

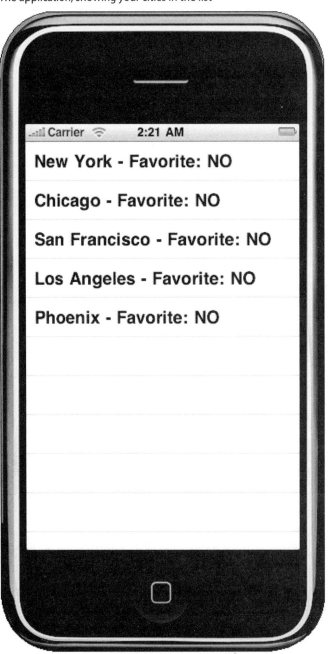

Taking action when a row is touched

Remember that the `UITableViewDelegate` provides you with the functions that enable you to customize how the `UITableView` behaves. So to add the ability for the user to touch a row and toggle the favorite status on a given element, you need to add and implement your `UITableViewDelegate` for this table.

Making the UITableViewDelegate

To add a `UITableViewDelegate`, again right-click on the classes group and choose Add ⇨ New File. Create an `NSObject` child class and name it `MyViewDelegate`. In its header file, specify that it implements the `UITableViewDelegate` protocol, as shown in Listing 6.10.

Listing 6.10

UITableViewDelegate Interface File

```
#import <Foundation/Foundation.h>
@interface MyViewDelegate : NSObject <UITableViewDelegate>
{
    NSMutableArray *favoriteCities;
}
@property (nonatomic, retain) NSMutableArray * favoriteCities;
-(id)initWithFavoriteCities:(NSMutableArray *)inFavoriteCities;
- (NSIndexPath *)tableView:(UITableView *)tableView
  willSelectRowAtIndexPath:(NSIndexPath *)indexPath;
- (void)tableView:(UITableView *)tableView
  didSelectRowAtIndexPath:(NSIndexPath *)indexPath;
@end
```

NOTE
As usual, be sure to declare your @synthesize for the `favoriteCities` property in your implementation.

Again, just like the `UITableViewDataSource`, you're going to hold a reference to your model, and so, naturally, that will need to be passed into the initializer method. This can be implemented just like you did for the data source in Listing 6.5, so I won't show it again here. I'll talk about the actual implementation of your delegate methods in just a moment, but for now, let's return to your `UITableViewController` to the `viewDidLoad` method and create an instance of your delegate and set it as the delegate on the `UITableView`. This is shown in Listing 6.11. You need to add an instance variable to hold the delegate, just like you did for the data source. This is left as an exercise for the reader If you get lost, look at the sample code for the chapter, or review the previous section and mimic what you did for the data source.

Listing 6.11

Updated viewDidLoad Method

```
-(void)viewDidLoad;
{
    [[self navigationItem] setRightBarButtonItem:[self editButtonItem]];
    [self setDataSource:[[MyViewDataSource alloc]
                        initWithFavoriteCities:favoriteCities]];
    [self setDelegate:[[MyViewDelegate alloc]
                      initWithFavoriteCities:favoriteCities]];
    [[super tableView] setDataSource:dataSource];
    [[super tableView] setDelegate:delegate];
    [super viewDidLoad];
}
```

CAUTION
Again, remember that to instantiate your delegate, you have to import it's header file in the MyViewController implementation.

Once you've done this, you will go back to your delegate class and begin implementing the methods that you're interested in.

Allowing the user to select a row

Recall that the `UITableViewDelegate` provides the method `tableView:willSelectRowAtIndexPath:` to enable you to take action as the user is selecting a row in the table. Specifically, this method gets called when the user has touched the row but before the selection has been confirmed. What you need to do here is to essentially tell the `UITableView` that it's okay for the user to select this row. Currently in your application, you're going to let the user select all of the rows. So all you need to do here is to simply return the index path that is passed into you. Therefore, the implementation for this method is very simple and is shown in Listing 6.12.

Listing 6.12

Implementation of the willSelectRowAtIndexPath Method

```
-  (NSIndexPath *)tableView:(UITableView *)tableView
   willSelectRowAtIndexPath:(NSIndexPath *)indexPath;
{
    return indexPath;
}
```

Updating the model when the row is touched

The method that actually does the heavy lifting of row selection is the method `tableView:didSelectRowAtIndexPath:`. It is from this method that you actually take whatever action is necessary based on the selection. This method is called after the `tableView:willSelectRowAtIndexPath:` method has been called and it has returned the index path that you actually want the user to select. In your case, you are going to use this method to toggle the value of "FAVORITE" for the selected city. You will then tell the table view to reload its data so that the updated status can be shown on the UI. Implementation of this method is shown in Listing 6.13.

Listing 6.13

The didSelectRowAtIndexPath Method Allowing You to Update Your Favorites

```
- (void)tableView:(UITableView *)tableView
didSelectRowAtIndexPath:(NSIndexPath *)indexPath;
{
    NSMutableDictionary *cityData =
            [favoriteCities objectAtIndex:[indexPath row]];
    if([[cityData objectForKey:@"FAVORITE"] isEqualToString:@"NO"])
        [cityData setObject:@"YES" forKey:@"FAVORITE"];
    else
        [cityData setObject:@"NO" forKey:@"FAVORITE"];

    [tableView reloadData];
    [tableView deselectRowAtIndexPath:indexPath animated:YES];
}
```

NOTE
The last line of this method explicitly deselects the selected row. This is always necessary in implementing this method. If you do not deselect the selected row, then the row will stay selected in the `UITableView`. Applications have been known to be rejected by Apple for this error.

Running the application and selecting favorites

If you've done everything correctly, and wired everything together correctly, then at this point you can run your application and have it display your list of cities. You can also scroll the list and choose items on the list to set them as favorites. Congratulations! You've implemented the basics of a `UITableView`.

Remember that in this particular case, you separated out the data source and the delegate from your `UITableViewController`. This makes it easy to see what methods are implemented in what portion of the code. You may find it more straightforward to simply make the `UITableViewController` adopt these protocols. This is perfectly acceptable, though you may find for a very complicated model that it makes your `UITableViewController` a bit more bloated. The point here is that you have flexibility to do it either way, so choose what is right for your code. You can continue the examples with these components separated for now.

Configuring a UITableView to Be Grouped

Something that might be fun to add to this particular application would be to group the cities by state. This also provides you with a convenient example of how to make a grouped table view.

To implement a grouped table view, the first thing you need to do is to open your table view nib (you can find it in the Resources group in Xcode; it should be called "MyView.xib") in Interface Builder again and select your table view. From the Properties window, select the first tab, the Attributes tab, and find the setting for Style. Click this drop-down list and choose Grouped instead of Plain.

If you now save your nib and run your application again, you see that you now have a grouped list. This is exciting, but it's really just one group right now. In order to implement multiple groups, you need to change your model and then change your data source. Let's do that now.

Adding the state to your model

To keep things simple, you're going to modify your model so that your top-level dictionaries will actually be for each state. Each state dictionary will then have an array within it that contains the cities in that state. The city dictionaries will be exactly the same as they were before, containing the name of the city and whether or not it's a favorite. The state dictionaries will have the state name and the city array.

The new implementation for your `makeModel` method is shown in Listing 6.14. It's a bit repetitive, but it's straightforward to read.

Listing 6.14
The Updated makeModel Method

```
-(NSMutableArray *)makeModel;
{
    NSMutableDictionary *state;
    NSMutableDictionary *city;
    NSMutableArray *cities;
    NSMutableArray *ret = [NSMutableArray array];
    // New York
    state = [NSMutableDictionary dictionary];
    [state setObject:@"New York" forKey:@"NAME"];
    city = [NSMutableDictionary dictionary];
    [city setObject:@"New York" forKey:@"NAME"];
    [city setObject:@"NO" forKey:@"FAVORITE"];
    cities = [NSMutableArray arrayWithObject:city];
    [state setObject:cities forKey:@"CITIES"];
    [ret addObject:state];

    // Chicago
    state = [NSMutableDictionary dictionary];
    [state setObject:@"Illinois" forKey:@"NAME"];
```

```
    city = [NSMutableDictionary dictionary];
    [city setObject:@"Chicago" forKey:@"NAME"];
    [city setObject:@"NO" forKey:@"FAVORITE"];
    cities = [NSMutableArray arrayWithObject:city];
    [state setObject:cities forKey:@"CITIES"];
    [ret addObject:state];

    // San Francisco and Los Angeles
    state = [NSMutableDictionary dictionary];
    [state setObject:@"California" forKey:@"NAME"];
    city = [NSMutableDictionary dictionary];
    [city setObject:@"San Francisco" forKey:@"NAME"];
    [city setObject:@"NO" forKey:@"FAVORITE"];
    cities = [NSMutableArray arrayWithObject:city];

    city = [NSMutableDictionary dictionary];
    [city setObject:@"Los Angeles" forKey:@"NAME"];
    [city setObject:@"NO" forKey:@"FAVORITE"];
    [cities addObject:city];
    [state setObject:cities forKey:@"CITIES"];
    [ret addObject:state];

    // Phoenix
    state = [NSMutableDictionary dictionary];
    [state setObject:@"Arizona" forKey:@"NAME"];
    city = [NSMutableDictionary dictionary];
    [city setObject:@"Phoenix" forKey:@"NAME"];
    [city setObject:@"NO" forKey:@"FAVORITE"];
    cities = [NSMutableArray arrayWithObject:city];
    [state setObject:cities forKey:@"CITIES"];
    [ret addObject:state];

    return ret;
}
```

Updating your UITableViewDataSource

Now that you've changed your model to support the additional data, you need to change your data source and your delegate so that rather than accessing the cities from the top level of your array, you'll dig into the state and then access the city from there.

Getting the group names from your model

Recall that the first method that is called from the UITableView when it needs to access the data in the model is the method numberOfSectionsInTableView:. Now that you want to have more than one section in your table view, you need to change this method so that it returns the number of states in your model. To do this, instead of just returning one, you will return the count from the top array in your model. This is shown in Listing 6.15.

Listing 6.15

The Updated numberOfSectionsInTableView Method

```
- (NSInteger)numberOfSectionsInTableView:(UITableView *)tableView
{
    return [favoriteCities count];
}
```

Next, you have to implement a new method that you didn't need before; that method is `tableView:titleForHeaderInSection:`. This method returns the title that you want to have show up at the top of each section; in your case, you want this to be the state name. The implementation for this is shown in Listing 6.16.

Listing 6.16

Getting a Title for Each Header

```
- (NSString *)tableView:(UITableView *)tableView
titleForHeaderInSection:(NSInteger)section;
{
    NSDictionary *state = [favoriteCities objectAtIndex:section];
    return [state objectForKey:@"NAME"];
}
```

Showing the correct value in the cells

The next thing you need to do is to update your `tableView:cellForRowAtIndexPath:` method to reflect the fact that you now need to look first at the section and then the row in your index path in order to access the appropriate item in the model. The new implementation of this method is shown in Listing 6.17.

Listing 6.17

Updated Code to Retrieve the Cells for Your Data

```
- (UITableViewCell *)tableView:(UITableView *)tableView
        cellForRowAtIndexPath:(NSIndexPath *)indexPath
{
    static NSString *CellIdentifier = @"Cell";

    UITableViewCell *cell =
            [tableView dequeueReusableCellWithIdentifier:CellIdentifier];
```

```
    if (cell == nil)
    {
        cell = [[[UITableViewCell alloc]
            initWithStyle:UITableViewCellStyleDefault
            reuseIdentifier:CellIdentifier] autorelease];
    }
    NSDictionary *state = [favoriteCities objectAtIndex:[indexPath section]];
    NSArray *cities = [state objectForKey:@"CITIES"];
    NSDictionary *city = [cities objectAtIndex:[indexPath row]];

    NSString *cityName = [city objectForKey:@"NAME"];
    NSString *favorite = [city objectForKey:@"FAVORITE"];
    NSString *cellText =
        [NSString stringWithFormat:@"%@ - Favorite: %@", cityName, favorite];
    [[cell textLabel] setText:cellText];
    return cell;
}
```

Updating the delegate

Finally, you need to update your delegate so that when you touch a city, you are updating the appropriate element in the model. The new implementation for this is shown in Listing 6.18.

Listing 6.18

Updated didSelectRowAtIndexPath Method

```
- (void)tableView:(UITableView *)tableView
didSelectRowAtIndexPath:(NSIndexPath *)indexPath;
{
    NSDictionary *state = [favoriteCities objectAtIndex:[indexPath section]];
    NSArray *cities = [state objectForKey:@"CITIES"];
    NSMutableDictionary *cityData = [cities objectAtIndex:[indexPath row]];
    if([[cityData objectForKey:@"FAVORITE"] isEqualToString:@"NO"])
        [cityData setObject:@"YES" forKey:@"FAVORITE"];
    else
        [cityData setObject:@"NO" forKey:@"FAVORITE"];

    [tableView reloadData];
    [tableView deselectRowAtIndexPath:indexPath animated:YES];
}
```

Once you've done all of this, the new version of your application should list your cities grouped by state, as shown in Figure 6.4.

Figure 6.4
The finished app with grouped table view

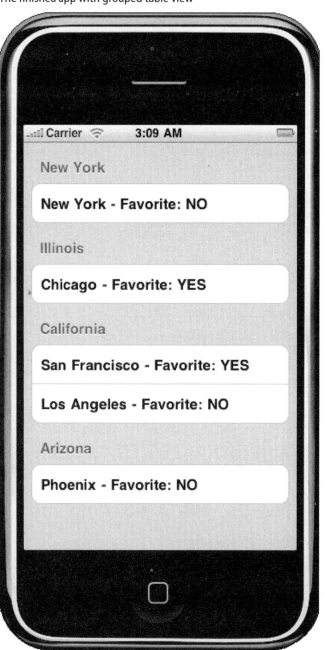

Doing Advanced UITableView Configuration

The greatest customization of a `UITableView` can usually be done by customizing the `UITableViewCells` that are used within it. You'll see how to do this in the next chapter, but first, I want to take a look at a couple of the things that can be customized on the list. Specifically, I'm going to show you how to add an alphabetical list to the side of your table. This makes it really handy for managing large quantities of data, as it gives the user the opportunity to jump to a particular section of the table view without having to scroll through the entire list. Secondly, you'll add a search box to the top of your table view so that the user can search for a particular element.

Adding an alphabetical list to the side

Adding an alphabetical index list to the side of `UITableView` is incredibly easy. You've probably seen these index lists in things like the address book.

To do this, you simply implement the UITableViewDataSource method `sectionIndex TitlesForTableView:`, returning an array of strings that will be used as the header labels for each of the sections. These will also be used as the alphabetical list on the side of the table view. You could, for example, return an array with the letters A to Z. In this case, you could return the first letter of each of your states, as shown in Listing 6.19.

Listing 6.19

sectionIndexTitlesForTableView Implementation

```
- (NSArray *)sectionIndexTitlesForTableView:(UITableView *)tableView;
{
    NSMutableArray *names = [NSMutableArray array];
    for(NSDictionary *state in favoriteCities)
    {
        NSString *stateName = [state objectForKey:@"NAME"];
        NSString *letter = [stateName substringToIndex:1];
        [names addObject:letter];
    }
    return names;
}
```

NOTE
Don't actually add this to your example application, because when you implement your swipe-to-delete support, the delete button can sometimes cause problems with the index list. So leave this out of your example code for now. I just wanted to show you how to do it.

Adding search

Adding search to `UITableView` is slightly less straightforward but still not particularly complicated. Essentially, what you need to do is to add a `UISearchBar` to your nib file, and then assign that to be the custom view for your header using the `UITableView` property `tableHeaderView`. This a special view displayed above the `UITableView`, such that it appears at the top of the list when you scroll to the top. You then configure your data source to be your `UISearchBarDelegate`. As you receive events from the UI search bar, you filter your data source model accordingly. The UI search bar delegate method that you need to implement is the `searchBar:textDidChange:` method, which is called as the user is editing the text.

The standard Apple applications also add an overlay over the initial table view when searching begins to prevent the user from touching the list while search is active. You can easily do this by adding an overlay view into your nib and displaying it over top of your `UITableView`. The method to implement in the UI search bar delegate to do this is the method `searchBarTextDidBeginEditing:`.

If you choose to do this, be sure to hide this overlay view when the user actually begins typing his search string so that he can choose an item from the filtered list. In addition to this, you may want to implement on your overlay view a handler for touch events such that if the user touches on the overlay view without having typed any search criteria, it clears the search and returns him to the main list. For more information about handling touch events, see Chapter 11.

Summary

In this chapter, you've taken an in-depth look at the `UITableView` class and its sister classes `UITableViewDataSource` and `UITableViewDelegate`. Using these classes, you can display lists of data in your applications efficiently and easily. In the next few chapters, you will see how to add further customization to your `UITableView` using custom `UITableViewCells`, and then you'll see how you can use your `UITableView` in conjunction with a `UINavigationController` to display and navigate large trees of hierarchical data. This chapter, however, has given you a solid foundation in how to utilize `UITableView`s in your application.

Working with UITableViewCells

In the last chapter, you looked at how you could add a `UITableView` to your application to display a list of data. The way that you used the `UITableView` was pretty basic. You didn't do anything fancy in terms of customizing how it looks or anything like that. Naturally, it is possible to add quite a bit of customization to `UITableView`s, and in particular, the cells that make up the individual rows in the table view. And, it is possible to add quite a bit of customization to `UITableView`. Most of that customization applies to the appearance and behavior of the cells that make up the rows in the `UITableView`. You do this by customizing the `UITableViewCell` objects that you return from the `UITableViewDataSource` when displaying the data. This is an incredibly powerful method for customizing your `UITableView` and will be the focus of this chapter.

You'll begin exploring the `UITableViewCell` class by looking at the parts that make up a `UITableViewCell` and how you can use those parts to provide a custom appearance. You will then see how you can use inheritance to create a custom `UITableViewCell` that enables you to do virtually anything that you want to.

Understanding the Parts of a UITableViewCell

Before you begin customizing a `UITableViewCell`, you need to get a feel for the lay of the land. That is to say, you need to know what the parts of a typical `UITableViewCell` are, how they behave, and where they are located in the cell. Figure 7.1 outlines the typical layout of the subviews of a `UITableViewCell`.

In This Chapter

Customizing a UITableView's appearance

Exploring options for laying out UITableViewCells

Optimizing UITableView performance by using a reuse identifier with cells

Exploring UITableViewCell inheritance

Figure 7.1

Parts of a `UITableViewCell`

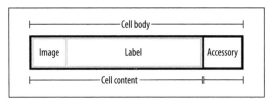

As Figure 7.1 shows, inside the `UITableViewCell` there are two primary subviews. On the far right is the accessory view. The accessory view is where a button is displayed if you configure the cell to have an accessory view. Typically this might be something like an indicator, which shows that there is more data to be displayed. When the table view cell is placed into editing mode, this accessory view is used to display the reordering control, or the delete button when a row is selected for deletion and awaiting confirmation.

The view to the left, which takes up the rest of the cell, is the content view. This is where the content of your cell goes, and where the majority of your customization should be done.

NOTE
The content view automatically resizes when necessary to make room for the accessory view. When customizing, you should take this into account, meaning that you should also resize any custom subviews you have created.

When you construct a `UITableViewCell`, you have the ability to pass a style parameter to the constructor. The possible values for this style parameter are listed in Table 7.1. The types of subviews contained within the content view vary, depending upon the type of cell that you have requested.

Table 7.1 Cell Styles

Value	Description
`UITableViewCellStyleDefault`	A simple cell with just a single text field and an image on the left side. In iPhone OS 2.x, this was the only style available.
`UITableViewCellStyleValue1`	This style of cell contains a label on the left side of the cell with normal black text, and a label on the right side of the cell with smaller blue text. This is similar to the types of cells used in the settings application.
`UITableViewCellStyleValue2`	This style of cell contains blue text on the left side and black text on the right side. The text on the right side is small. This is similar to how the contacts application works.
`UITableViewCellStyleSubtitle`	This style of cell contains a label across the top and another label below it in smaller gray text.

NOTE
The ability to choose different styles of `UITableViewCell` is new in iPhone OS 3. Prior to this release, only the default cell style was supported.

Each of the subviews shown in the cells in Table 7.1 can be fully customized to change various parameters, including their background color, their text color, and their font. This provides you with a fair amount of configurability without doing any inheritance whatsoever.

In addition to simply changing the properties of the existing views that are already part of the `UITableViewCell`, you can even add your own custom subviews directly to the content view. This is handy for when you want to do things such as add a progress bar to a `UITableViewCell`. To do this, you simply create your additional custom view and add it as a subview to the content view.

CAUTION
When adding custom subviews, be very careful about where you construct your subviews. You don't want to be re-creating the subview for every cell. Try to reuse your subviews as much as possible. Object creation can be surprisingly hard on performance on the iPhone.

As you may have noticed in Table 7.1, in a default `UITableViewCell` there is an optional `UIImageView`. This is enabled by setting the image property on the `UITableViewCell`. In the next section, you will use this `UIImageView` to set a star next to cities that are your favorites in your favorite cities application.

Adding Images to a UITableViewCell

Up to now, in your cities application, to indicate a favorite city you've simply added the word *favorite* next to the city name in your display. Wouldn't it be great if, instead, you could just add a graphical element to each row in your list that displayed a star. When you've marked a favorite city, perhaps the star can be filled in with color. When a city is not a favorite, the star can just be an outline. Let's take a look at how you would implement this behavior.

CROSS-REF
The images that you will be using in this example can be found on the book's Web site in the example code for this chapter. In order to use these images in your project, be sure to add them to your Xcode project. A good place for them is under the Resources group.

In the previous section, when you looked at the default style of `UITableViewCell`, there was an optional `UIImageView`, which displays on the left side of the cell if the image property has been set. This is what you will use to display your star. Listing 7.1 shows your previous implementation of the method `tableView:cellForRowAtIndexPath`, which simply returns each row with a text value listing the city name and then whether or not it's a favorite.

Listing 7.1

Method tableView:cellForRowAtIndexPath: Implementation

```
- (UITableViewCell *)tableView:(UITableView *)tableView
        cellForRowAtIndexPath:(NSIndexPath *)indexPath
{
    static NSString *CellIdentifier = @"Cell";

    UITableViewCell *cell = [tableView
                    dequeueReusableCellWithIdentifier:CellIdentifier];
    if (cell == nil)
    {
        cell = [[[UITableViewCell alloc]
                initWithStyle:UITableViewCellStyleDefault
                reuseIdentifier:CellIdentifier] autorelease];
    }
    NSDictionary *state = [favoriteCities objectAtIndex:
                                        [indexPath section]];
    NSArray *cities = [state objectForKey:@"CITIES"];
    NSDictionary *city = [cities objectAtIndex:[indexPath row]];

    NSString *cityName = [city objectForKey:@"NAME"];
    NSString *favorite = [city objectForKey:@"FAVORITE"];
    NSString *cellText =
      [NSString stringWithFormat:@"%@ - Favorite: %@", cityName, favorite];
    [[cell textLabel] setText:cellText];
    return cell;
}
```

In your new version, you are simply going to change the cell text to only show the city name, and then you are going to set the cell's image property to the appropriate image based on whether or not the city has been selected as a favorite. Listing 7.2 shows how to do this.

NOTE

The `+[UIImage imageNamed:]` method automatically caches the images it loads so that subsequent calls to it do not have to go to the disk, but instead return the image from memory if an image has been previously loaded. Because of this, it's acceptable to use it even in such performance-critical areas of code like this.

Listing 7.2

Method tableView:cellForRowAtIndexPath: Implementation

```
- (UITableViewCell *)tableView:(UITableView *)tableView
        cellForRowAtIndexPath:(NSIndexPath *)indexPath
{
    static NSString *CellIdentifier = @"Cell";

    UITableViewCell *cell = [tableView
                    dequeueReusableCellWithIdentifier:CellIdentifier];
    if (cell == nil)
    {
        cell = [[[UITableViewCell alloc]
                    initWithStyle:UITableViewCellStyleDefault
                    reuseIdentifier:CellIdentifier] autorelease];
    }
    NSDictionary *state =
            [favoriteCities objectAtIndex:[indexPath section]];
    NSArray *cities = [state objectForKey:@"CITIES"];
    NSDictionary *city = [cities objectAtIndex:[indexPath row]];

    NSString *cityName = [city objectForKey:@"NAME"];
    [[cell textLabel] setText:cityName];
    if([[city objectForKey:@"FAVORITE"] isEqualToString:@"YES"])
        [cell setImage:[UIImage imageNamed:@"FullStar.png"]];
    else
        [cell setImage:[UIImage imageNamed:@"EmptyStar.png"]];

    return cell;
}
```

> **NOTE**
> iPhone OS 3 added several new views to the default `UITableViewCell` implementation to allow for much more configuration than was possible in OS 2.x. In particular, the `backgroundView`, `selectedBackgroundView`, `textColor`, `selectedTextColor`, `selectedImage`, and `detailTextLabel` properties give you a lot of possibilities.

Once you've updated the code to this new version, you should be able to run the application and see a full star next to your favorites and an empty star next to the ones that are not your favorites, as shown in Figure 7.2.

Figure 7.2
The finished favorite cities app with stars next to the cities.

Performing Deeper Customization of UITableViewCells

You've seen how you can do some basic customization of UITableViewCells using the existing views that are part of UITableViewCell and also by adding your own. But what if you want to go even further? What if you want to introduce completely new behavior, for example, when you want the cell to behave differently than normal when it is touched. It is here that you enter the realm of using inheritance to customize your UITableViewCells.

To create your own custom cell that inherits from `UITableViewCell`, simply create a new class and declare it as a child class of `UITableViewCell`, as shown in Listing 7.3.

Listing 7.3

CustomUITableViewCell Header

```
#import <UIKit/UIKit.h>

@interface CustomUITableViewCell : UITableViewCell
{
    UILabel *customLabelOne;
    UILabel *customLabelTwo;
    UIImageView *customImage;
}
@property (nonatomic, retain) UILabel * customLabelOne;
@property (nonatomic, retain) UILabel * customLabelTwo;
@property (nonatomic, retain) UIImageView * customImage;
@end
```

An important thing to realize when inheriting from `UITableViewCell` is that, although you can construct your subviews in your designated initializer — which for the `UITableViewCell` class is `initWithStyle:reuseIdentifier:reuseIdentifier` — when you do this, you should probably pass `CGRectZero` as the frame for your view. This initializes your subview with an empty frame. You should then subsequently adjust the frame of your view in the `layoutSubviews` method. This enables you to use the content view as the basis for how to size your own views. Doing this will ensure that your subviews adjust their sizes accordingly when accessory views are displayed.

Listing 7.4 shows a typical implementation of a custom `UITableViewCell` with its designated initializer and its `layoutSubviews` method. In this example, you're making a custom cell with a colored background and some additional graphics and text in it.

Listing 7.4

CustomUITableViewCell Implementation

```
#import "CustomUITableViewCell.h"
@implementation CustomUITableViewCell
@synthesize customLabelOne;
@synthesize customLabelTwo;
@synthesize customImage;
(id)initWithFrame:(CGRect)frame
    reuseIdentifier:(NSString *)reuseIdentifier
{
```

continued

Listing 7.4   *(continued)*

```objc
        if (self = [super initWithFrame:frame reuseIdentifier:reuseIdentifier])
        {
            customLabelOne = [[UILabel alloc] initWithFrame:CGRectZero];
            [customLabelOne setBackgroundColor:[UIColor lightGrayColor]];
            [customLabelOne setFont:[UIFont boldSystemFontOfSize:
                                    [UIFont labelFontSize]]];
            [customLabelOne setLineBreakMode:UILineBreakModeWordWrap];
            [customLabelOne setTextAlignment:UITextAlignmentLeft];
            [[self contentView] addSubview:customLabelOne];

            customLabelTwo = [[UILabel alloc] initWithFrame:CGRectZero];
            [customLabelTwo setBackgroundColor:[UIColor lightGrayColor]];
            [customLabelTwo setFont:[UIFont boldSystemFontOfSize:
                                    [UIFont labelFontSize]-8]];
            [customLabelTwo setLineBreakMode:UILineBreakModeWordWrap];
            [customLabelTwo setTextAlignment:UITextAlignmentLeft];
            [[self contentView] addSubview:customLabelTwo];

            customImage = [[UIImageView alloc] initWithImage:
                            [UIImage imageNamed:@"SomeImage.png"]];
            [customImage setBackgroundColor:[UIColor lightGrayColor]];
            [[self contentView] addSubview:customImage];

            [[self contentView] setBackgroundColor:[UIColor lightGrayColor]];
        }
        return self;
    }
    - (void)dealloc
    {
        [self setCustomLabelOne:nil];
        [self setCustomLabelTwo:nil];
        [self setCustomImage:nil];
        [super dealloc];
    }
    -(void)layoutSubviews;
    {
        [super layoutSubviews];
        [customLabelOne setFrame:CGRectMake(0, 0, 160, 40)];
        [customLabelTwo setFrame:CGRectMake(0, 0, 160, 40)];
        [customImage setFrame:CGRectMake(300, 0, 20, 20)];
    }
    - (void)setSelected:(BOOL)selected animated:(BOOL)animated {
        [super setSelected:selected animated:animated];
        // Configure the view for the selected state
    }
    @end
```

Once you have your custom `UITableViewCell`, you can use it in your table view by simply allocating an instance of it when necessary, instead of the default `UITableViewCell`. An example of this is shown in Listing 7.5.

Listing 7.5

Updated tableView:cellForRowAtIndexPath: Method

```objc
- (UITableViewCell *)tableView:(UITableView *)tableView
        cellForRowAtIndexPath:(NSIndexPath *)indexPath
{
    static NSString *CellIdentifier = @"Cell";

    CustomUITableViewCell *cell =
            [tableView dequeueReusableCellWithIdentifier:CellIdentifier];
    if (cell == nil)
    {
        cell = [[[CustomUITableViewCell alloc]
                    initWithStyle:UITableViewCellStyleDefault
                  reuseIdentifier:CellIdentifier] autorelease];
    }

    [[cell customLabelOne] setText:@"label one"];
    [[cell customLabelTwo] setText:@"label two"];
    return cell;
}
```

As you can see, this is all very straightforward.

In some cases, it may be necessary to have the appearance of the cell change when the cell is selected. To do this, simply override the method `setSelected:animated:` to take whatever action is necessary to display the cell differently for the selection. In Listing 7.6, you're going to set your background color to a different color when the cell is selected.

Listing 7.6

Method setSelected:animated: Implementation

```objc
- (void)setSelected:(BOOL)selected animated:(BOOL)animated {
    [super setSelected:selected animated:animated];
    if(selected)
    {
        [[self contentView] setBackgroundColor:[UIColor grayColor]];
        [[self customLabelOne] setBackgroundColor:[UIColor grayColor]];
```

continued

Listing 7.6 *(continued)*

```
        [[self customLabelTwo] setBackgroundColor:[UIColor grayColor]];
    }
}
```

Using the inheritance method of customizing `UITableViewCell` doesn't mean that you can't also use the same existing UI labels and other views that you used previously in working with `UITableViewCells`. You can use them both, or either one. It's completely up to you.

Thinking about Performance in Custom Cells

The last thing I want to talk about with regards to customizing `UITableViewCells` is performance. The default `UITableViewCell` implementation is highly optimized for high-performance in scrolling of your `UITableView`. It's very important, when making custom `UITableViewCells`, that you also take similar precautions to make sure that your cells can be drawn quickly. If you don't, the scrolling performance of your `UITableView` can be drastically impacted. There's nothing that kills a great user interface more than having a table view that, when you flick it, sputters and staggers and doesn't feel smooth in its scrolling.

When thinking about performance of `UITableViewCells`, there are really three concepts that you need to keep in mind: reuse, efficiency, and opacity.

The first concept, reuse, means that object allocation can be very expensive on the iPhone, particularly when done repeatedly in a tight loop, such as when scrolling a `UITableView`. This means that the first rule that you have to keep in mind when customizing your `UITableViewCells` is to utilize the cell queuing mechanism and reuse your cells whenever possible. This also spills over into how you handle updating those reused cells. Specifically, you shouldn't remove and add new subviews to your cells whenever you update one of the properties of that cell. For example, if you have a custom cell with an image in it, don't remove the image and then add it back in when reusing that cell. Instead, simply reset the value of that property. Listing 7.7 shows the proper way to handle this.

Listing 7.7

Proper Handling of a Custom Cell

```
- (UITableViewCell *)tableView:(UITableView *)tableView
        cellForRowAtIndexPath:(NSIndexPath *)indexPath
{
```

```
static NSString *CellIdentifier = @"Cell";

UITableViewCell *cell = [tableView
            dequeueReusableCellWithIdentifier:CellIdentifier];
UILabel *customLabel;
if (cell == nil)
{
    cell = [[[UITableViewCell alloc]
                initWithStyle:UITableViewCellStyleDefault
                reuseIdentifier:CellIdentifier] autorelease];
    customLabel =
      [[UILabel alloc] initWithFrame:CGRectMake(0, 0, 50, 50)];
    [customLabel setTag:1];
    [[cell contentView] addSubview:customLabel];
}
else
{
    customLabel = (UILabel *)[[cell contentView] viewWithTag:1];
    [customLabel setText:@""];
}
[customLabel setText:@"new value"];

return cell;
}
```

NOTE
This code utilizes the UIView Tag feature to enable you to retrieve a view from another view's hierarchy using a tag. This is a great way to get views from complex view configurations.

The second concept, efficiency, means that whenever you are doing any sort of drawing, you should always strive to do it in the most efficient manner possible. This means that, for example, if you have a custom view that implements a custom `drawRect` method, you should utilize the tools at your disposal to ensure that you are only drawing as much of that view as is absolutely required. In cases where part of your view is covered by something else, this may mean that you do not have to redraw the entire view; you may only need to draw the portion of the view that is actually exposed. Utilize these tools to be as conservative and as efficient in your drawing operations as possible.

Finally, there is the concept of opacity. Alpha transparency on the iPhone, while useful, is also quite expensive. You should avoid transparency whenever possible. In some cases, this is essentially a free performance gain. For example, if you have text with a white background, over the top of another view that is also white, there's really no need to set the background to be transparent. Instead, set the background to be opaque and set it to be the same color as the view behind it. The user will never know the difference and the GPU of the iPhone will be able to draw the cell significantly faster. If you can imagine a complex cell with a lot of subviews, being drawn with a lot of transparency, the cost of drawing those cells can become very high. You can set your custom views to opaque using the opaque property.

Reusing UITableViewCells

One of the important concepts that I touched on in the previous chapter but that I didn't really go into in a lot of detail was the concept of UITableViewCell reuse.

I'll say it here and I'll say it again later: object allocation on the iPhone is expensive. Particularly, object allocation done in tight loops, such as when scrolling a table view. Additionally, when you have a table view with a lot of elements in it, if you allocated a cell for every one of those elements in the entire data set, your memory use would balloon significantly and quickly. This is why Apple introduced the idea of cell reuse.

The idea behind cell reuse is relatively simple. Essentially, when a cell that is in the UITableView scrolls off of the screen, it's placed into a queue. When you need another cell to display on the screen, you can then ask the queue for a cell. If one is available, you're given that same cell that was placed in the queue when it scrolled off the screen. This means that you only need to allocate enough cells to actually display the current content on the screen. You don't need to have additional cells that aren't being displayed taking up memory for all of the elements in your data set.

This is a powerful concept, and best of all, it's incredibly easy to use in your applications. In fact, all of the examples that I've shown you so far using UITableViews already utilize this capability. You can see it in Listing 7.8 where you first attempt to dequeue a cell from the queue. If you don't get one, then you allocate one, but if you do, you simply reuse the one that you get.

Listing 7.8

Cell Reuse

```
- (UITableViewCell *)tableView:(UITableView *)tableView cellForRowAtIndexPath:(NS
  IndexPath *)indexPath
{
    static NSString *CellIdentifier = @"Cell";

    UITableViewCell *cell =
        [tableView dequeueReusableCellWithIdentifier:CellIdentifier];
    if (cell == nil)
    {
        cell = [[[UITableViewCell alloc]
                    initWithStyle:UITableViewCellStyleDefault
                    reuseIdentifier:CellIdentifier] autorelease];
    }
```

```
NSDictionary *state = 
    [favoriteCities objectAtIndex:[indexPath section]];
NSArray *cities = [state objectForKey:@"CITIES"];
NSDictionary *city = [cities objectAtIndex:[indexPath row]];

NSString *cityName = [city objectForKey:@"NAME"];
[[cell textLabel] setText:cityName];

if([[city objectForKey:@"FAVORITE"] isEqualToString:@"YES"])
    [cell setImage:[UIImage imageNamed:@"FullStar.png"]];
else
    [cell setImage:[UIImage imageNamed:@"EmptyStar.png"]];

return cell;
}
```

CAUTION

When you get that cell and are attempting to reuse it, it may still contain the data from your previous use of the cell. Because of this, you should make sure that you clear any labels that you won't be using prior to setting your values.

Understanding the reuse identifier

In the examples that I've shown where you are reusing `UITableViewCells`, you may have noticed the reuse identifier string. This is how you tell the difference between different kinds of cells in your table. For example, you may have initialized a table of style so-and-so. You would set the `reuseIdentifier` on this cell to be something descriptive about that particular cell type — for example "cell number one." Perhaps the second cell in your table is a different kind of cell, "cell number two." By using different reuse identifiers, you can dequeue a cell of the correct type for reuse.

The reuse identifier is simply a string that is used to determine what kind of cell you want to dequeue. That's all it is. In practice, it's probably a good idea to define this as a constant in your code so that you can get code completion on your identifiers. However, for simple use, you can set it to any kind of string that you want.

Using different types of cells in the same UITableViYou've seen how you can customize `UITableViewCells` and use them in place of the standard cells. But what if you want to mix and match your cells? What if you want to use one type of cell in one section and another kind of cell in another section? Well, the good news is that this is really easy to do.

To use multiple types of cells in a single table view, simply follow the same pattern that you would use if you were allocating one kind of cell, but for the rows where you want to use a different kind of cell, you can allocate the new cell.

You should still use the `dequeueReusableCellWithIdentifier:` method to see if you have a cell of the correct type that can be reused. But if you don't, then simply allocate one using the appropriate `reuseIdentifier`, and return it.

Listing 7.9 shows how you can use two different kinds of cells: a standard cell if you are in section 0 in your table and a custom cell if you are in section 1.

Listing 7.9

Using Different Types of Cells

```
- (UITableViewCell *)tableView:(UITableView *)tableView
        cellForRowAtIndexPath:(NSIndexPath *)indexPath
{
    UITableViewCell *cell = nil;
    if([indexPath section] == 0)
    {
        static NSString *CellIdentifier = @"CellOne";
        UITableViewCell *cell =
          [tableView dequeueReusableCellWithIdentifier:CellIdentifier];
        if (cell == nil)
        {
            cell = [[[UITableViewCell alloc]
                        initWithStyle:UITableViewCellStyleDefault
                        reuseIdentifier:CellIdentifier] autorelease];
        }
    }
    else if([indexPath section] == 1)
    {
        static NSString *CellIdentifier = @"CellTwo";
        UITableViewCell *cell =
          [tableView dequeueReusableCellWithIdentifier:CellIdentifier];
        if (cell == nil)
        {
            cell = [[[CustomCell alloc]
                        initWithStyle:UITableViewCellStyleDefault
                        reuseIdentifier:CellIdentifier] autorelease];
        }
    }
    // ...

    return cell;
}
```

Summary

In this chapter, you have seen how you can customize the appearance of your `UITableView Cells` using both existing `UITableViewCells` and composition and also using inheritance. These two methods give you all the tools necessary to make table views that can represent virtually any kind of data you can throw at them.

Next up, you'll take a look at how you can tie in the `UITableView` into a `UINavigation Controller` to build an application that handles trees of hierarchical data.

Working within the UINavigationController Model

In this chapter, you're going to take an in-depth look at how the `UINavigationController` navigation model works and how to make your `UIViewControllers` fit within that navigational model. You'll begin by exploring the details of the `UINavigationController`. You will then take a look at how to use your `UIViewController` with the `UINavigationController` to navigate your data. When finished with this chapter, you should have a solid, basic understanding of how this fundamental set of classes works and how you can utilize them in your code.

Understanding the UINavigationController Navigational Model

The `UINavigationController` provides an architecture for managing a stack of `UIViewControllers`. Typically, this is used to manage lists or trees of data where you are choosing an item from a list and then subsequently you need to bring in an additional view to edit that item or show detailed information for that item. This is one of the most commonly used GUI tools in Cocoa Touch.

Figure 8.1 shows the `UINavigationController` architecture at work. Specifically, it shows how the `UINavigationController` provides sort of a moving window from view controller to view controller.

In This Chapter

Displaying hierarchical lists of data using UINavigationController

Configuring the UINavigationBar and the buttons that control its navigation

Pushing and popping UIViewControllers on a UINavigationController

Allowing adding, editing, and deleting of rows

Figure 8.1

How a `UINavigationController` works

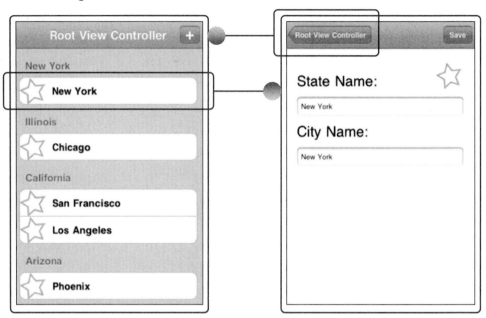

To navigate between the view controllers displaying the data, you push and pop the view controllers onto and off of the `UINavigationController`. When you push a `UIViewController` onto a `UINavigationController`, the `UINavigationController` then displays that `UIViewController` as its primary view. Similarly, when you pop a `UIViewController` off of the `UINavigationController`, it then removes the top `UIViewController` and brings in whatever `UIViewController` was pushed off when that `UIViewController` was pushed on originally.

It may seem complicated, but in fact it's actually a reasonably simple and intuitive interface to use.

Creating a UINavigationController

Xcode provides a template for creating an application with a top-level `UINavigationController`. This is a perfectly acceptable way to create an application with the `UINavigationController`. However, when you create your application for the first time, you don't always know exactly what it is that you want your UI to do. So it's helpful if you learn how to manually create the `UINavigationController` and add it to an existing project. Fortunately, this is extremely easy.

Chapter 8: Working within the UINavigationController Model

Probably the easiest way to do this is to open your `MainWindow.xib` file and drag-and-drop a `UINavigationController` object from the objects palette to the objects browser window. This creates an instance of a `UINavigationController` in your nib. You then create an `IBOutlet` in your application delegate to connect this navigation controller to.

The `UINavigationController` will include a default root `UIViewController` that is prebuilt into it. To make it use your `UIViewController`, you need to configure it to pull your `UIViewController` from its nib.

To do this, you create a separate nib for your `UIViewController`. Typically, this `UIViewController` is stored in a separate nib, not in your main window nib. To add this `UIViewController` to the `UINavigationController`, edit the Nib Name property on the default root view controller object and set it to the name of the nib containing your `UIViewController`. You also need to edit the Class property of the root view controller and set it to be the class of your view controller. The root view controller is a subobject of the navigation controller, as shown in Figure 8.2. In Figure 8.2, it's called "`MyViewController`." With this object selected in the object browser, the NIB name property can be found on the first tab of the info panel, and the Class property can be found on the last tab.

Figure 8.2

Configuring the `UINavigationController`

Once you have created the `UINavigationController`, you need to add the `UINavigation Controller`'s view as a subview of your window so that it is displayed. This can be done in your `applicationDidFinishLaunching` method on your application delegate. A typical implementation is shown in Listing 8.1.

Listing 8.1

Implementation of the applicationDidFinishLaunching Method

```
- (void)applicationDidFinishLaunching:(UIApplication *)application
{
    // Override point for customization after app launch

    [window addSubview:[navigationController view]];
    [window makeKeyAndVisible];
}
```

This code adds the navigation controller as a root view in your application. There are, of course, situations where you don't want this to be the case. For example, if you have a game whose primary interface is a `UIView`, then you probably don't want that game's main view to be wrapped in a `UINavigationController`, but you may want to use a `UINavigation Controller` somewhere deeper in the game, for example, for settings.

To do this, this same pattern can be used anywhere else in your application as well. The view property of the `UINavigationController` is the view that is used to display the `UINavigationController` itself, and can be used anywhere you would use any other UIView. To display your `UINavigationController` in an arbitrary place in your app, add your `UINavigationController` to the nib file where it's going to be displayed, and then add the navigation controller's view as a subview of whatever view you want it to display over the top of.

Typically, you may want to animate the display of the `UINavigationController`. The navigation controller's view is just like any other view and can be animated using the `UIView` animation methods.

The `UINavigationController` even inherits from `UIViewController`. This means, theoretically, that you can also use a `UINavigationController` anywhere that you might use a `UIViewController` instead. This includes, for example, using the `UIViewController` method `presentModalViewController:animated:` to display the `UINavigation Controller` modally.

Configuring a UINavigationController

It's useful here to take a small side trip to discuss the pieces that make up the `UINavigation Controller` and how you configure those pieces.

Figure 8.3

A typical `UINavigationController`

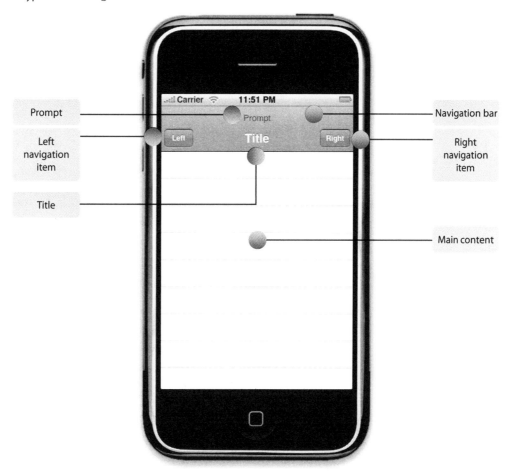

Figure 8.3 shows a typical `UINavigationController`, including the locations of the title, the right and left navigation buttons, and the prompt. Your user content is displayed in the

middle. When a `UIViewController` is pushed onto the `UINavigationController`, a `UINavigationItem` is created as a property of the `UIViewController` and gives access to the elements displayed in the navigation bar. This makes it very convenient to control what elements are displayed in the navigation bar, depending upon the `UIViewController` that is currently being displayed. The `UINavigationController` will interrogate the `UINavigationItem` on the `UIViewController` to find out things like the title, the navigation buttons, and the prompt. If these properties have been set, they will override any settings set on the `UINavigationController`. Table 8.1 shows the properties on the `UINavigationItem` used to configure these settings.

Table 8.1 UINavigationItem Properties

Property Name	Purpose
`Title`	Used to specify the title displayed in the center of the `UINavigationController` navigation bar when this `UIViewController` is displayed
`Prompt`	Specifies a supertitle over the title in the `UINavigationController` navigation bar
`leftBarButtonItem`	Specifies the `UIBarButtonItem` displayed on the left side of the `UINavigationController` navigation bar
`rightBarButtonItem`	Specifies the `UIBarButtonItem` displayed on the right side of the `UINavigationController` navigation bar

The buttons displayed on the `UINavigationBar` are of type `UIBarButtonItem` and are created just like normal buttons, with a target, an action, and so on. `UINavigationController` automatically provides the "back" button that's displayed on the left side of the `UINavigationBar` when the stack of `UIViewControllers` that are being managed by this `UINavigationController` contains more than one `UIViewController`. This enables the user to have a consistent user experience wherein they navigate a list of data, select an item from that list, view the detailed information for that item, and then have a "back" button to take them back to the list.

Finally, you can also choose to show or hide the navigation bar, if you need to.

Pushing and Popping UIViewControllers

Now that you have configured the `UINavigationController`, you come to its raison d'être, which is managing `UIViewController`s.

To cause a `UINavigationController` to display a `UIViewController`, you use the method `pushViewController:animated:`, passing your `UIViewController` as the first argument. This causes the `UINavigationController` to slide the current view controller off to the left and bring in the new view controller from the right to display it as the primary view. Similarly, to remove a view controller from the `UINavigationController`, you use the method `popViewControllerAnimated:`, which removes the top view controller from the stack and slides in whatever view controller was pushed off in order to display that view controller.

If you have multiple levels of view controllers being displayed, the method `popToViewController:animated:` continues popping view controllers off the stack until it finds the view controller you have passed in.

Finally, if you want to remove all of the view controllers except for the root view controller and return to the top level of your stack, the method `popToRootViewControllerAnimated:` does exactly that.

Adding a UIToolbar

With the introduction of iPhone OS 3, Apple added the ability to easily manage a toolbar displayed at the bottom of a `UINavigationController`. By default, the toolbar is hidden, so to display it, all you need to do is call the method `setToolbarHidden:animated:` on the `navigationController` object. The items displayed in the toolbar are retrieved from the currently active `UIViewController` by accessing the `toolbarItems` property. The toolbar items property should contain an array of `UIBarButtonItem` objects and can be set using the method `setToolbarItems:animated:`.

Allowing Editing of the Rows

Now that you've seen the basics of how to work with a `UINavigationController`, let's extend the favorite cities application to add the ability to edit your model. In doing so, you will use the `UINavigationController` as your top-level view, enabling you to select items from your list of cities, and edit them in a subview that you will push onto the `UINavigationController`.

Moving MyView into a UINavigationController

The first thing you have to do in order to add a `UINavigationController` to your example project is to remove your `UIViewController` from your `MainWindow` nib and replace it with a `UINavigationController`.

To do this, open the `MainWindow` nib in Interface Builder, find your `UIViewController` that you added for your `MyView` object, and delete it. Then drag-and-drop a `UINavigationController` from the palette on to the object browser window. Configure the `UINavigationController` so that the root `UIViewController` is now your `MyView`. You also have to configure the Class type on the root `UIViewController` after you have set the root view controller on the `UINavigationController`. Do this using the Class property under the Class Identity settings in the properties window after you have selected the root view controller.

You should set it to be the class of your main UIViewController (if you are following along, this will be MyViewController).

You need to change your `IBOutlet` in your app delegate to now be an instance of `UINavigationController` instead of `MyViewController`. Remember that you still need to add the view for the `UINavigationController` to your main Windows view.

Listing 8.2 shows the completed interface, and Listing 8.3 shows the implementation of the `applicationDidFinishLoading` method.

Listing 8.2

New Application Delegate Interface

```
#import <UIKit/UIKit.h>
@interface FaveCitiesListAppDelegate : NSObject <UIApplicationDelegate>
{
    UIWindow *window;
    UINavigationController *navController;
}
@property (retain, nonatomic) IBOutlet UIWindow *window;
@property (retain, nonatomic) IBOutlet
            UINavigationController * navController;
@end
```

Listing 8.3

Updated Implementation of the ApplicationDidFinishLaunching Method

```
- (void)applicationDidFinishLaunching:(UIApplication *)application
{
    [window addSubview:[navController view]];
    [window makeKeyAndVisible];
}
```

NOTE

To make the model more easily accessible to the view controller, you will need to move the creation of the model data to MyViewController. In this code listing, that's already been done. Be sure to do it in your code as well.

Once you've done all of this, you're ready to implement editing. You can even run the application at this point and see that your lists now show up in a `UINavigationController`.

Making Your EditViewController

The first thing that you need to do is to create a new view that will be used for editing the item that you select from the `UITableView`. You've already seen how to create a `UIView` and how to create a `UIViewController`, so I'm just going to show you a screen shot of what I want the user interface to look like. Use Interface Builder to create a new view in a new nib that looks just like this. Create a `UIViewController` to use with this new view. The `UIViewController` should have outlets for the city name and a favorite star. If you make the favorite star a `UIButton` and configure it to be a custom button, using the star images as the button image, you can make it so that when the user touches the star it toggles the status of whether or not that item is a favorite. Remember to connect your buttons to actions on your `UIViewController`.

If you need help, you can reference the sample code for this chapter. Figure 8.4 shows the completed view for editing a particular city.

The only tricky things I'm doing in the `EditViewController` code surround the favorites button and how you notify the parent view controller whether the user has actually edited the values or if they canceled out. This code is shown in Listing 8.4.

Figure 8.4

The `EditView` Interface

Listing 8.4

EditViewController Implementation Code

```
-(void)updateStar;
{
    if(favorite)
        [starButton setImage:
            [UIImage imageNamed:@"FullStar.png"]
            forState:UIControlStateNormal];
    else
        [starButton setImage:
            [UIImage imageNamed:@"EmptyStar.png"]
            forState:UIControlStateNormal];
}
-(IBAction)starTouched:(id)sender;
{
    [self setFavorite:!favorite];
    [self updateStar];
}
-(IBAction)doneTouched:(id)sender;
{
    [self setStateInfo:[stateNameField text]];
    [self setCityInfo:[cityNameField text]];
    [delegate editViewFinishedEditing:self];
    [[self navigationController] popViewController:YES];
}
-(void)viewDidLoad;
{
    [self updateStar];

    [cityNameField setText:cityInfo];
    [stateNameField setText:stateInfo];

    UIBarButtonItem *doneButton =
        [[UIBarButtonItem alloc]
         initWithBarButtonSystemItem:UIBarButtonSystemItemSave
         target:self action:@selector(doneTouched:)];
    [[self navigationItem] setRightBarButtonItem:doneButton];
    [doneButton release];
}
```

Going down these items in order, the `updateStar` method is really just a convenience method used in multiple locations to set the star image to be either the empty or full star, depending on the state of the favorite value. It's used both in the starTouched: method and in the `viewDidLoad` method. This just prevents you from duplicating this code.

The starTouched: method is an IB action that's used in conjunction with the star button on the `UIView`. This toggles the state of the favorite value and then updates the star button with the appropriate image.

The `doneTouched:` method is called when the user hits the Save button in the upper-right corner of the view controller. This method takes the values from the text fields and puts them into the member variables and then notifies the delegate that you finished editing the values. Finally, it pops the top view controller (yours) off the navigation controller's stack, thus returning you to the parent view.

NOTE
In the `doneTouched:` method, you call the method `editViewFinishedEditing:` on the delegate. This is only called when you're done editing. If the user uses the back button to leave the edit screen, it isn't called. This is a convenient way to tell whether the user has canceled or committed his edits.

Finally the `viewDidLoad` method first updates the star, then updates the text fields, and then finally creates the `UIBarButtonItem` that will be used to display the "Save" button in the upper-right corner of the `UINavigationController`. Remember that you access the `navigationItem` off of the current `UIViewController` to set the right or left bar button items to whatever kind of button you would like to have displayed in those particular locations.

NOTE
`UIViewController` provides the `setEditing:animated:` method for setting and unsetting a view in "Edit Mode". I don't use it here because I want to demonstrate the use of the `UINavigationController` and the `UINavigationItem`, as well as adding buttons to it. However, for future reference, overriding this method might also be a way to implement your Save button.

Editing rows

Now that you have your edit view designed and built, you can actually use it to edit your rows.

To make your life easier, you should move the `UITableViewDelegate` code from the `MyViewDelegate` into `MyViewController`. This is a common way of handling the delegate. Recall that in the previous sections, I broke it out into another file to make the distinctions between the components clearer. Now is the time to bring them back together. Do this before proceeding by just cutting the methods from the delegate and data source and pasting them into MyViewController. Also connect the delegate and data source outlets in Interface Builder from your UITableView to your MyViewController instance.

Now pull up your `MyViewController` (which is now also the table view delegate), and edit the `tableView:didSelectRowAtIndexPath:` method. Previously, when the user selected a row, all you did was toggle the favorite status. Now you're going to move the toggling of the favorite into the edit window, and so what you need to do here is to allocate the new view controller for the edit view and push that onto the navigation controller. The new implementation of this method is shown in Listing 8.5.

Listing 8.5

Updated tableView:didSelectRowAtIndexPath: Method

```
- (void)tableView:(UITableView *)tableView
didSelectRowAtIndexPath:(NSIndexPath *)indexPath;
{
    NSDictionary *state =
        [favoriteCities objectAtIndex:[indexPath section]];
    NSArray *cities = [state objectForKey:@"CITIES"];
    NSMutableDictionary *city = [cities objectAtIndex:[indexPath row]];

    NSString *stateName = [state objectForKey:@"NAME"];
    NSString *cityName = [city objectForKey:@"NAME"];
    BOOL isFavorite =
        [[city objectForKey:@"FAVORITE"] isEqualToString:@"YES"];

    EditViewController *controller = [[EditViewController alloc]
                    initWithCity:cityName
                           state:stateName
                        favorite:isFavorite
                        delegate:self];
    [[self navigationController] pushViewController:controller
                                           animated:YES];
    [controller release];

    [self setEditedItem:indexPath];
}
```

This code uses the `indexPath` to find the appropriate row in the model. You then use that row to initialize the `EditViewController`. You then push the `EditViewController` onto the navigation controller and release it. The initializer for EditViewController simply sets the parameters in its instance variables. You should be able to write that method yourself. Because the navigation controller retains the `EditViewController`, you don't need to keep a copy of it retained. Finally, you're going to hold onto the `indexPath` that was passed in so that when the `EditViewController` is finished editing, you can find the row that was edited to update its values.

The method that updates the values when the editing is complete is the delegate method from the `EditViewController`. That method is shown in Listing 8.6.

Listing 8.6
Delegate Method for the EditViewController

```
-(void)editViewFinishedEditing:(EditViewController *)inEditView;
{
    NSString *favorite = nil;
    if([inEditView favorite])
        favorite = @"YES";
    else
        favorite = @"NO";

    NSString *stateName = [inEditView stateInfo];
    NSString *cityName = [inEditView cityInfo];
    if([self editedItem]) // we're editing an existing item
    {
        NSMutableDictionary *origState =
            [favoriteCities objectAtIndex:[editedItem section]];

        NSString *origStateName = [origState objectForKey:@"NAME"];
        // first we check to see if the state name changed...
        if(![origStateName isEqualToString:stateName])
        {
            // if it did, we need to move the city to the new state.
            NSMutableDictionary *newState = nil;
            for(NSMutableDictionary *state in favoriteCities)
            {
                if([[state objectForKey:@"NAME"]
                        isEqualToString:stateName])
                    newState = state;
            }
            if(!newState)
            {
                newState = [NSMutableDictionary dictionary];
                [newState setObject:stateName forKey:@"NAME"];
                [newState setObject:
                        [NSMutableArray array] forKey:@"CITIES"];
                [favoriteCities addObject:newState];
            }
            NSMutableDictionary *newCity =
                    [NSMutableDictionary dictionary];
            [newCity setObject:cityName forKey:@"NAME"];
            [newCity setObject:favorite forKey:@"FAVORITE"];
            [[newState objectForKey:@"CITIES"] addObject:newCity];
            // now remove the old city and optionally state
            // if the state has no more cities.
            [[origState objectForKey:@"CITIES"]
```

continued

Listing 8.6 *(continued)*

```
                    removeObjectAtIndex:[editedItem row]];
            if([[origState objectForKey:@"CITIES"] count] == 0)
                [favoriteCities removeObjectAtIndex:[editedItem section]];
        }
        else
        {
            // the state didn't change, we can just update the old one.
            NSMutableArray *cities = [origState objectForKey:@"CITIES"];
            NSMutableDictionary *city = [cities objectAtIndex:[editedItem row]];
            [city setObject:cityName forKey:@"NAME"];
            [city setObject:favorite forKey:@"FAVORITE"];
        }
    }
    else // it was a new add
    {
        NSMutableDictionary *newState = nil;
        for(NSMutableDictionary *state in favoriteCities)
        {
            if([[state objectForKey:@"NAME"] isEqualToString:stateName])
                newState = state;
        }
        if(!newState)
        {
            newState = [NSMutableDictionary dictionary];
            [newState setObject:stateName forKey:@"NAME"];
            [newState setObject:[NSMutableArray array] forKey:@"CITIES"];
            [favoriteCities addObject:newState];
        }
        NSMutableDictionary *newCity = [NSMutableDictionary dictionary];
        [newCity setObject:cityName forKey:@"NAME"];
        [newCity setObject:favorite forKey:@"FAVORITE"];
        [[newState objectForKey:@"CITIES"] addObject:newCity];
    }
    [[self tableView] reloadData];
}
```

This is a pretty hefty chunk of code, but the important thing to realize is that the majority of this is simply manipulating your model. Arrays of dictionaries are not exactly an intuitive model to work with. All you really need to know in looking at this is that you are updating your model, and that's all you're doing. At the top you handle an edit, and at the bottom you handle an add. You determine whether it is an edit or an add, based on whether or not you have previously set the `editedItem` property, which is the index path of the item that you are editing. If this value is nil, then you assume that it is an add.

Next, you'll see how you actually handle adding rows.

Adding rows

To allow for adding rows, you're going to add a button to the upper-right corner that is an add button. This shows up as a little plus icon on your `UINavigationController`. You'll connect the action from that button to an `IBAction` in your `MyViewController` class. The line that you need to add to your `viewDidLoad` method to insert your add button is shown in Listing 8.7.

Listing 8.7

Code Required to Display the Add Button on the EditView

```
[[self navigationItem] setRightBarButtonItem:
 [[[UIBarButtonItem alloc]
   initWithBarButtonSystemItem:UIBarButtonSystemItemAdd
   target:self
     action:@selector(addCity:)] autorelease]];
```

The `addCity` method is shown in Listing 8.8.

Listing 8.8

The Implementation of the addCity Method

```
-(IBAction)addCity:(id)sender;
{
    [self setEditedItem:nil];
    EditViewController *controller =
        [[EditViewController alloc] initWithCity:@""
                                          state:@""
                                       favorite:NO
                                       delegate:self];
    [[self navigationController] pushViewController:controller
                                           animated:YES];
    [controller release];
}
```

As you can see, this is relatively similar to how you handled editing rows, except that you are simply passing empty strings for the city and state, and you're making sure that the edited item is set to nil.

Deleting rows

The last item that I want to talk about with regard to editing of rows is deleting rows.

Again, like you did with the delegate, if you haven't already done so, now is a good time to move the data source methods from the separate module into the `MyViewController` class. This makes it simpler for you to manipulate the model when you go to make your deletions.

To add deletion, you need to implement two methods in your data source. Those methods are `tableView:canEditRowAtIndexPath:` and `tableView:commitEditingStyle:forRowAtIndexPath:`. These methods allow you to control whether the row at the specified index path is editable at all, and if it is, to take action when the row has been deleted.

To perform a delete on a `UITableView`, you use a swipe gesture across the row you want to delete. When you perform this gesture, the `UITableView` first asks your data source if it is okay to edit the row that you have swiped. This is when it calls the `tableView:canEditRowAtIndexPath:` method. If this method returns YES, then it displays a delete button in the accessory view of the row for confirmation. If the user then presses the delete button, then the method `tableView:commitEditingStyle:forRowAtIndexPath:` is called to perform the actual deletion.

Summary

In this chapter, you've seen how to use a `UINavigationController` to manipulate and display lists of data. In the example, you displayed the list of your favorite cities, and then the detailed view for each of those cities. Another way this could have been used would be to display a list of states, and then the list of cities for each of the states as the detail view for the state.

The important thing to take away from this chapter is that the `UINavigationController` is a really fundamental and powerful tool in building iPhone applications. It's also surprisingly easy to use. Using it enables you to build an application that already follows the UI paradigms that the built-in applications follow. This means that your applications will be easier for your customers to learn and use.

Understanding the UIApplicationDelegate

Thus far, every application that you've written has contained an application delegate class. Typically, these are generated by the template that you use to create the project in the first place. Therefore, it's logical that I should take a moment and discuss the application delegate and what it does for you.

I won't go into tremendous amounts of detail in this chapter; the goal here is to become familiar with the architecture and some of the messages that the UIApplicationDelegate can receive.

Exploring the Purpose of the UIApplicationDelegate

Every application written for the iPhone has an application delegate. In some ways, you can think of the application delegate as your main class for your application. It's the first place that your application is handed control from the operating system when it's launched, and it is also the last place that the application executes from as it is exiting. This makes it the perfect place to handle tasks such as loading and saving of persistent state. In addition to this, it also handles certain operating system messages that are sent to the application as it is running. These include handling memory warnings, or changes in the size of the status bar.

The way that you utilize a UIApplicationDelegate is by creating a class that implements the UIApplicationDelegate protocol and then creating an instance of that class in your MainWindow nib file. Usually there is no reason for you to think about any of this because the project templates do all this work for you. All you really need to know is to look in your project for the class that's called something along the lines of <YourProjectName>AppDelegate. This is the application delegate that is provided by your template. All you need to do is implement the methods that you want to utilize in this class, and everything else will fall into place.

Let's take a look at some of the methods that you may want to use in your applications.

In This Chapter

Learning the ins and outs of the UIApplicationDelegate

Handling startup and shutdown of your application

Receiving memory warnings and status information

Taking action to respond to notifications

Receiving network notifications and handling opened URLs

Handling Startup and Shutdown

The first and foremost purpose of the `UIApplicationDelegate` is to handle the startup and shutdown of your application. It does this through the implementation of several methods, which are listed in Table 9.1.

Table 9.1 Methods Used for Startup and Shutdown	
Method Signature	**Description**
`- (void)applicationDidFinish Launching:(UIApplication *) application`	This is the first method in your application that is called when your application finishes launching from the iPhone OS. This is a good place to perform operations such as restoring saved state or setting up global variables.
`- (BOOL)application: (UIApplication *)application did FinishLaunchingWithOptions: (NSDictionary *)launchOptions`	This is an alternative method to the `application DidFinishLaunching:` method, and if your application delegate implements it, then it is called in favor of that method. This method is used when the iPhone OS receives a remote notification for your application or if another application has issued an open URL request for a URL type that your application handles.
`- (void)applicationWillTerminate: (UIApplication *)application`	This is the last method in your application delegate that is called as your application is exiting.

Understanding launch methods

The methods used in launch, `applicationDidFinishLaunching:` and `application: didFinishLaunchingWithOptions:`, are called when your application is finished launching. The first method, `applicationDidFinishLaunching:`, was the only one available in iPhone OS 2.0 and is generally what you will see in most applications. Usually this method is used for application initialization and adding your initial views to your main window.

The second method, `application:didFinishLaunchingWithOptions:`, is new with iPhone OS 3, and if implemented on your `UIApplicationDelegate` is called in favor of the older method. It was added so that an application could more easily handle tasks such as remote notifications and opening URLs. When the iPhone OS receives a remote notification, it examines the notification to determine what application handles that type of notification. It then gives the user the option to launch the application to view the notification. If the user chooses to view the notification, then the application is launched and the options parameter for this method includes a dictionary containing data from the notification. You will look at this method in more depth in Chapter 20.

Another case where this method is called is when another application has requested that the operating system handle a URL of a type that your application has previously registered as being one that it can open. In this case, the `launchOptions` parameter contains an object representing the URL and the bundle ID of the application that sent the request.

NOTE
The `UIApplicationDelegate` also includes the optional method `application:handleOpenURL:` for handling situations where your application is launched by opening a URL.

In the case where the application is simply launched from the iPhone OS via its application icon, the `launchOptions` parameter contains nil.

Table 9.2 shows a list of the keys that can be used to access the elements of the launch options dictionary. These keys are actually used in other methods as well, but I present them all here for your information.

Table 9.2 Keys for the launchOptions Dictionary

Value	Description
`UIApplicationStatusBarOrientationUserInfoKey`	Accesses a value used to indicate the current orientation of the status bar. This value is encoded in an `NSNumber`.
`UIApplicationStatusBarFrameUserInfoKey`	Accesses a `CGRect` value to indicate the new location and frame of the status bar. This value is encoded as an `NSNumber`.
`UIApplicationLaunchOptionsURLKey`	Accesses an `NSURL` that has been initialized with the URL used to launch the application.
`UIApplicationLaunchOptionsSourceApplicationKey`	Accesses a string containing the bundle ID of the application that was used with a URL to launch the application.
`UIApplicationLaunchOptionsRemoteNotificationKey`	Accesses a dictionary representing the payload from the remote notification used to launch the application.

Understanding the applicationWillTerminate method

When the user presses the home button on the iPhone, they are requesting that your application be terminated. The application then sends the `applicationWillTerminate:` message to your application delegate. You are expected to handle whatever cleanup is necessary for your application in a timely manner. This means persisting any objects to disk, freeing resources, or doing other typical shutdown operations.

NOTE
When I talk about persisting data, I'm talking about saving it either in the application's Documents folder or into the user's defaults. I talk more about saving data in Chapters 17 and 18.

I mention that you need to do this in a timely manner because if you don't, the operating system force-quits your application. It does this regardless of whether or not you have completed your save operation.

Receiving Notifications while Your Application Is Running

In addition to handling startup and shutdown of your application, the application delegate also handles messages sent to the application from the operating system while your application is running. Usually these kinds of messages include warnings about resources (memory, or time changes) and remote notifications. Some of these you may be safe to ignore; others, you ignore at your peril.

Table 9.3 shows each of the methods that you can implement to receive these messages. I will discuss each of these in turn.

Table 9.3 Runtime Notification Messages

Methods	Description
`-(void)applicationDidReceiveMemoryWarning:(UIApplication *)application`	Received when the iPhone is running low on memory.
`-(void)applicationSignificantTimeChange:(UIApplication *)application`	Received when a significant time event has occurred.

`-(void)application:(UIApplication *)application willChangeStatusBarOrientation:(UIInterfaceOrientation)newStatusBarOrientation duration:(NSTimeInterval)duration,`
 	
`-(void)application:(UIApplication *)application didChangeStatusBarOrientation:(UIInterfaceOrientation)oldStatusBarOrientation,`
 	
`-(void)application:(UIApplication *)application willChangeStatusBarFrame:(CGRect)newStatusBarFrame`
 	
and
 	
`-(void)application:(UIApplication *)application didChangeStatusBarFrame:(CGRect)oldStatusBarFrame`
	Used to indicate changes to the size and orientation of the status bar.

Methods	Description

`(void)applicationWillResignActive:(UIApplication *)application`
 	
and
 	
`(void)applicationDidBecomeActive:(UIApplication *)application`
	These notifications are received when the iPhone is put to sleep or when a phone call is received.

`(void)application:(UIApplication *)application didReceiveRemoteNotification:(NSDictionary *)userInfo,`
 	
`- (void)application:(UIApplication *)application didRegisterForRemoteNotificationsWithDeviceToken:(NSData *)deviceToken`
 	
and
 	
`- (void)application:(UIApplication *)application didFailToRegisterForRemoteNotificationsWithError:(NSError *)error`
	Used to signal your application that a remote notification for your application has been received by the operating system.

Handling resource alerts

The first category of messages that you might receive while your application is running are resource alerts. The most important of these is the `didReceiveMemoryWarning:` message, which you receive if the iPhone's memory is running low.

Remember that even though only one user application is allowed to run at a time, there are still system applications that are running in the background on the iPhone all the time, and so the memory being used may not necessarily be strictly due to your application. Nonetheless, you are still expected to reduce your memory footprint when you have received this warning. Furthermore, if you do not reduce your memory footprint, and the operating system sends another low-memory warning, your application can be force-terminated in order to free up memory for the operating system.

Your view controllers also receive this message when there is a low-memory warning, and so you typically free memory in those objects themselves. However, this method is provided on the application delegate itself for any other global data that you can purge. Usually, the ways that you can find to free up data include purging caches or deleting temporary data. You can also flush data out to disk, persisting it in a way that enables you to retrieve it again later when you may need it.

Another resource message that you may receive is the `applicationSignificantTimeChange:`. This message is received when a significant time event occurs. Examples of a significant time event are the changing of daylight savings time and the arrival of midnight. The purpose of this message is to allow you to update parts of your application that utilize time and to handle the time change appropriately. When writing applications for the iPhone, it's important to realize that the iPhone can be used at all times of day and every day of the year. Many applications have suffered from improper handling of times and dates. For example, the leap year and daylight savings time make handling dates and times in a constant 99.9 percent uptime environment such as the iPhone particularly challenging.

Handling phone calls and sleep

Some of the more important status messages that you can receive via the `UIApplication Delegate` are those that indicate that the phone has been put to sleep by the lock button, or that an incoming phone call has temporarily interrupted the application. Games in particular probably want to pause the action during these events. You may also want to save your game or application state. The notifications that are sent in these circumstances are the messages `applicationWillResignActive:` and `applicationDidBecomeActive:`, which are called when the application is interrupted, and when control returns to the application after the interruption, respectively.

In the event that the interruption has been caused by an incoming phone call, the user is presented with a dialog giving them the option to take the phone call. If they do, control temporarily returns to the application, and the application terminates just as if the user had pressed the home button. This means that the `applicationWillTerminate:` method is called as well.

Changing status

The final status messages that you might receive are notifications related to changing the size and orientation of the status bar. For example, when a phone call is received, the status bar can grow in size to indicate that the user is on a phone call. Your application should be aware of the possibility of these changes and be prepared to adjust your UI to allow for them.

The first of these notifications concerns changes in the orientation of the status bar. These are received using the methods `application:willChangeStatusBarOrientation: duration:` and `application:didChangeStatusBarOrientation:`, which are sent before and after the status bar has changed its orientation, respectively. In both cases, you receive the new status bar orientation as a parameter to this method.

The next of these notifications, which are probably a bit more useful, are the notifications surrounding changes in the frame of the status bar. These are the `application:willChange StatusBarFrame:` and `application:didChangeStatusBarFrame:` methods. Like the status bar orientation messages, you receive information about the status bar's frame before and after the change, and you should use this information to adjust your own UI to make room for the new status bar size.

NOTE
You may be asking yourself "When would I see the status bar ever change size?" Well, the answer is, when a user is on a phone call and using your application. The status bar grows in size with a message that indicates the user can touch it to return to their phone call.

Handling remote notifications

iPhone OS 3 added the ability to handle remote notifications. I will discuss notifications in much greater depth in Chapter 20. However, it is instructive to at least be familiar with the methods that are called on the application delegate when a notification for your application has been received.

Firstly, when your application registers to receive a remote notification of a particular type, the method `application:didRegisterForRemoteNotificationsWithDeviceToken:` is received, indicating that registration for the remote notification was successful. If an error during registration occurs, the method `application:didFailToRegisterForRemoteNotificationsWithError:` is called.

Finally, when your application receives a notification, the method `application:didReceiveRemoteNotification:` is called. When this method is called, it is passed a dictionary containing the payload from the remote notification. You can use this payload to take some action in your application based on that notification. It contains components such as an alert message, a badge quantity, and a sound to play as the alert is displayed. Additionally, this dictionary can also have custom data defined by you when you set up your notification supplier.

Summary

In this chapter, you've taken a brief look at the `UIApplicationDelegate` and the methods that you can implement on your application delegate to receive information from the operating system about the device, how it was launched, and so on. While this is an extremely simple protocol, it's one that most applications have to interact with at least in some minimal way. Understanding the proper use of each of these methods will help you to develop an application that performs well on the iPhone OS. I will revisit many of these methods in future chapters.

Applying Custom UIViews

In this chapter, we're going to return to the Model, View, Controller design pattern, and focus on one particular component of this pattern, views. Every graphical component that you use in building your applications inherits from `UIView`, the Cocoa Touch class that represents the View part of the Model, View, Controller pattern. This means that understanding this fundamental component is vital to your success on the platform.

Doing custom drawing is one of the most common requirements when creating custom `UIViews`. One of the most powerful things that Cocoa Touch inherits from its Cocoa lineage is its drawing API. When it was originally developed, this drawing API, known as Core Graphics, was unparalleled for its ease of use and capabilities. Even now, through its acceleration with OpenGL, it provides an excellent way to draw custom 2-D graphics on the screen.

In addition to its outstanding abilities when it comes to drawing graphics, `UIView` also provides some simplified mechanisms for doing basic animation using Core Animation.

In this chapter, you are going to take a look at how to add custom `UIViews` to your projects and then how to draw graphics in those views.

In This Chapter

Working within the Cocoa Touch view geometry

Adding a custom view to an interface

Exploring Core Graphics

Using basic drawing commands

Understanding Cocoa Touch View Geometry

The first thing to understand about using `UIViews` is how the Cocoa Touch view geometry works. Figure 10.1 shows a typical iPhone screen and the coordinate system that is used on that screen. Note that the coordinate system has its origin in the upper-left corner, which is different from the default coordinate system used by Cocoa on Mac OS X. On Mac OS X, the origin of the coordinate system is in the lower-left corner. This means that when drawing graphics on the iPhone screen, your points and lines must work within this coordinate system.

Figure 10.1
The iPhone screen coordinate system

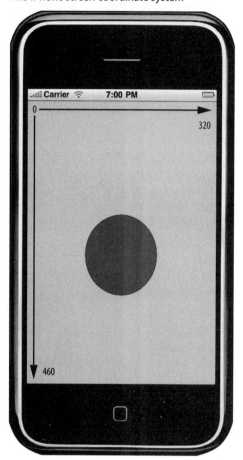

Adding Custom Views to a Project

Adding a custom view to your project is relatively simple. The first thing that you need to do is to create a class that inherits from `UIView`. To do this, add a new file to your project, and from the final template selection dialog, choose a new `UIView` subclass. This creates a `UIView` subclass for you with the skeletons of its most essential methods.

Once you've done this, open Interface Builder to the nib that you want to add the custom view to. If you don't already have a view object on your UI that you want to use as your custom view, scroll through the object palette until you find a view object. Drag and drop this object onto your UI wherever you want to have the custom UIView display. Next, with the UIView object in your interface selected, find the Class Identity property in the information window for the object, and set this property to your newly created custom view class.

NOTE
Another way of adding a custom view to your view hierarchy is to do it programmatically by simply adding it as a subview of one of your existing views.

Now, wherever you placed the instance of your view in your UI, your application will create an instance of your custom view and display it when your UI displays.

At this point, you can return to Xcode and begin writing the code that actually implements your custom view.

Implementing the Custom View Code

When inheriting from UIView, there are three core methods that you need to be concerned about. The first is the designated initializer, which is initWithFrame:. This method will be called to construct your custom view. You should expect that the parameter passed to this method, a CGRect indicating the frame for the view, will probably initially be CGRectZero, and you should plan to do your layout in the method layoutSubviews, which is specifically intended for this purpose. Use the initializer instead to initialize data that you will later need for drawing your view.

The second method to be concerned about when inheriting from UIView, as alluded to in the previous paragraph, is the layoutSubviews method. This method is called by any super views to your view whenever layoutIfNeeded is called. You should use this method to lay out any subviews or elements within your view. The default implementation of this method does nothing.

The final method that you need to be concerned about is the drawRect: method. This is the method where all the magic happens. It is where your custom view actually does the drawing to display your data. It is important to note that you should not manually draw your subviews in your drawRect: method. The view system will send a drawRect: call to each of your subviews in turn. They are responsible for drawing themselves. You do not draw them. You need only be concerned with drawing your own content.

Typically, you use Core Graphics calls to actually do the physical drawing of your view. However, keeping in mind what I just said about some views drawing themselves, you should remember that your custom view, even though it's custom, is still a view and still fits into the UIView hierarchy. This means that not only can it be a subview of another view, but it can also contain subviews. The importance of this is that, for example, rather than compositing an image into your view in your drawRect: method, it may actually be more efficient to use a UIImage subview.

Similarly, rather than drawing text directly into your view using the Core Graphics function calls, it may be more efficient instead to use a UILabel. You should remember this when you're designing your view and try not to duplicate functionality that is already there. Leverage the tools that Cocoa Touch gives you.

Working with Core Graphics

Before I get too involved with looking at an actual implementation of a `drawRect:` method I feel it's important to give a short overview of some key Core Graphics concepts.

A comprehensive coverage of Core Graphics is beyond the scope of this book, but the few tools that I will expose you to here should give you at least the basic knowledge that you'll need to do some simple drawings. From here, referencing the Apple documentation and some of the excellent books on Core Graphics available for the desktop will help you to take your work to the next level.

Working with basic Core Graphics types

The first thing to understand is that Core Graphics is a C API. As such, it utilizes a mixture of C structures and C functions to do its work. While it tries to emulate objects using references, it is not a traditional Object Oriented API, and does not utilize Objective-C classes.

The types that you will be working with in this chapter mostly consist of those listed in Table 10.1.

Table 10.1 Basic Core Graphics Types

Type	Purpose
`CGFloat`	A float value specifically for use in Core Graphics functions.
`CGPoint`	C structure indicating a single point in the `UIView` coordinate system or a Core Graphics context. This consists of two `CGFloat`s indicating X and Y coordinates.
`CGSize`	C structure defining dimensions. This consists of two `CGFloat`s indicating height and width.
`CGRect`	C structure defining a rectangle in a `UIView` or Core Graphics context. This consists of a `CGPoint` to indicate the origin of the `CGRect` and a `CGSize` to indicate its dimensions.
`CGImageRef`	A reference to a `CGImage`, which is the Core Graphics version of a UIImage.
`CGAffineTransform`	C structure defining an affine transform matrix for use in translations, scaling, and rotations.
`CGPathRef`	A reference to a Core Graphics path, which can be used to draw lines or objects.
`CGContextRef`	A reference to a Core Graphics context.

Because it is a C API, I should take a few moments to discuss how to manage the memory that will be used to allocate each of these objects.

Understanding the core frameworks memory management model

Because Core Graphics is a core framework, it utilizes the Core Foundation memory management system. The Core Foundation memory management system consists of a combination of reference counting, method naming conventions, and specialized allocator functions. For the purposes of this Core Graphics discussion, you need only be concerned with the reference counting and naming conventions. You will not look at allocator functions, as they are not needed for anything you will be working with.

It should be noted that although the Core Foundation memory management system may seem similar to the Cocoa memory management system (Cocoa also utilizes reference counting for its memory management), they are not compatible and cannot be used interchangeably. Whenever you work with Core Foundation objects, you must always use the Core Foundation system, and when using Cocoa Touch objects, you must always use the Cocoa system.

Understanding Core Foundation object ownership

Like Cocoa, Core Foundation utilizes reference counting for its primary memory management architecture. Because it's a C-based API, however, it would have been onerous to force developers to use additional calls on every object created to increment the reference count. Instead, Core Foundation utilizes a naming convention and a set of rules to govern the ownership of memory allocated.

There are three rules governing the ownership of memory allocated. The first rule is that if you create an object (from a Create or Copy function), you own it.

The second rule is that if you get an object (from a Get function), you do not own the memory, and if you want to keep it around, you must retain it using a call to `CFRetain`. You own memory that you have called `CFRetain` on.

The third and final rule is that if you own the memory, either through creation or `CFRetain`, when you finish using it you must release it using `CFRelease`.

Core Foundation and Core Graphics contain several basic types that are defined as C structures. These types, such as `CGRect` and `CGPoint`, which you have already seen, do not need to be retained or released. As a naming convention, objects that need to be retained or released are named with a type name containing the word *Ref*, for example, `CGImageRef`. The only types of this nature that you will work with in this chapter are `CGImageRef` for images and `CGPathRef` for paths.

Using CFRetain and CFRelease

The functions used to increment and decrement the retain count on blocks of memory are `CFRetain` and `CFRelease`. `CFRetain` increments the retain count, and `CFRelease` decrements the retain count. When the retain count on a block of memory reaches 0, that block of memory is freed. There is no `autorelease` in Core Foundation.

Getting the graphics context

In order to draw using Core Graphics, you have to first get the graphics context for the view into which you want to draw. Cocoa Touch provides a function that allows you to acquire the graphics context for your current view. That function is `UIGraphicsGetCurrentContext`, and it returns a `CGContextRef`. Remember, because the method contains "Get," the object returned from this function is not owned by you, and if you expect to keep it around beyond the scope of the current method you should retain it using `CFRetain`.

The Core Graphics context provides something akin to the drawing board in which you can perform drawing operations. Graphics contexts can be acquired, saved, and restored. When you have acquired a graphics context, you are able to set your drawing parameters on it, things such as your stroke color, your fill color, and transformations. Any of these settings specified on a graphics context are used for all subsequent drawing on that graphics context.

Contexts can also be saved and restored. For example, you can set up a graphics context, configuring your drawing environment, then save that graphics context, changing some parameters, do some additional drawing, and then restore back to your previously saved graphics context, thus restoring your previously configured drawing environment. In this way, the graphics context saving and restoring function works a bit like a stack in that you can push a graphics context onto the stack and then pop it back off. In this way, multiple calls to save and restore your graphics context can be nested so that you can have multiple states to use for drawing. The method used to save a graphics context is `CGContextSaveGState()`, and the method used to restore a graphics context is `CGContextRestoreGState()`. Listing 10.1 shows an example where the stroke color is set first to blue then to black, and then back to blue again.

Listing 10.1

Saving and Restoring a CGContextRef

```
{
    CGContextRef context = UIGraphicsGetCurrentContext();
    CGContextSetStrokeColorWithColor(context,
                            [[UIColor blueColor] CGColor]);
    // … do some drawing with the blue color …
    CGContextSaveGState(context);
    CGContextSetStrokeColorWithColor(context,
                            [[UIColor blackColor] CGColor]);
    // … now do some drawing with black …
    CGContextRestoreGState(context);
    // … now the drawing color is back to blue …
}
```

Think of this as if you are pushing and popping your graphics contexts onto and off of a stack. As you set up your graphics environment, you can push a new graphics context onto the stack, which saves your old context and its settings. Later, you can pop the top-level graphics context off the stack, removing any new settings that you had put into the graphics context and restoring the settings that you had in your old graphics context before you pushed the new one onto the stack.

Doing basic drawing operations

Once you have a Core Graphics context, you can begin drawing in it. The first thing you'll probably want to do with your new graphics context is to set your drawing colors. There are two functions that you can use to set these colors; the first is `CGContextSetStrokeColorWithColor()`, which allows you to configure the stroke color for drawing, and the second is `CGContextSetFillColorWithColor()`, which allows you to configure the fill color. In addition to these methods, there are other methods that enable you to configure settings such as patterns, shadows, glyphs, line caps, and a variety of other settings to help you with your drawing. For these simple examples, however, I'll stick to the basics.

The stroke color controls the color that is used to draw lines and to draw the outlines of paths and objects. The fill color determines the color that is used to fill in enclosed objects, such as circles, squares, and paths. For example, if you draw an ellipse, the inside of the ellipse is drawn with the fill color and the outside edge is drawn with the stroke color.

NOTE
The stroke color is also used when drawing lines and text.

Once you've configured your colors and your other drawing parameters, you can begin actually doing your drawing. A summary of the most often used drawing commands is shown in Table 10.2.

Table 10.2 Frequently Used Core Graphics Drawing Functions

Function Signature	Purpose
`void CGContextClearRect(CGContextRef c, CGRect rect);`	Clears the given rectangle in the context
`void CGContextFillPath(CGContextRef c);`	Fills the current path in the given context using the current fill color
`void CGContextFillRect(CGContextRef c, CGRect rect);`	Fills the given `CGRect` with the current fill color

continued

Table 10.2 Continued

Function Signature	Purpose
`void CGContextStrokePath (CGContextRef c);`	Strokes the current path in the given context using the stroke color
`void CGContextStrokeRect (CGContextRef c, CGRect rect);`	Strokes the given `CGRect` using the current stroke color
`void CGContextStrokeEllipseInRect (CGContextRef context, CGRect rect);`	Strokes an ellipse that fits within the `CGRect` passed using the current stroke color

Several of the methods in Table 10.2 utilize the current path as the object that they are going to draw. The Core Graphics path system provides a mechanism that enables you to draw complex shapes by drawing segments of the path and joining the segments together into a single shape. Table 10.3 shows the methods used to define a path within the Core Graphics context.

Table 10.3 Methods for Defining Paths in Core Graphics

Function Signature	Purpose
`void CGContextMoveToPoint (CGContextRef c, CGFloat x, CGFloat y);`	Creates a new path in the given context and the location you specify
`void CGContextAddLineToPoint (CGContextRef c, CGFloat x, CGFloat y);`	Adds a straight line from the current path endpoint to the specified location
`void CGContextAddArcToPoint (CGContextRef c, CGFloat x1, CGFloat y1, CGFloat x2, CGFloat y2, CGFloat radius);`	Adds a curved line from the current path endpoint to the specified location using the given settings
`void CGContextClosePath (CGContextRef c);`	Closes the current path by drawing a line from the current location to the origin of the path

This is by no means a comprehensive listing of the methods related to paths. This is simply to give you an idea of what's available. Consult the Apple documentation for more information.

Doing transforms

Transformations are one of the more powerful capabilities of Core Graphics. Using them, you can rotate, scale, and translate your points, sizes, and rectangles.

To create a transform, you use one of the `CGAffineTransformMake` methods, which are listed in Table 10.4.

Once you have a transform, you can apply it to any of the Core Graphics primitive types that I previously discussed, for example, `CGPoint`, `CGRect`, or `CGSize`. The end result is the value transformed accordingly.

Table 10.4 CGAffine Transform Creation Methods

Function	Purpose
`CGAffineTransform CGAffineTransformMakeRotation (CGFloat angle);`	Creates a transform initialized to rotate to the specified angle
`CGAffineTransform CGAffineTransformMakeScale (CGFloat sx, CGFloat sy);`	Creates a transform initialized to scale to the specified size
`CGAffineTransform CGAffineTransformMakeTranslation (CGFloat tx, CGFloat ty);`	Creates a transform initialized to translate the specified distance in x and y space

Again, there is a lot more power to be found here. My goal here is simply to expose you to some of the more commonly used methods. I will not be using any transforms in the example code for this chapter.

Implementing drawRect

Now that you've gone through a brief overview of some of the basics of Core Graphics, let's take a look at how you actually implement a custom view and how you actually write the code that you use to display your graphics. In this example, you will just do a very simple custom view that draws a circle in the middle of it.

I'm not going to bore you with the details of how to add a new file to your project or how to create a new project for this example. If you've been working through the tutorials so far, you already know how to do that and it doesn't need to be repeated here. If necessary, you can reference the sample code that's provided on the Web site.

The purpose of this exercise is to create a new project that is a view-based project. You're then going to add a new file to your project that is a subclass of `UIView`. In Interface Builder, you need to set your view in your project to have its class be an instance of your custom view class, so be sure to do that. Remember, you do this by selecting your `UIView` instance in Interface Builder, and then set the Class property to your custom `UIView` class.

Your designated initializer doesn't need to do anything special for this, nor will you have any subviews that you will be managing. So you don't need to write either the `initWithFrame:` or the `layoutSubviews:` methods yourself; you'll simply rely on the inherited versions.

The only method you actually need to concern yourself with in this particular case is the `drawRect:` method.

When you create this new `UIView` subclass, its header should look like Listing 10.2.

Listing 10.2

UIView Subclass Interface

```
#import <UIKit/UIKit.h>

@interface MyCustomView : UIView
{
}
@end
```

You don't need to implement the `initWithFrame:` method, but you can leave the default implementation if you used the `UIView` subclass template. Let's jump right to `drawRect:`, which is shown in Listing 10.3.

Listing 10.3

The Implementation of the drawRect: Method

```
- (void)drawRect:(CGRect)rect
{
    // acquire the current context
    CGContextRef ctxt = UIGraphicsGetCurrentContext();
    // set the stroke color to black
    CGContextSetStrokeColorWithColor(ctxt, [[UIColor blackColor] CGColor]);
    // set the fill color to blue
    CGContextSetFillColorWithColor(ctxt, [[UIColor blueColor] CGColor]);
    // get a rect that's inset from the frame
    CGRect circleRect = CGRectInset([self frame], 100, 165);
    // draw the outline of the circle
    CGContextStrokeEllipseInRect(ctxt, circleRect);
    // fill the circle
    CGContextFillEllipseInRect(ctxt, circleRect);
}
```

CAUTION

The `initWithFrame` designated initializer is not called when you instantiate a view from within a nib. Instead, you can use the `awakeFromNib` method on the view, or the `viewDidLoad` method on the view controller.

What you're doing here is getting your context, setting your stroke and fill colors, and then just drawing the circle in your rect. It's that simple! When you run this application, you see something like Figure 10.2.

Figure 10.2
The completed custom view

Summary

In this chapter, you've seen how you can create a custom `UIView` and add it to your projects. Remember that even custom `UIViews` are still views and still work within the view hierarchy system, which means that you can even insert these views as subviews of existing views when needed.

You also saw some basic Core Graphics drawing code. Core Graphics is a tremendously powerful drawing system for 2-D graphics. For the most part, it also functions very similarly between the iPhone and Mac OS X, with the notable exception that the coordinate system origin for the two platforms differs. That is, on Cocoa Touch, the origin is in the upper left, and in Cocoa, it's in the lower left. That said, most books about the subject covering the desktop will be applicable to iPhone development.

Keep the example code from this chapter handy, because in the next chapter you're going to see how you handle touch events using this custom view. You're going to expand your little circle view so that the user can use his finger to move the circle around and resize it.

Handling Touch Events

A defining feature of the iPhone is its use of multi-touch for its primary input. There is nothing more unique to iPhone than this capability. When working with custom views, it's important to utilize multi-touch to its greatest advantage. Fortunately, Cocoa Touch makes implementing a multi-touch interface on your custom views incredibly easy. All you have to do is override a few methods to handle the touch events to unlock the potential of this innovative capability.

In this chapter, you're going to take a look at how you implement multi-touch on a custom view. First, you will take a look at what you need to do to support multi-touch, and then you will implement multi-touch on your custom `UIView` example code from the last chapter.

Handling Touch Events in a Custom UIView

When the user's finger touches the iPhone screen, the view that the touch occurs within receives a set of messages. These messages are translated into calls to specific methods that are defined as part of the `UIResponder` class. `UIView`, as a subclass of `UIResponder`, also has the option to override these methods so that it can take action on these events. The methods that are called are `touchesBegan:withEvent:`, `touchesMoved:withEvent:`, and `touchesEnded:withEvent:`. These methods must be overridden in your custom `UIView` to handle the touch events. The signatures for these methods are shown in Table 11.1.

In This Chapter

Handling touch events in your custom UIViews

Implementing the methods needed to handle touch events

Learning about the UITouch class and how to use it

Configuring a UIView to receive multi-touch events

Extending the custom UIView sample code to handle touches and multi-touches

Table 11.1 Touch Event Methods

Method	Purpose
- (void) touchesBegan:(NSSet*) touches withEvent:(UIEvent*)event	Called when the user's finger initially touches the screen
- (void) touchesMoved:(NSSet*) touches withEvent:(UIEvent*)event	Called as the user drags his finger across the screen
- (void) touchesEnded:(NSSet*) touches withEvent:(UIEvent*)event	Called when the user lifts his finger from the screen

As you can see from the descriptions, the chain of events that occurs when a touch event occurs essentially consists of events associated with the touch beginning, moving, and finally ending. I'll describe each of these in detail.

NOTE
The iPhone screen is capable of detecting up to four touches simultaneously, and no more.

Implementing touchesBegan:withEvent:

The first of these methods, the `touchesBegan:withEvent:` method, is called when the user initially touches the screen. This is a good place for you to do things such as detect where the user is placing her finger to determine if she wants to drag an object. Listing 11.1 shows a typical implementation of a `touchesBegan:withEvent:` method.

Listing 11.1

A typical touchesBegan:withEvent: Method

```
- (void) touchesBegan:(NSSet*)touches withEvent:(UIEvent*)event
{
    UITouch *touch = [touches anyObject];
    if([touch tapCount] == 2)
    {
        // ... take some action on a double tap ...
    }
    else
    {
        // .. do something with the touch, or store it. ...
    }
}
```

The first parameter to the `touchesBegan:withEvent:` method is an `NSSet`, which contains the touches that triggered this event. If it is a multi-touch event (more than one finger touching the screen at a time), then this contains multiple touches, one for each finger. If it is a single touch event, then this only contains one touch.

In this example, you are checking to see if the touch has a tap count of two. This is to allow you to determine if the user is double tapping versus single tapping to drag.

Another action you might consider taking in this method is to store the initial touch so that you can then compare its location to the touch location in the `touchesEnded:withEvent:` method to detect things such as swipes.

As the user drags his finger across the screen, `touchesMoved:withEvent:` is called repeatedly to update the position of the touch.

When the user lifts his finger, the `touchesEnded:withEvent:` method is called.

Working with the touches NSSet

Each of these methods takes as its first parameter an `NSSet` containing instances of the `UITouch` class to represent the touches that triggered the event. The `UITouch` class has several methods available on it that enable you to determine things such as where the touch occurred, the last location of this touch, how many times the user tapped the screen when beginning this touch, and so on. These methods are shown in Table 11.2.

Table 11.2 UITouch Methods

Method	Description
`- (CGPoint)locationInView:(UIView *) view`	Returns a `CGPoint` indicating the location of this touch in the provided `UIView`
`- (CGPoint)previousLocationInView: (UIView *)view`	Returns a `CGPoint` indicating the previous location of this touch in the provided `UIView`
`@property(nonatomic, readonly) NSUInteger tapCount`	Returns the number of taps used to initiate this touch event
`@property(nonatomic, readonly) NSTimeInterval timestamp`	Returns the time this touch was last updated

The most commonly used of these methods is the `locationInView:` method, which gives you the location of this touch in the provided `UIView`. This is perfect for determining where in your user interface the touch occurred.

Implementing touchesMoved:withEvent:

The next method is the `touchesMoved:withEvent:` method. This method is called as the user drags her finger across the screen. It is called constantly, over and over, to update the location as accurately as possible. This is where you would typically do things such as updating the location of an object being dragged.

Implementing touchesEnded:withEvent:

The final method is the `touchesEnded:withEvent:` method, which is called when the user lifts her finger from the screen. It should be noted that this method is not called if the user simply pauses her finger but does not lift it. Typically, in this method you would do things such as reset any state that you may have set in `touchesBegan:withEvent:`.

Handling multi-touch events

A multi-touch event differs from a single-touch event primarily because the touches `NSSet` that you receive on each of these methods contains more than one touch. To enable multi-touch on a view, you need to set the property `multiTouchEnabled` either through code or through Interface Builder. Once you've done this, to detect a multi-touch event, all that's necessary is to check the count of objects in the touches `NSSet`. If it is more than one, then you have a multi-touch event.

CAUTION
In practice, users rarely place both fingers on the screen at the same instant; therefore, do not rely on your `touchesBegan:withEvent:` method to detect a two-finger touch on its first invocation. Instead, be prepared to handle a transition from a single touch to a multi-touch event as you receive additional calls to touchesBegan:withEvent and touchesMoved:withEvent:. The simulator can be deceiving in this case, because the simulator always does its multi-touch events simultaneously.

Updating Your Custom View with Touch Events

Now that you've seen what you need to do to implement touch events in theory, let's see how you actually do it in practice by updating your custom view example project with touch events.

First, you will take your custom `UIView` example code and update it so that if the user touches inside the circle with his finger, he can drag the circle around the screen. If the user touches outside the circle, nothing happens.

After you've done this, you will update the example code one more time to add the ability to scale the circle, making it larger and smaller by using a pinch gesture with multi-touch.

Moving the circle with a touch

In order to update your example code to allow users to move the circle around the screen, the first thing you have to do is store the circle's rectangle as a member variable. Listing 11.2 shows the updated interface for your custom view, which now includes a member variable for this purpose.

Listing 11.2
The updated Custom UIView Interface

```
#import <UIKit/UIKit.h>

@interface MyCustomView : UIView
{
    CGRect circleRect;
    BOOL moving;
}
@end
```

I've also added a Boolean value, which you will be using to track whether or not you are currently moving the circle, based on whether the user's initial touch is inside the circle.

Because of this change to how you're storing `circleRect`, you also need to update your `drawRect:` method to make it so that you are just using your member variable instead of using a hard-coded `CGRect`. The updated version of this method is shown in Listing 11.3.

Listing 11.3
The Updated drawRect: Method

```
- (void)drawRect:(CGRect)rect
{
    // acquire the current context
    CGContextRef ctxt = UIGraphicsGetCurrentContext();
    // set the stroke color to black
    CGContextSetStrokeColorWithColor(ctxt, [[UIColor blackColor] CGColor]);
    // set the fill color to blue
    CGContextSetFillColorWithColor(ctxt, [[UIColor blueColor] CGColor]);
    // draw the outline of the circle
    CGContextStrokeEllipseInRect(ctxt, circleRect);
    // fill the circle
    CGContextFillEllipseInRect(ctxt, circleRect);
}
```

In addition to updating your `drawRect:` method, you also need to initialize these member variables. For this example, your custom `UIView` is being instantiated from within a nib file. This means that your designated initializer, `initWithFrame:`, is not called. Instead, you need to override the method `awakeFromNib`, which is called when your view is instantiated by the nib and is the appropriate place to initialize member variables. The code for this method is shown in Listing 11.4.

Listing 11.4
Implementation of the awakeFromNib Method

```
-(void)awakeFromNib;
{
    circleRect = CGRectInset([self frame], 100, 165);
    moving = NO;
    [super awakeFromNib];
}
```

> **NOTE**
> You are still initializing your member variable version by using the view frame to set your initial size. Once it's set, though, you are free to update it at any time before you redraw it.

In order to make it easy to set the new location of your circle, I've added a small utility function that sets the origin on your circle at an offset from the given point so that the center of your circle tracks your touches. This method is called `setCircleCenter:` and is shown in Listing 11.5. Add this method to your implementation as well so you can use it later.

Listing 11.5
Implementation of the setCircleCenter: Method

```
-(void)setCircleCenter:(CGPoint)centerPoint;
{
    CGSize centerSize;
    centerSize.width = circleRect.size.width/2;
    centerSize.height = circleRect.size.height/2;
    circleRect.origin.x = centerPoint.x - centerSize.width;
    circleRect.origin.y = centerPoint.y - centerSize.height;
}
```

Finally, you get to the implementation of your touches methods. The first one, `touchesBegan:withEvent:`, checks to see if the touch event is multi-touch. If it is, then for now you just ignore it. If it's not multi-touch, then you check to see if the location of the touch is inside your circle. If it is, then you set your moving member variable to YES, and begin updating the location of your `circleRect`. Implementation of this method is shown in Listing 11.6.

Listing 11.6
Implementation of the touchesBegan:withEvent: Method

```
- (void) touchesBegan:(NSSet*)touches withEvent:(UIEvent*)event
{
    if([touches count] == 1)
    {
        UITouch *touch = [touches anyObject];
        CGPoint touchPoint = [touch locationInView:self];
        if(CGRectContainsPoint(circleRect, touchPoint))
        {
            moving = YES;
            [self setCircleCenter:touchPoint];
            [self setNeedsDisplay];
            return;
        }
    }
}
```

The `touchesMoved` method is also simple. In it, again, you check to see if this is a multi-touch event; if it is, then you ignore it for now. If it isn't, and you set the moving variable to YES, then you update the location of the `circleRect` so that its center is where your new touch is located. Listing 11.7 shows how you do this.

Listing 11.7
Implementation of the touchesMoved:withEvent: Method

```
- (void) touchesMoved:(NSSet*)touches withEvent:(UIEvent*)event
{
    if(moving && [touches count] == 1)
    {
        UITouch *touch = [touches anyObject];
        CGPoint centerPoint = [touch locationInView:self];
        [self setCircleCenter:centerPoint];
        [self setNeedsDisplay];
    }
}
```

Last, but not least, you implement the `touchesEnded:withEvent:` method. In it, you check one last time to see if you're moving and to see if it's a single touch; if so, then you update your location one last time and then set your moving variable back to `NO`. In this way, you have ended your dragging of the object. This is shown in Listing 11.8.

Listing 11.8

Implementation of the touchesEnded:withEvent: Method

```
- (void) touchesEnded:(NSSet*)touches withEvent:(UIEvent*)event
{
    if(moving && [touches count] == 1)
    {
        UITouch *touch = [touches anyObject];
        CGPoint centerPoint = [touch locationInView:self];
        [self setCircleCenter:centerPoint];
        [self setNeedsDisplay];
    }
    moving = NO;
}
```

If you compile this code and run the application, you should now be able to touch inside the circle and drag it around the screen. If you touch outside the circle, or if you use a multi-touch with multiple fingers, nothing happens.

Now let's take this one step further and add the ability to scale the circle using a pinch gesture.

Adding scaling

To demonstrate how to handle touch events in custom views, let's extend this example app to include scaling your circle.

To do this, you first need to add another utility function to your custom view, which does the actual scaling. In this case, you're passing your two touches to the method and then setting the `circleRect` using those touches. Listing 11.9 shows that utility method.

CAUTION

Remember that you need to enable multi-touch on a view to receive multi-touch events. In this case, I did it in Interface Builder.

Listing 11.9

Implementation of the setCircleSizeFromTouches: Method

```
-(void)setCircleSizeFromTouches:(NSSet *)touches;
{
    // get the two points from the set.
    NSArray *touchesArray = [touches allObjects];
    UITouch *touchOne = [touchesArray objectAtIndex:0];
    UITouch *touchTwo = [touchesArray objectAtIndex:1];
    CGPoint pointOne = [touchOne locationInView:self];
    CGPoint pointTwo = [touchTwo locationInView:self];

    CGFloat x = MIN(pointOne.x, pointTwo.x);
    CGFloat y = MIN(pointOne.y, pointTwo.y);
    CGFloat width = MAX(pointOne.x, pointTwo.x) - x;
    CGFloat height = MAX(pointOne.y, pointTwo.y) - y;

    circleRect = CGRectMake(x, y, width, height);
}
```

Now that you have the method that you'll use to actually do the scaling, let's modify your touch methods.

You don't need to do anything to `touchesBegan:withEvent:`, but you do need to update `touchesMoved:withEvent:`. Listing 11.10 shows your updated version of that method.

Listing 11.10

The updated touchesMoved:withEvent:

```
- (void) touchesMoved:(NSSet*)touches withEvent:(UIEvent*)event
{
    if(moving && [touches count] == 1)
    {
        UITouch *touch = [touches anyObject];
        CGPoint centerPoint = [touch locationInView:self];
        [self setCircleCenter:centerPoint];
        [self setNeedsDisplay];
    }
    if([touches count] == 2)
    {
        [self setCircleSizeFromTouches:touches];
        [self setNeedsDisplay];
    }
}
```

Basically, you just check to see if there are two touches included in the set, and if so, you call your scale function. It's as easy as that! The `touchesEnded:withEvent:` method does virtually the same thing as shown in Listing 11.11.

Listing 11.11

The updated touchesEnded:withEvent: Method

```
- (void) touchesEnded:(NSSet*)touches withEvent:(UIEvent*)event
{
    if(moving && [touches count] == 1)
    {
        UITouch *touch = [touches anyObject];
        CGPoint centerPoint = [touch locationInView:self];
        [self setCircleCenter:centerPoint];
        [self setNeedsDisplay];
    }
    if([touches count] == 2)
    {
        [self setCircleSizeFromTouches:touches];
    }
    moving = NO;
}
```

That's it! Now if you run this app and use a pinch gesture, the circle scales appropriately.

Summary

In this chapter, I've demonstrated how to extend your custom `UIView` to support touch events and then to support multi-touch. This is a defining feature of the iPhone and one that you can leverage in your apps.

Although this feature is often used in games, you should also try to think about how it can be used in your apps to provide a more intuitive and natural user experience.

Working with Other Controls

You've seen how to create your own custom UIViews. You've also seen how to work with some of the basic controls. In this chapter, you'll take a look at some of the other controls that are available in Cocoa Touch and then see how you can use some of those controls in your application.

The important thing to remember is that Cocoa tends to use the same design patterns over and over again. This means that if you learn how to work with a few controls, it makes working with all of the controls easier. In particular, you'll notice extensive use of the delegate pattern and the data source pattern in most of these controls.

This is one of the things that make Cocoa and Cocoa Touch great, the fact that the designers of these APIs used well-known design patterns to guide them and reuse those design patterns over and over again for different types of components. If you learn these design patterns thoroughly, then you'll have no problem whatsoever working with any of the components available in Cocoa Touch.

Finding What Other Controls Are Available in Cocoa Touch

Cocoa Touch provides a rich library of user interface (UI) elements to use in your applications. The components available to you include switches, sliders, buttons, progress indicators, busy indicators, tab bars, segmented controls, and much more. I could probably fill an entire book just by talking about the different types of UI components available and how to use them. The collection that I'm going to talk about here is just a small sampling that represents three different styles of programming interfaces.

The first component you'll look at is the UISlider. I am using this as an example because it represents a basic control. You place it on your parent view, set its properties, and then forget it. If you need more information about it, you can simply access it from one of your IBOutlets.

In This Chapter

Working with different kinds of components in your UI

Using a UISlider to take variable value input from the user

Using a UITabBar to control a set of view controllers

Configuring a UIPickerView

The second component that you're going to look at is a `UITabBar`. I chose this one because it's used in conjunction with other views to control which view is the currently displayed one. It's a pretty complex component but shares some similarities with features such as navigation controllers and so on.

The third and final component that you're going to look at is a `UIPickerView`. The interesting thing about the `UIPickerView` is that it provides both a delegate interface and a data source interface to configure its components and to enable you to take action when the component is used. This is a really common pattern in Cocoa Touch. You've already seen it in use with `UITableView` and you will see it in use elsewhere.

Again, this is not a comprehensive list of all of the components in Cocoa Touch. The goal here is to look at a set of representative components and allow you to take the knowledge that you gain from working with them and apply it to other components. To really explore all of the different components available in Cocoa Touch, I suggest you look through the Interface Builder object selection palette and drag-and-drop components onto a test window to see how they look. Then look through the Apple documentation on the component's class. The documentation will tell you exactly what is necessary to utilize that component.

Working with a UISlider

The `UISlider` is the most basic of the components that you will be looking at in this chapter. It's really very simple to work with.

The purpose of the `UISlider` is to provide the user with an interface for setting a variable numeric value via a sliding control. The range of values that can be represented by the `UISlider` is entirely up to you and is configured when you set up the `UISlider`. To set the initial value of the `UISlider`, there are some simple methods that you use via an `IBOutlet` from your controller or by directly accessing them on your instance variable. Figure 12.1 shows what a `UISlider` looks like on a typical iPhone application screen.

Configuring a UISlider through Interface Builder

To add a `UISlider` to your application, you simply drag-and-drop it from the object selection palette onto whatever view you want to place it in. As with all other GUI components in Cocoa Touch, `UISlider` is a view, and so this means that you can also add it programmatically by simply creating a `UISlider` and calling `addSubview:` on the parent view.

The most important properties to configure on a `UISlider` are its Minimum value, Maximum value, and Initial value. Figure 12.2 shows the properties for a `UISlider` in Interface Builder. The Continuous property determines whether or not the `UISlider` sends continuous updates of its value as the user is sliding it. If you set this property to true, then as the user is adjusting the slider, you receive continuous updates. If you set this property to false, then you only receive an update when the user lifts her finger from the slider after setting it to its final position.

Figure 12.1
A `UISlider` on an iPhone application screen

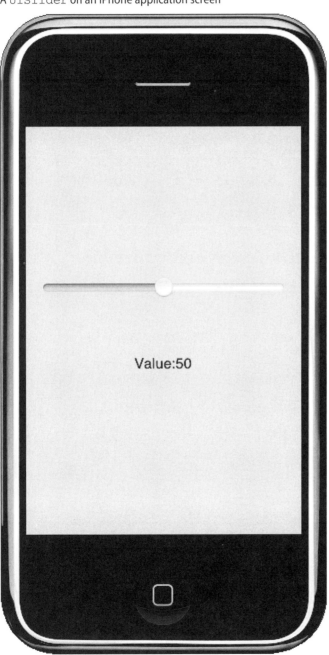

Figure 12.2
`UISlider` properties

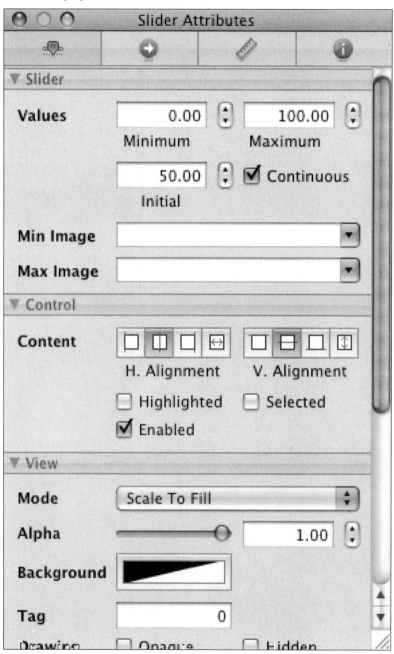

The Minimum and Maximum values indicate the minimum and maximum values that the slider can represent. The Initial value property in Interface Builder maps to the value property on the class and represents the initial value that the slider is set to. To set this programmatically, you would simply use the value property.

When customizing the appearance of UISlider, there are several properties that you can use to alter its appearance. Perhaps the most commonly used are the minimum and maximum value images. These images are placed underneath the slider at the minimum and maximum locations of the slider and can be used to display icons to represent the values at either of those locations. For example, if it's a volume slider, you might have a small speaker at one end and a large speaker at the other. You can also customize the appearance of the knob that's used to actually adjust the value. This knob is called the *thumb*, and it can be customized using the thumbImageForState property.

NOTE
You can set the thumb and track images for various states on the control. These states represent things such as selected state, highlighted state, and normal state.

Finally, you can also customize the appearance of the track itself by setting the property track ImageForState. When customizing the track image, you can customize either the minimum track image or the maximum track image, which correspond to the portion of track on the lower value side of the thumb versus the higher value side of the thumb. Remember that the track image must be able to stretch to fill in the region between the endpoints of the track and the current location of the thumb, so be sure your image will stretch appropriately.

Updating the status of your UISlider

When the user adjusts the UISlider it sends a message to an IBAction that you define. It sends these messages constantly when the user is sliding the control if you have selected continuous updates, or when the user is finished updating the control and he lifts his finger in the case of noncontinuous updates. You can use this IBAction to read the value of the slider and set whatever property you have assigned that slider.

Using UITabBar

The UITabBar component is interesting to look at because it is similar in functionality to the UINavigationBar. That is to say, it is more than just a graphical component that you put on your UI. It also manages a navigation metaphor for you by abstracting a navigational UI from your view controllers, which are subsequently controlling views that the UITabBar displays. Figure 12.3 shows a typical UITabBar in a UI.

Figure 12.3
A typical `UITabBar`

Configuring a UITabBar through Interface Builder

Like `UINavigationBar`, you should actually interact with the `UITabBar` primarily through a `UITabBarController`. To add a `UITabBarController` to your interface, in Interface Builder, you can drag-and-drop a new `UITabBarController` (shown in Figure 12.4) into your object management window. This adds a new `UITabBarController`, which includes a view containing a `UITabBar` that you can use to add to your main window as a subview. `UITabBarController` is a subclass of `UIViewController`, and so, alternatively, you can also instantiate a `UITabBarController` programmatically in the same way that you would a `UIViewController`.

> **NOTE**
> `UITabBarController` is intended to be used as-is and not subclassed.

Adding UITabBarItems to a UITabBar

As `UINavigationController` does with its `UINavigationItems`, `UITabBar Controllers` have `UITabBarItems` that are added to the `UITabBarController` by adding new `UIViewControllers` via the `UITabBarController`'s `viewControllers` property. There is also a method available on the `UITabBarController` called `setView Controllers:animated:`, which enables you to set the view controllers property while passing a parameter to determine whether or not the addition of the new view controllers will be animated.

> **NOTE**
> If you set more `UIViewControllers` on your `UITabBarController` than can be displayed, the `UITabBarController` displays the standard "More" `UITabBarItem`, allowing the user to customize which `UITabBarItems` to display. This behavior can be customized using the `customizableView Controllers` property on the `UITabBarController`.

By default, the `UITabBarController` creates a `UITabBarItem` for each `UIView Controller` that is added to the `UITabBarController`. This default `UITabBar` item doesn't contain an image but shows a title based on the title property of the `UIView Controller` being managed.

In order to configure images for the `UITabBarController`, your `UIViewController` should create a custom `UITabBarItem` with your desired image, and store it in its `tabBar Item` property.

> **NOTE**
> The image displayed on a `UITabBar` should be approximately 30x30 pixels. The image is displayed without color, using the image's alpha channel to create a silhouette effect on the `UITabBar`.

Figure 12.4
UITabBarController in Interface Builder

Once fully configured, the UITabBarController displays a UITabBar at the bottom of your user interface. When the user touches individual items on the UITabBar, it displays the root view from the view controller associated with that particular UITabBar item.

Doing advanced configuration of UITabBars

You can further customize the behavior of the UITabBarController by utilizing a UITabBarControllerDelegate. The UITabBarControllerDelegate protocol defines several methods that enable you to determine whether or not a user should be able to select a particular tab bar item or to take action when a tab bar item has been selected. Additionally, there are also events defined on the UITabBarControllerDelegate protocol that are triggered when the user begins customizing the UITabBarItems via the "More" UITabBarItem.

See the class documentation on UITabBarControllerDelegate for more information on this behavior.

Using UIPickerView

The last component that you are going to look at in this chapter is the UIPickerView. The UIPickerView is an instructive component to examine because it has both a delegate interface and a data source interface. This is a very common pattern that's used in Cocoa and Cocoa Touch. The UIPickerView component enables the user to make a single selection from a list of possible choices.

CAUTION

The UIDatePicker class looks very similar to UIPickerView, but it is not a subclass, nor does it allow for the same kind of customization that a UIPickerView does. The UIDatePicker does a great job selecting dates, but the UIPickerView is designed to be a more generic and customizable component.

Configuring UIPickerView through Interface Builder

Just like the other components that I've talked about, you can drag-and-drop a UIPickerView component onto your UI wherever you want to display it. Typically you want to connect an IBOutlet to the UIPickerView, and connect the UIPickerView's delegate and data source outlets to objects that handle those responsibilities.

The only real customizable property in Interface Builder is to determine whether or not the UIPickerView shows a selection indicator. Instead, the majority of customization for UIPickerView comes from the delegate and the data source.

Creating the UIPickerViewDataSource

A `UIPickerView` appears in your UI to be a barrel-shaped dial that you can spin to select a particular value. The barrel is broken up into parts called components. In this way, you can allow the user to configure several different values at once using the different components. For example, you can have a component to represent each of the hundreds, tens, and ones places in a number selection. This is much easier on the user than, for example, displaying a single component with the numbers 1 through 1,000.

The `UIPickerViewDataSource` works a little differently from data sources that you've seen so far in that it doesn't actually present the data displayed in the `UIPickerView` per se, but instead configures only the components and row counts for the `UIPickerView`. The actual data displayed in the `UIPickerView` is retrieved from the delegate.

To configure the number of components on your `UIPickerView`, implement the method `numberOfComponentsInPickerView:` on your `UIPickerViewDelegate`. Again, this method should return the number of components that you want displayed. So, following your example earlier where you wanted to display a component for each of the 100s, 10s, and 1s places in a number, the number of components would be three.

To configure the number of rows in a given component, implement the method `pickerView: numberOfRowsInComponent:`. Still following the previous example, to display ten digits, you would return 10 (0-9) as the number of rows in each of the components.

Listing 12.1 shows implementations of both of these methods.

Listing 12.1

Implementation of a UIPickerViewDataSource

```
- (NSInteger)numberOfComponentsInPickerView:(UIPickerView *)pickerView
{
    return 3;
}
- (NSInteger)pickerView:(UIPickerView *)pickerView
numberOfRowsInComponent:(NSInteger)component
{
    return 10;
}
```

These are incredibly simple examples, but I include them here so that you can see the complete method signatures.

Creating a UIPickerView delegate

In a `UIPickerView`, the delegate is the one that provides the actual data that's displayed in the rows. You can implement either the method `pickerView:titleForRow:forComponent:` or the method `pickerView:viewForRow:forComponent:reusingView:`; these return the text to display in a given row for a given component, or an actual complete view to display for that row for that component, respectively.

Following my earlier example, you might implement the method `pickerView:titleForRow:forComponent:` such that it returns the numbers zero through nine, depending on the row that's being requested. This is shown in Listing 12.2.

Listing 12.2

UIPickerViewDelegate titleForRow code

```
- (NSString *)pickerView:(UIPickerView *)pickerView
         titleForRow:(NSInteger)row
         forComponent:(NSInteger)component
{
    return [NSString stringWithFormat:@"   %ld", row];
}
```

An important thing to take away here is that the `UIPickerView` asks the data source for the number of rows in the number of components. It then asks the delegate what title to display in each one of those rows.

Further customization, such as the width and height of components and rows, can also be done by implementing the methods `pickerView:widthForComponent:` and `pickerView:titleForRow:forComponent:` on the delegate.

Finally, when the user selects an item from the `UIPickerView`, the method `pickerView:didSelectRow:inComponent:` is called on your delegate, giving you the opportunity to take the selected row and set it in your model or take some other action.

Summary

By looking at these three different types of components that can be used in your iPhone application interfaces, you've seen three different models of UI component design.

The first, demonstrated by use of the `UISlider`, is the most basic. Things such as buttons, labels, switches, and so on all follow the same model. You place them on your view, you configure them, you might attach them to an `IBAction`, or you might have an `IBOutlet` that enables you to set their value or read their value later.

The second, demonstrated by the use of `UITabBar`, demonstrates a view and its accompanying view controller that control a set of child view controllers and provide a consistent navigational model to view each of those child view controllers.

And finally, the `UIPickerView` utilizes a delegate and data source to allow you to configure the appearance and behavior of the component.

Between these three different types of components, you should be able to take the knowledge that you've learned here and apply it to other components that you can find in Interface Builder.

Handling Text Input

Most applications allow some kind of textual input to be used with them. Whether it's text, telephone numbers, or whatever, more than likely during your career as an iPhone app developer you're going to need to work with text input.

Another of the unique features of the iPhone is its on-screen keyboard. Using it, the iPhone can be adapted to display different kinds of alphabets, symbols, and so forth. The text system that provides this keyboard has a variety of capabilities that can be utilized by developers, allowing them to select the kind of capitalization, the kind of letters displayed, and the type of auto correction to use, among other things.

In this chapter, you're going to take a look at how you can configure the keyboard, both through Interface Builder and through code. You're also going to look at how you handle events that occur with the keyboard, such as when the text changes or when the user is finished editing. When you're finished, you will have a solid grasp of how to work with text with your iPhone applications.

Configuring the Keyboard through Interface Builder

There are two UI components in Cocoa Touch that are designed specifically for handling text entry: `UITextField` and `UITextView`. `UITextField` is designed to handle a single-line entry of text and `UITextView` is designed to handle multiple lines of text. Both of these components adhere to the `UITextInputTraits` protocol, which means that they're able to configure the type of keyboard that is used to enter text into them.

I'll talk more about `UITextInputTraits` when I talk about configuring text behavior through code. For this section, you're just going to look at how you configure the keyboard on text fields through Interface Builder. You don't necessarily need to know why it works yet; you just need to know how it works.

In This Chapter

Learning how to work with text input and text fields

Configuring the on-screen keyboard

Understanding UITextInputTraits

Handling keyboard events with a delegate

> **NOTE**
> To demonstrate setting up text fields through Interface Builder, most of the screenshots that I'm going to show you are using `UITextFields`. You should be aware that everything that I'm going to show you with regard to configuring the text behavior is equally applicable to a `UITextView`.

> **NOTE**
> `UIWebView` does not support setting of `UITextInputTraits`, but you can set some of the `UITextInputTraits` properties within the HTML used in the `UIWebView` by using properties on the input fields in the HTML code.

Figure 13.1 shows a view with a text field. It's nothing special, but that's what you're going to use.

Figure 13.1

Interface Builder with a `UITextField`

The items you're interested in configuring are shown in the properties window when the `UITextField` is selected. They are the Text Input Traits values.

Configuring capitalization

When you look at the text-input properties configuration items in Interface Builder, highlighted in Figure 13.2, you can see what you're going to be configuring with regard to text input. The first thing I'm going to talk about is capitalization.

Figure 13.2

The text input traits properties in Interface Builder

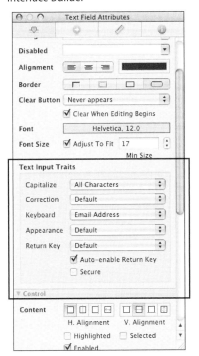

Capitalization on text entry can be configured to capitalize words, sentences, none, or everything. In the case of capitalizing words, the first letter of each word in the sentence is capitalized. In the case of capitalizing sentences, the first letter of each sentence is capitalized. And naturally, in the case of everything or none, this setting controls whether all of the text in the field is completely capitalized or if nothing is capitalized. The user can, of course, override these settings for the most part by pressing Shift on the keyboard.

Enabling and disabling auto correction

The next setting to look at is auto correction. You can enable or disable auto correction by setting the auto correction property. When auto correction is enabled, words are suggested to the user as she is typing, and if she doesn't want the suggested correction to occur, then she needs to touch the screen where the word is to remove the suggested correction. If you disable the auto correction feature, no words are suggested to the user as she is typing, and the input box does not attempt to complete the words that she is typing. This is a good idea for fields where you know that the words are typically not going to be in the iPhone dictionary, terms such as URLs for Web sites and so on.

CAUTION
If the user has chosen to disable auto correction in settings, this will override what you set here. This is important to know, in the event you enable auto correction, but you don't see it working. This might be why.

Setting the keyboard type

The next setting is for the keyboard type. Figure 13.3 shows several of the available keyboard types that you can choose from.

Figure 13.3
Different types of keyboards in iPhone OS

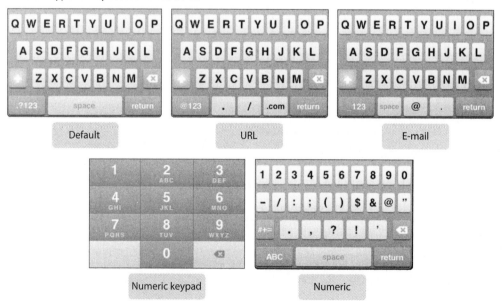

Because a picture is worth a thousand words, I don't think there's much to be added here in terms of what kinds of keyboards are available with what kinds of settings. However, I will say that generally speaking, the default keyboard works for most applications. If, however, you are working with, for example, URLs or e-mail addresses, then you may find that the keyboards specifically designed for those purposes result in a better user experience. They have strategically placed "@" keys and the ability to bring up extended text by touching and holding the ".com" key.

Setting the behavior of the Return key

The text displayed in the Return key can be customized as well. It can be set to anything from "Done" to "Go," and so on. You should configure this text according to what would be appropriate for your application. The property to set this, defined on UITextInputTraits, is returnKeyType.

Looking at other settings

Finally, let's take a look at the last few settings that are available. If you set a text field to be secure, then instead of displaying the text as you're typing, the text field displays a circle for each of the letters as they are typed. This is intended for things such as passwords so that you don't compromise the user's password while he is typing it.

You have the ability to automatically enable and disable the Return key. When you enable this check box, if the user has not typed anything into the text field, then he is unable to press the Return key. Once he has entered something into the text field, he is able to press Return. This is useful for restricting input to ensure that the user enters a particular value versus leaving it blank.

You have the ability to set the general "appearance" setting, which essentially designates whether the keyboard is to be used on an alert message or on normal input. Alert messages are displayed in a different color on the iPhone from a normal input screen; as a result, you may want your keyboard to take on this appearance when using it.

Working with a Keyboard through Code

You've seen how you work with text input properties through Interface Builder. You've also seen how simply dropping a text field onto a form is enough to get the text keyboard to display once the user touches inside the field. Now let's take a look at how you force the keyboard to display through code and also how you can manipulate UITextInputTraits through code.

Making the keyboard appear

By now, you've probably seen that if you drop a UITextField onto an interface and allow the user to touch inside that UITextField, it automatically displays the keyboard. But what if you want to force the keyboard to display as soon as your view displays, or as a result of the user touching some other place in the UI? The way to do this is simply by setting your text field to be the first responder.

Remember that Cocoa and Cocoa Touch incorporate a concept of a responder chain. The responder chain is a list of user interface elements that handle input from the user. The first responder is the `UIView` that currently has focus. When a `UITextView` becomes the first responder, it creates and configures a keyboard based on its `UITextInputTraits` and displays that keyboard. So the way to force the `UITextView`'s keyboard to display is to simply call the method `becomeFirstResponder` on the `UITextView`. Listing 13.1 shows an example of how to do this.

Listing 13.1

Showing the Keyboard

```
-(void)showKeyboard
{
    [myTextField becomeFirstResponder];
}
```

Similarly, to dismiss the keyboard, you simply resign the first responder using the `resignFirstResponder` method, as shown in Listing 13.2.

Listing 13.2

Dismissing the Keyboard

```
-(void)dismissKeyboard
{
    [myTextField resignFirstResponder];
}
```

This works with either `UITextField` or `UITextView`.

Understanding UITextInputTraits

So far, all the work that you've done has been related to using `UITextField` and `UITextView`, and all of the items that you configured through Interface Builder can also be configured using properties on those classes. The reason that these classes have these properties is because they both adhere to the `UITextInputTraits` protocol and implement the properties specified by it.

If you need to create a custom view that supports text input, you should also implement this protocol. Table 13.1 shows the properties defined by this protocol.

Table 13.1 Properties Defined on the UITextInputTraits Protocol

Property Definition	Purpose
`@property(nonatomic) UITextAutocorrectionType autocorrectionType`	Configures the auto correction type
`@property(nonatomic) UITextAutocapitalizationType autocapitalizationType`	Configures the auto capitalization type
`@property(nonatomic) BOOL enablesReturnKeyAutomatically`	Defines whether or not the Return key automatically enables and disables
`@property(nonatomic) UIKeyboardAppearance keyboardAppearance`	Configures the keyboard appearance
`@property(nonatomic) UIKeyboardType keyboardType`	Configures the keyboard type
`@property(nonatomic) UIReturnKeyType returnKeyType`	Configures the Return key type
`@property(nonatomic, getter=isSecureTextEntry) BOOL secureTextEntry`	Defines whether or not the text field is secure

Setting these properties on any field implementing the `UITextInputTraits` protocol should enable you to configure these values as shown.

Handling Events for the Keyboard

The keyboard provides a variety of events that your application can monitor to take action in particular conditions. For example, you can be notified when the keyboard displays so that you can move your view out of the way of the keyboard or be notified when the user has finished editing so that you can take the field data and do something useful with it. Additionally, `UITextField` supports configuring a `UITextFieldDelegate`, which enables you to customize the behavior of the input to control things such as whether or not the user should be able to begin editing, finish editing, or change particular characters in the field.

Let's take a look at how you work with these notifications and delegate methods.

Creating a UITextFieldDelegate

Because most applications want to be notified when the user has finished editing the text, you will probably want to implement a `UITextFieldDelegate` or `UITextViewDelegate` in your application. The implementation between the two of them varies slightly, but for demonstration purposes, I'll be focusing on the `UITextFieldDelegate`.

Knowing when the text changes

Typically, you set your view controller to be the `UITextFieldDelegate` for your `UIText Fields` on your view. The most important methods that you'll probably be interested in implementing are `textFieldShouldEndEditing` and `textFieldDidEndEditing`. These methods are called just before and after the text field is asked to resign the first responder. The method `textFieldShouldEndEditing:` provides an excellent hook for your application to do input validation, to determine whether or not you want to allow the user to dismiss the keyboard and continue working. The `textFieldDidEndEditing:` method provides an excellent location for you to take the data from the text field and update your model.

Table 13.2 shows some of the other methods defined on the `UITextFieldDelegate` protocol.

Table 13.2 Methods Defined in the UITextFieldDelegate Protocol

Method Signature	Purpose
`- (BOOL) textField: (UITextField *) textField shouldChangeCharactersIn Range: (NSRange) range replacement String: (NSString *) string`	This method is called whenever the user types a new character into the text field and gives you the opportunity to prevent that character from being typed.
`- (void) textFieldDidBeginEditing: (UITextField *) textField`	This method is called after the text field has become the first responder.
`- (void) textFieldDidEndEditing: (UITextField *) textField`	This method is called after the text field has resigned the first responder.
`- (BOOL) textFieldShouldBeginEditing: (UITextField *) textField`	This method is called before the text field becomes the first responder.
`- (BOOL) textFieldShouldClear: (UITextField *) textField`	This method is called in response to pressing the built-in clear button on the `UITextField`.
`- (BOOL) textFieldShouldEndEditing: (UITextField *) textField`	This method is called before the text field resigns the first responder.
`- (BOOL) textFieldShouldReturn: (UITextField *) textField`	This method is called when the user has pressed the Return key and gives you the opportunity to disallow her from doing so.

Taking action when the user has finished editing

Again, perhaps the most important of these methods is the `textFieldDidEndEditing:` method, which allows you to be notified when the user has finished editing the text field and thus take the value in the text field and set it on one of your model objects.

Listing 13.3 shows how to handle a text field that has resigned the first responder and how to take its value and put it into a model.

Listing 13.3

Handling the User Finishing Editing

```
- (void)textFieldDidEndEditing:(UITextField *)textField
{
    [[self setTextValue:[textField text]];
}
```

Moving your view in response to the keyboard appearance

When the keyboard slides in and out of view, it posts several different types of notifications to the `NSNotificationCenter`. These notifications are shown in Table 13.3.

Table 13.3 Notifications Sent When the Keyboard Displays and Hides

Notification	Purpose
`UIKeyboardWillShowNotification`	This notification is sent before the keyboard appears.
`UIKeyboardDidShowNotification`	This notification is sent after the keyboard appears.
`UIKeyboardWillHideNotification`	This notification is sent before the keyboard disappears.
`UIKeyboardDidHideNotification`	This notification is sent after the keyboard disappears.

One of the most common tasks that needs to be done when working with text fields is to move your view to a more appropriate location when the keyboard is sliding into view.

To do this, configure your view to listen for the `UIKeyboardDidShowNotification`, and when it occurs, slide your view such that your text field is still visible over the top of the keyboard.

In portrait mode, the keyboard is 216 pixels tall. In landscape mode, the keyboard is 162 pixels tall. When the keyboard show and hide notifications are posted, the user info dictionary that is included with the notification includes information about the size of the keyboard that will be displayed. You can use this information to adjust your content as needed. Listing 13.4 shows an example of how to handle this event.

Listing 13.4

Handling Moving Your View to Make Room for the Keyboard

```
-(void)viewDidLoad;
{
    // .. other initialization ...
    NSNotificationCenter *ctr = [NSNotificationCenter defaultCenter];
    [ctr addObserver:self selector:@selector(handleShowKeyboard:)
                name:UIKeyboardDidShowNotification object:nil];
    [ctr addObserver:self selector:@selector(handleHideKeyboard:)
                name:UIKeyboardDidHideNotification object:nil];
    [super viewDidLoad];
}
-(void)handleShowKeyboard:(NSNotification *)inNotification;
{
    NSNotification *userInfo = [inNotification userInfo];
    NSValue *kbBounds = [userInfo objectForKey:UIKeyboardBoundsUserInfoKey];

    [UIView beginAnimations:@"MOVING FOR KB" context:nil];
    CGRect currentFrame = [myTextField frame];
    currentFrame.origin.y -= [kbBounds CGRectValue].size.height;
    [myTextField setFrame:currentFrame];
    [UIView commitAnimations];
}
-(void)handleHideKeyboard:(NSNotification *)inNotification;
{
    NSNotification *userInfo = [inNotification userInfo];
    NSValue *kbBounds = [userInfo objectForKey:UIKeyboardBoundsUserInfoKey];

    [UIView beginAnimations:@"MOVING FOR KB" context:nil];
    CGRect currentFrame = [myTextField frame];
    currentFrame.origin.y += [kbBounds CGRectValue].size.height;
    [myTextField setFrame:currentFrame];
    [UIView commitAnimations];
}
```

Summary

In this chapter, you've taken a wide-ranging tour of how text input is handled on the iPhone. You saw how to customize a keyboard for different kinds of text input. You learned how to display and dismiss the keyboard. You saw how to handle events related to the keyboard so that you could adjust your UI appropriately when the keyboard is displayed and so on. You also saw how to implement a `UITextFieldDelegate` so that you can handle the input from the user, and how to validate input and prevent the user from exiting out of text entry fields, if needed.

Using these capabilities, you can build applications that efficiently utilize text input to make it easier for users to enter data into your applications.

Building Cinematic UIs with Core Animation

One of the cornerstones of the iPhone UI is its use of animation to build what Apple calls a "Cinematic User Experience." A cinematic user experience is essentially one where the user elements don't simply pop into view; instead, they slide, fade, and otherwise transition into position using animation. The judicious use of animation provides what can only be described as a certain *weight* to the user interface elements. They *feel* more realistic because of the use of animation. This results in a UI that is more intuitive, and easier to use.

Because of the ubiquity of animation in iPhone OS, Apple has provided you with several very convenient APIs for using animation within your own applications.

In this chapter, I'll talk about each of these techniques, and see how you can apply them in your own applications.

Using the UIViews Animation Methods

The first, and simplest, method for doing animation in iPhone OS 3 is by using the capabilities built right into the `UIView` class.

The `UIView` class contains several class methods that you can use to do simple animations of your views. Because all user interface elements inherit from `UIView`, this means that this capability is already built into all of your apps; you need only utilize it, and utilizing it is incredibly simple.

`UIView` Core Animation methods work by setting up animation blocks within your code using the `UIView` class methods `begin Animations:context:` and `commitAnimations`. Then, any changes you make to `UIView`s within that block, such as moving, resizing, or changing their color, are automatically animated for you. It's as simple as that! This is incredibly powerful and easy to use when you are animating the repositioning of UI elements in your applications.

In This Chapter

Using UIView class methods to animate user interface elements

Using CALayer to do more complex keyframe animation

Using UIImageView animated images to display rendered animations

Because code is often easier to understand than a description, let's take a look at an example, where you want to move a button from one location to another, but you want to animate that motion. Listing 14.1 shows how this is done.

Listing 14.1

Moving a Button with Animation

```
-(void)moveButton;
{
    [UIView beginAnimations:@"BUTTONMOVING" context:nil];
    [UIView setAnimationDuration:3.0];
    [UIView setAnimationDelegate:self];
    [UIView setAnimationDidStopSelector:@selector(animation:didFinish:context:)];

    [button setCenter:CGPointMake(160, 240)];

    [UIView commitAnimations];

}
```

NOTE
The @"BUTTONMOVING" parameter in Listing 14.1 is a unique identifier for this animation. It can be anything you want, and can be used to later identify a particular animation in the animation delegate methods.

In this case, the animation that you are doing is very simple. Basically, this code is saying, "animate moving the button to the location 160, 240, over 3 seconds. When the animation completes, call the method on your `self` object called `animation:didFinish:context:`."

The important thing to understand about this code is that an animation block is defined by the `begin` animations call to the `commit` animations call. Any changes to properties of `UIViews` within those two calls will be animated. In this particular case, the only thing you're doing is changing the center location of your button. However, you could be doing things as complicated as changing the frame, or changing the alpha transparency to cause a fade.

You set the properties of the animation itself using class methods on the `UIView` class. The methods available on the `UIView` class for configuring animation are shown in Table 14.1.

Table 14.1 UIView Animation Methods

Method	Purpose
`+ (void)beginAnimations:(NSString *) animationID context:(void *)context`	A call to this method begins an animation block. Any calls made subsequent to this call before the `commitAnimations:` call are applied to the current animation block.
`+ (void)commitAnimations`	A call to this method closes an animation block and begins the animation.
`+ (void)setAnimationStartDate: (NSDate *)startTime`	This method is used to set an animation to start at some point in the future, as opposed to starting immediately as soon as the commit animations call is made.
`+ (void)setAnimationsEnabled:(BOOL) enabled`	This method enables or disables animations.
`+ (void)setAnimationDelegate:(id) delegate`	This method sets the animation delegate, which will receive events when the animation completes.
`+ (void)setAnimationWillStart Selector:(SEL)selector`	This method sets the selector that will be called on the delegate prior to the animation beginning.
`+ (void)setAnimationDidStopSelector: (SEL)selector`	This method sets the selector that will be called on the delegate when the animation completes.
`+ (void)setAnimationDuration: (NSTimeInterval)duration`	This method sets the duration of the animation.
`+ (void)setAnimationDelay:(NSTime Interval)delay`	This method configures the animation to start after the given delay, as opposed to immediately.
`+ (void)setAnimationCurve:(UIView AnimationCurve)curve`	This method configures the desired animation curve.
`+ (void)setAnimationRepeatCount: (float)repeatCount`	This method configures the animation to repeat for the given number of times.
`+ (void)setAnimationRepeat Autoreverses:(BOOL)repeat Autoreverses`	This method configures the animation to automatically reverse when it repeats.
`+ (void)setAnimationBeginsFrom CurrentState:(BOOL)fromCurrentState`	This method configures the animation to begin from the current `UIView` state. The default value of this is NO. This option only relates to when you are beginning an animation while another animation is currently executing. If you configure this for YES, the beginning location of the new animation that you are configuring will be wherever the currently executing animation is at. The default value allows the prior animation to complete before beginning the new animation.

continued

Table 14.1 Continued

Method	Purpose
`+ (void) setAnimationTransition: (UIViewAnimationTransition) transition forView: (UIView *) view cache: (BOOL) cache`	This method allows you to animate a transition from one view to another using one of the standard view transition animations, such as flipping from left to right or vice versa.
`+ (BOOL) areAnimationsEnabled`	This method returns whether or not animations are enabled.

Most of these methods are reasonably straightforward, but let's discuss a few of them in more detail. The method `setAnimationCurve:` is particularly interesting, because it enables you to configure the behavior of the animation itself. An animation curve defines the speed of the animation as it runs through its course. The default type of animation curve is `UIViewAnimationCurveEaseInOut`, which causes the animation to start off slowly, accelerate during the middle of the animation, and then slow down again as it reaches its completion. This may or may not be appropriate for your particular UI needs; as a result, Apple provides you with a variety of settings for this particular property, as shown in Table 14.2.

Table 14.2 Animation Curve Types

Type	Purpose	Diagram
`UIViewAnimationCurveEaseInOut`	This is the default animation curve. This curve is defined by starting off slowly, accelerating to full speed during the middle of the animation, and then slowing down again at the end.	
`UIViewAnimationCurveEaseIn`	This animation curve is defined by starting off slowly, then speeding up during the middle, and maintaining its speed until the end.	

Type	Purpose	Diagram
`UIViewAnimationCurveEaseOut`	This animation curve is defined by starting off at full speed, and then gradually slowing down as it reaches the end of the animation.	Start → End (Time)
`UIViewAnimationCurveLinear`	This animation curve is defined as having a steady speed throughout its entire duration.	Start → End (Time)

Another task that's often required when working with animations is related to knowing when an animation completes. The animations themselves take place in the run loop of the application asynchronously. This means that you don't simply tell it to animate and then wait for the animation to complete. The animation is running while your application is doing other things. This means that you need to know when the animation has completed so that you can do any finishing tasks, such as launching another animation or removing a view that was faded out.

To do this, you configure the delegate for the animation block to be the object that reacts to these events. You then configure a selector on that delegate to be called when the animation has completed. You do this using the calls `setAnimationDelegate:` and `setAnimationDidStopSelector:`.

Let's take a look at an example where you want the button to fly off the screen, and then be removed. In this case, you're going to have the animation start off slowly, accelerate to full speed, and stay at full speed until it's finished. In this way, it will appear that the button has weight — that it needs to gain momentum in the beginning as it's beginning to move, and once it's going, it's gone. Listing 14.2 shows this code in action.

Listing 14.2

Animating Removal of a Button

```
-(void)removeButton;
{
    [UIView beginAnimations:@"BUTTONMOVING" context:nil];
    [UIView setAnimationDelegate:self];
    [UIView setAnimationDidStopSelector:@selector(animation:didFinish:context:)];

    // this is setting the curve
    [UIView setAnimationCurve:UIViewAnimationCurveEaseIn];

    // set the button position off screen somewhere.
    [button setCenter:CGPointMake(480, 720)];

    [UIView commitAnimations];
}
```

Now the interesting thing here is that because you are removing this button from the screen, you also want to remove it from the view hierarchy when the animation is completed. If you remove the button from the view hierarchy here in this method, you don't see the animation, because the removal of the button takes place before the animation has actually started. Remember, the animation starts asynchronously in your run loop after you've exited this method. Therefore, you need to be notified that the animation has completed. Then, in that method you remove the button from the view hierarchy.

To be notified when the animation completes, you set the animation "did stop selector" using the method `setAnimationDidStopSelector:`, which takes as an argument a selector, on your delegate, that will be called when the animation stops. You then implement this method on your delegate, and within this implementation you remove the button from its super view. Listing 14.3 shows how you do this.

Listing 14.3

Implementation of the Animation "Did Stop Selector"

```
- (void)animation:(NSString *)inAnimationID
       didFinish:(NSNumber *)inFinished
         context:(void *)inContext
{
    // the animation finished, so remove the button
    [button removeFromSuperview];
}
```

Finally, the `UIView` animation blocks can be used when you want to animate the flipping of the user interface to transition from one `UIView` to another; you can use the method `set AnimationTransition:forView:cache:` to configure the view to transition from one view to another. This was a very common requirement under iPhone OS 2.0, in cases where you wanted to display the flip side of the user interface for configuring settings. iPhone OS 3 introduces the `UIViewController` property `modalTransitionStyle`, which makes this less necessary today, but it's still a valuable technique to know, both for working with older devices and for doing flip transitions on views that don't necessarily have a view controller associated with them.

To use an animation transition to toggle between two `UIView`s involves several steps. First, you have to create a container view that will actually do the transition. You add your currently displaying `UIView` to the container view, and display it. When you want to transition the views, you begin your animation block using the `beginAnimations:context:` class method. Within this animation block is where the real action happens. You set the animation transition using the method `setAnimationTransition:forView:` to the type of transition you want (flip left, flip right, and so on). You then remove the current subview from the container view and add your new subview. When you commit the animations, your transition takes place. Listing 14.4 shows this in action.

Listing 14.4
Animating a Flip Transition from One View to Another

```
-(void)flipTransition;
{
    [UIView beginAnimations:@"FLIP" context:nil];
    [UIView setAnimationTransition:UIViewAnimationTransitionFlipFromRight
                           forView:containerView
                             cache:YES];

    [currentSubview removeFromSuperview];
    [containerView addSubview:newSubview];
    [UIView commitAnimations];
}
```

As you can see, this makes animating basic UI transitions very simple. Now, what if you want something a little bit more complex? The animations that I've shown you here are all very basic. However, the kinds of animation that you may want to use involve things like keyframes, complex motion, and so forth. The class methods on `UIView` are not well suited to this. To do this, you need to go a little bit deeper and at a slightly lower level and work with the core animation API itself.

Using Advanced Core Animation with CALayer

As I said before, the `UIView` animation block code makes working with simple animations trivial. But what if you want something with a little bit more control? For example, what if you want to animate moving a graphic over a particular path? You see this done in the mail application when you send messages to folders: You see a small icon that moves in an arc and plunks down into the folder. Doing this using the `UIView` animation blocks would be onerous. Ideally, what you want to be able to do is simply define a path along which your image then moves. Fortunately for you, there is a low-level API that enables you to do exactly that.

Core animation consists of a collection of classes that work together to enable you to do complex animations. The most central of those classes is the `CALayer` class, which provides an abstraction for a view-like hierarchy of layers that can be animated. All `UIViews` on iPhone OS have an associated `CALayer` property that can be accessed and operated on using the other core animation classes. In the following example, you will use this capability to animate an image view over a particular path.

To access a view's `CALayer`, you simply call the method layer on a `UIView` instance. The path that the image you will travel along is simply a core graphics path. You create it using core graphics calls, and then you create a `CAKeyframeAnimation` using the path. You then add this animation to your layer.

Listing 14.5 shows the creation of your animation path. In it, you want to move from a location in the upper-left corner to the lower-right corner. You're going to do it in an arc, so that it's like you're dropping it into a bucket.

Listing 14.5

Creation of the Animation Path

```
CGPoint startPoint = CGPointMake(20, 20);
CGPoint controlPoint1 = CGPointMake(280, 100);
CGPoint controlPoint2 = CGPointMake(280, 100);
CGPoint endPoint = CGPointMake(300, 440);

CGMutablePathRef path = CGPathCreateMutable();
CGPathMoveToPoint(path, NULL, startPoint.x, startPoint.y);
CGPathAddCurveToPoint(path, NULL,
                      controlPoint1.x, controlPoint1.y,
                      controlPoint2.x, controlPoint2.y,
                      endPoint.x, endPoint.y);
```

Figure 14.1 shows the path you want to see exhibited.

Figure 14.1
The path of the animation

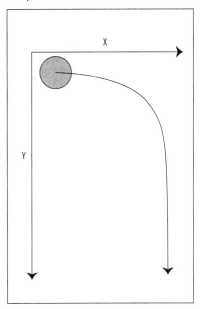

Once you have the path created, you're going to create a keyframe animation. Now when you create this animation, you have to associate the animation with a particular property on your layer so that when that property is modified, your animation will fire. In this particular case, you're going to use the "position" property. This enables you, when the position of the layer is changed, to have the animation run. This is a little bit counterintuitive, in that you create the animation itself with the property that you want to observe in mind. This is shown in Listing 14.6.

Listing 14.6
Creating the Keyframe Animation

```
CAKeyframeAnimation *keyFrameAnimation =
[CAKeyframeAnimation animationWithKeyPath:@"position"];
[keyFrameAnimation setDuration:1.0];
[keyFrameAnimation setPath:path];
[keyFrameAnimation setTimingFunction:
[CAMediaTimingFunction functionWithName:kCAMediaTimingFunctionEaseIn]];
```

NOTE
The property shown here is the KVC (Key Value Coding) description of the property on the layer that you want associated with this animation, in this case, @"position". When that property is modified, the animation fires.

Note that you set the path for the animation here. You're also setting a timing function so that again, the animation runs slowly to begin with and then speeds up until it finally finishes.

Finally, you get the layer for your image view that you want to animate, and you add the animation to that layer. To have the animation actually fire, you call the method `setPosition`, which is the key path for the property that you specified the animation should be associated with. In this case, you are setting your position to be the endpoint of your animation.

Listing 14.7 shows the final implementation of this entire method.

Listing 14.7
Animating an Image View Dropping to the Bottom of the Screen

```
-(void)kerPlunk;
{
    CGPoint startPoint = CGPointMake(20, 20);
    CGPoint controlPoint1 = CGPointMake(280, 100);
    CGPoint controlPoint2 = CGPointMake(280, 100);
    CGPoint endPoint = CGPointMake(300, 440);

    CGMutablePathRef path = CGPathCreateMutable();
    CGPathMoveToPoint(path, NULL, startPoint.x, startPoint.y);
    CGPathAddCurveToPoint(path, NULL,
                          controlPoint1.x, controlPoint1.y,
                          controlPoint2.x, controlPoint2.y,
                              endPoint.x, endPoint.y);

    CAKeyframeAnimation *keyFrameAnimation =
    [CAKeyframeAnimation animationWithKeyPath:@"position"];
    [keyFrameAnimation setDuration:1.0];
    [keyFrameAnimation setPath:path];
    [keyFrameAnimation setTimingFunction:
     [CAMediaTimingFunction functionWithName:kCAMediaTimingFunctionEaseIn]];

    CALayer *layer = [imageView layer];
    [layer addAnimation:keyFrameAnimation forKey:@"KERPLUNK"];
    [layer setPosition:endPoint];
    CFRelease(path);
}
```

Core animation is relatively complex subject. It provides a lot of power to the application developer to build complex animations for animating user interface elements. As a result, you could probably fill a book this size completely with core animation tips and tricks and documentation.

What I've shown you here is really just a taste of the capabilities that you can unleash by digging just a little bit deeper into core animation. For more information, I recommend looking at the core animation documentation, which also includes a cookbook for developers who want to work with core animation.

Now before you finish this chapter, I'd like to talk about one more aspect of animation in Cocoa Touch: using animation in `UIImageView`s.

Animating UIImageViews

So now you've seen how to do basic animations using properties directly on `UIView`s, and when that doesn't get you what you're looking for, you've seen that you can dig deeper into the API to directly use the core animation classes to do more complicated animation. But what if the animation that you want to do involves something that would be easier to render ahead of time and play in your user interface? For example, you may want to have a rendered animation of a mailbox slamming shut and its mail-waiting flag popping up. An animation such as this would be relatively complex to do using core animation and transforms. It would be much easier to have an artist render an animation such as this and then simply display it as a little movie on your UI. It's not possible to use the media playing capabilities of the iPhone to display a movie on a button. So then how do you do this?

The answer is relatively simple. `UIImageView` has the capability of animating itself using cells that are provided for that purpose. The `UIImageView` instance method `animationImages` allows you to set an array of images that make up the animation. The method `startAnimating` then causes the `UIImageView` to display the cells provided in `animationImages`, one after another, as if it were displaying a movie. You can set the duration of the animation, as well as whether or not it loops. When you want the animation to stop, you can simply call `stopAnimating`.

NOTE
You can configure both the main image of the image view and the `animationImages` property. The main image displays when it's not animating. In most cases, the animated images can be a much lower quality and resolution than the main image of the image view. Because it's animating, the user's eye compensates for the difference in quality. This can help you to conserve resources.

Listing 14.8 shows an image view being loaded with cells for an animation and then animating. In this case, the animation will loop each second, showing 15 frames per second.

Listing 14.8
Showing an Animation in a UIImageView

```
-(void)animatingImageView;
{
    NSMutableArray *cells = [NSMutableArray array];
    for(int n = 0; n < 15; ++n)
    {
        NSString *cellName = [NSString stringWithFormat:@"cell%ld.png", n];
        UIImage *cell = [UIImage imageNamed:cellName];
        [cells addObject:cell];
    }
    [imageView setAnimationImages:cells];
    [imageView setAnimationDuration:1.0];
    [imageView startAnimating];
}
```

Summary

iPhone OS has an extraordinary set of tools that enable you to build animated user interfaces. Utilizing these tools gives you the ability to build applications that allow the user to have the illusion of direct manipulation of real-world objects when interacting with your application. Careful use of these capabilities can enhance an application's usability significantly. You should be cautious, however, not to use them gratuitously. By using them subtly, you provide the user with an additional level of detail in your application that they wouldn't otherwise have.

Using OpenGL ES

One of the things that excite application developers about the iPhone is that it ships with an excellent implementation of OpenGL ES. This means that developers have available to them the full power of OpenGL ES for developing 2-D and 3-D games. Some of the most amazing applications developed for the iPhone have been developed using OpenGL.

Unfortunately, OpenGL ES is a huge topic that I can't cover in depth in this book. It would simply be impossible for me to cover it with any level of competency in one or two chapters.

That said, however, I want to provide at least some information about OpenGL ES so that you have a basic understanding of how it works on the iPhone, and then to tell you where to get more information about OpenGL ES.

So, in order to do this, you'll first take a look at the implementation of OpenGL ES on the iPhone and then look at its features and limitations. You'll then take a look at the OpenGL ES template project and see what it gives you and how you can use it as a launching point for your own applications.

In This Chapter

Learning about the iPhone OpenGL ES implementation

Exploring the default OpenGL ES-based application

Interfacing your OpenGL ES code with the iPhone graphics hardware

Understanding the Capabilities of iPhone OpenGL ES

The first thing that you need to know about using OpenGL ES on the iPhone is that OpenGL ES is not OpenGL. While they address the same problem, the ability to draw two-dimensional and three-dimensional graphics on a computer screen in the API for OpenGL ES is different enough from the API for OpenGL that getting a book for OpenGL will not teach you anything about OpenGL ES. It will only lead to frustration. This doesn't mean they don't share similarities, of course. They are based on the same low-level functionality in the same way that Cocoa and Cocoa Touch are. However, they are different enough that when you look for resources on OpenGL ES, make sure you are looking at the ES version and not vanilla OpenGL.

The next thing to understand is that the OpenGL ES implementation of the iPhone has several limitations, even beyond just being an ES implementation. The first limitation is that the iPhone only supports triangular polygons, so when building your models, you need to make sure that your models are only using triangles. Polygons of any other shape do not work.

The second limitation is that the iPhone uses a shared memory model between the graphics and CPU. This means that although the iPhone OpenGL ES implementation supports vertex buffer objects (VBOs), you actually derive no significant performance improvement by using them. This also means that your application memory and your texture memory are the same; they come from the same memory pool. This means that when developing for any device other than an iPhone 3GS, your application and textures must not take up more than approximately 24MB of memory. As a result, you should also consider compressing your textures using the PVRTC format. This is the native texture format for the iPhone GPU.

Another limitation to be aware of is that the iPhone uses a technique called tile-based deferred rendering for its rendering system. This is an optimized rendering system that helps to keep memory use low by caching updates before uploading them to the GPU and only doing updates when it has a significant chunk of data to send. This means you should create and load all of your textures before rendering, and avoid operations that update or modify existing textures.

While not necessarily a limitation per se, you should try to avoid mixing OpenGL rendering with UIKit and core graphics rendering. OpenGL is optimized for applications that require high-performance animated gaming graphics, while UIKit is optimized for on-demand updates and precision-drawing operations. Don't try to mix the two.

Creating a Very Basic OpenGL View

Now let's take a look at how you create an application that uses OpenGL ES by looking at the template-generated code from Xcode when you choose to create an OpenGL ES-based application.

Figure 15.1 shows Xcode after you've created an OpenGL ES-based application. It contains the default files that are included as part of the project template.

Figure 15.1

An OpenGL ES application in Xcode

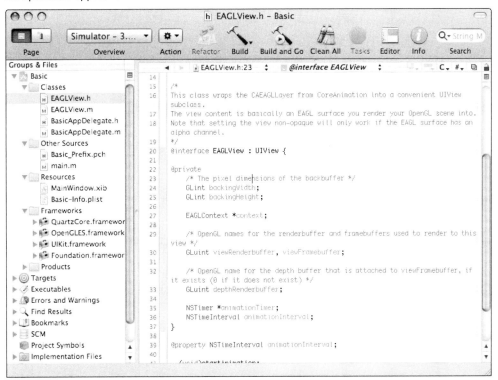

When you create an application using the OpenGL ES template, you get the usual application delegate, your main window nib, and so on. But you also get a custom `UIView`, which is a wrapper around an OpenGL view. This view is contained in the files `EAGLView.h` and `EAGLView.m`. These act as a wrapper around an OpenGL layer, to give you a convenient `UIView` that you can just drop into your application. You can just as easily take this chunk of code out of this project and put it into another of your projects and use it as a `UIView`. (Remember, though, that you shouldn't mix OpenGL views with `UIKit` views.)

Inside the `MainWindow.xib` file in the project, you see that you have a `UIWindow` and your app delegate. If you switch the instance browser window to list mode, you see that the `UIWindow` contains a `UIView` subview. The class identity of the `UIView` is your `EAGLView`, as shown in Figure 15.2.

Figure 15.2
Interface Builder with your view loaded

Looking now at the implementation of the `EAGLView`, let's see how it's implemented and where the actual OpenGL code goes.

Looking in the `EAGLView` header (shown partially in Listing 15.1), you see that you have an `EAGLContext` defined as a member variable. Most of the work surrounding how you interface OpenGL ES with the iPhone GUI involves working with the `EAGLContext` object.

Listing 15.1
The EAGLView Interface (partial)

```
@interface EAGLView : UIView {

@private
    /* The pixel dimensions of the backbuffer */
    GLint backingWidth;
    GLint backingHeight;
```

```
    EAGLContext *context;

    /* OpenGL names for the renderbuffer and framebuffers used to render to this
    view */
    GLuint viewRenderbuffer, viewFramebuffer;

    /* OpenGL name for the depth buffer that is attached to viewFramebuffer, if
    it exists (0 if it does not exist) */
    GLuint depthRenderbuffer;

    NSTimer *animationTimer;
    NSTimeInterval animationInterval;
}
@end
```

When you actually render the view, all OpenGL function calls must be made in relation to an `EAGLContext`. You set the current `EAGLContext` by calling `setCurrentContext:`, which is a static class method on `EAGLContext`. To initialize the `EAGLContext` object, you call `initWithAPI:`.

NOTE

`initWithAPI:` must be called with a system constant to tell the API which version of OpenGL ES you want to use. To use v1.0, pass `kEAGLRenderingAPIOpenGLES1`, which is the only version of OpenGL ES available on older generation iPhones and iPod Touches. iPhone 3GS added support for OpenGL ES 2.0. To use it, pass `kEAGLRenderingAPIOpenGLES2` as your parameter here.

In order to actually draw polygons, you need to use a render buffer. The call to `renderbuffer Storage:fromDrawable:` replaces the traditional OpenGL call to `glRenderbuffer Storage` and allocates the storage for a render buffer. After drawing your content into the render buffer, you can present it on-screen with a call to `presentRenderbuffer:`.

In the example code, you see the actual allocation of the render buffer in the method `create Framebuffer`. Then, the actual drawing of the view occurs in the `drawView` method. At the end of the `drawView` method, you can see that you call `presentRenderbuffer:` to send the data to the screen.

If you're wondering where to put your OpenGL code, the `drawView` method is where it goes. This is where you actually do the drawing of the OpenGL content, prior to presenting it on-screen.

If all you want is a static image displayed on an OpenGL screen, then you're done. You've just displayed that image. But usually you want to have animation and motion when you're working with OpenGL. The example code handles this by running a timer every 1/60 second. That timer simply calls the `drawView` method. This is set up in the `startAnimation` method.

So the important thing here is that in order to interface with the OpenGL system, you need to store and present your rendering buffers through the `EAGLContext`'s currently active context. Once you are familiar with these calls, it's just a matter of using standard OpenGL ES code to render your graphics.

The built-in template gives you almost everything you need for a bare-bones OpenGL ES game or application. It gives you an event loop for your rendering and updating of game logic, and it gives you a place to put your OpenGL ES code to render on-screen. It does all this and wraps it in a nice UIView to be used in your application. All you need to do is provide the game engine and the Open GL code.

Summary

In this chapter, you've taken a look at the APIs that enable you to interface with OpenGL ES from Cocoa Touch. The APIs that you focused on are specific to iPhone development and provide the bridge between OpenGL ES and the iPhone view system.

Again, you couldn't really focus too deeply on the OpenGL ES code. If you want more information about how to write OpenGL ES code, you should look for a book specifically on that subject.

Integrating Safari

Not only is the iPhone a great phone and great multimedia device, but it also has superb Internet capabilities. Among those capabilities is one of the best Web browsers available on any mobile platform. That Web browser is Safari. When developing Safari for the iPhone, Apple leveraged its previous technology from Safari on the desktop and developed a Web browser that not only renders pages excellently, but also provides a wonderful mobile user experience. It provides this experience by effectively utilizing the touch screen on the iPhone so that interactions with the Web browser are incredibly intuitive. An example is the Zoom feature, which uses a pinch gesture.

From a developer's point of view, Apple has also provided you with an excellent API for working with Safari. You can open URLs from within your application that cause Safari to load and display Web content. You can also embed a Safari browser right into your applications so that you can use it both to display Web content by downloading URLs and also to display local content, such as files included in your application bundles.

In this chapter, you're going to take a look at how to do both of these things. the code is very straightforward, and so there's not much to it. However, by leveraging what you learn in this chapter, you can provide a richer experience to your users.

In This Chapter

Opening URLs in Safari

Integrating a Web browser into your app with UIWebView

Displaying HTML files from your application bundle

Opening URLs Using the iPhone Web Browser

The first way to integrate Safari with your application is to allow users to open URLs within your application and then launch the iPhone Web browser externally with the URL loaded.

The advantage of this technique from a user's point of view is that the user can then stay in the browser, navigate the Web site, and so on. Essentially, you are handing off the user to your Web site.

There are also several disadvantages with this technique. The first is that because it launches Safari, it exits your application. This means that any state that the user had going at the time that she launched the URL is probably going to be lost. Additionally, it can be quite a jarring experience for the user to be sent out of your application and over to Safari.

When considering whether to use this technique for displaying Web content, be sure to carefully weigh these advantages and disadvantages. A good example of this kind of interaction might be if you have a link inside your application that takes the user to more of your company's products. In this case, the user is probably expecting to exit your application and go to your Web site.

To utilize Safari in this way, you first need an NSURL containing the URL you want to access. You can create this using the `NSURL` initializer, `initWithString:` or the factory method `URLWithString:`. Next, you use the `UIApplication sharedApplication` method to get the `UIApplication` singleton for your application. Finally, using the `UIApplication` singleton, you call the method `openURL:`, passing the URL that you created. Listing 16.1 shows how this is done.

Listing 16.1

Opening a URL in Safari

```
-(void)loadCompanyHomepage
{
    NSURL *url = [NSURL URLWithString:@"http://my.company.com/"];
    UIApplication *app = [UIApplication sharedApplication];
    [app openURL:url];
    // here your application will terminate and
    // safari will load with your content displayed
}
```

NOTE
You can also open URLs that are handled by other applications. If an application has registered to handle a particular URL type, opening its URL in this way exits your application and launches that application. You can pass any sort of data you want on the URL that you open. This can provide an interesting way of doing inter-process communication.

Using UIWebView

The next way to leverage Safari in your application provides for a much more integrated user experience. Specifically, I'm talking about using a `UIWebView` in your application. By using a `UIWebView`, you can actually embed a Web browser directly into your application. This enables you not only to view Web content inside of your application but also to load things like HTML files and the like from your application bundle. It is a tremendously powerful capability, and the API it provides is very easy to use.

Adding a UIWebView to your application

To add the `UIWebView` to your application, simply create a new view and its accompanying view controller, drag-and-drop a `UIWebView` from the object selection palette, and place it as a subview of the main view for your view controller. Connect it to an `IBOutlet` on your view controller and set your view controller to be its delegate. You may also want to configure the options that are available, but usually the default settings are suitable for most applications.

Of course, as usual, `UIWebView` is just a view and can also be added as a subview of another view programmatically.

Loading a URL

Once you've got your view controller with your `UIWebView` attached to it, you can ask it to actually load content. When using `UIWebView`, it requires that you provide it an `NSURLRequest`, configured with the URL that you want to load so that the Web view can use the URL request to load the content asynchronously. In the case where you simply want to display one page in your application, and you are loading this `UIWebView` specifically to display that page, a good place to set up this `NSURLRequest` is in the `viewDidLoad` method. Listing 16.2 shows an implementation of a `viewDidLoad` method where I do exactly that.

Listing 16.2

Loading a Web Page into a UIWebView

```
-(void)viewDidLoad
{
    NSURL *url = [NSURL URLWithString:@"http://my.company.com/help"];
    NSURLRequest *req = [NSURLRequest requestWithURL:url];
    [webView loadRequest:req];
    [self setTitle:@"Loading..."];
    [super viewDidLoad];
}
```

Now in this case, you are loading a help page off of your Web site. While loading the page, you want the title of the navigation bar to show the text, "Loading...." Once the URL has loaded, you change the title to be something else. To change the title, you simply set the title on your view controller to whatever you want displayed. The navigation item displays this in the navigation bar.

Implementing a UIWebViewDelegate

So, now that the page is loading, you want to be notified when the page is finished loading so that you can update your title.

To do this, you implement the `UIWebViewDelegate` method, `webViewDidFinishLoad:`, which is called when the request has completed and the content is loaded into the UIWebView. Listing 16.3 shows the implementation of this method.

Listing 16.3

Content Loading Has Completed.

```
- (void)webViewDidFinishLoad:(UIWebView *)webView
{
    [self setTitle:@"Help"];
}
```

If you want to give the user the ability to navigate to additional links, you can provide him with buttons that are connected to the `goBack:` and `goForward:` actions provided on the `UIWebView`. These actions are available both programmatically and through Interface Builder.

You simply connect the action from whatever button you have provided to the actions available on the `UIWebView` object. Similarly, you can also provide Stop and Reload buttons.

Finally, you probably want to implement the delegate method `webView:didFailLoad WithError:`, which is called when an error occurs during loading of the content. In this method, you can display an error message to tell the user what happened.

CAUTION

Apple routinely rejects applications that do not handle network errors cleanly. Be sure to do your error handling appropriately. Don't fail silently.

Listing 16.4 shows an implementation of this method.

Listing 16.4

Handling an Error

```
- (void)webView:(UIWebView *)webView didFailLoadWithError:(NSError *)error
{
    UIAlertView *alert = [[UIAlertView alloc]
```

```
                initWithTitle:@"Network Error"
                      message:[error localizedDescription]
                     delegate:nil
            cancelButtonTitle:@"OK"
            otherButtonTitles:nil];
    [alert show];
    [alert release];
}
```

Loading HTML content from the application bundle

Recall that I said that you could also use an embedded `UIWebView` to load pages from your application bundle. Doing this is also easy. Instead of using the load request method with an `NSURLRequest`, you use either of the methods `loadData:MIMEType:textEncodingName:baseURL:` or `loadHTMLString:baseURL:`, which load data from an `NSData` or an `NSString`, respectively. Listing 16.5 shows your help file being loaded from the app bundle instead of the Internet.

Listing 16.5

Loading from a File

```
-(void)viewDidLoad
{
    NSBundle *mainBundle = [NSBundle mainBundle];
    NSString *helpFilePath = [mainBundle pathForResource:@"help"
                                                  ofType:@"html"];
    NSString *helpContent = [NSString
                    stringWithContentsOfFile:helpFilePath];
    [webView loadHTMLString:helpContent baseURL:nil];
    [self setTitle:@"Loading..."];
    [super viewDidLoad];
}
```

Summary

In this chapter, you've seen how to leverage Safari both by launching an external copy of Safari and by embedding a browser right into your application. It's important to remember that the iPhone is always an Internet-connected device, and allowing it to access HTML content on the Internet is a powerful additional tool that you can use in your applications.

In addition to that, embedding a UI Web view in your application can be an effective way to display content that would normally be too time-consuming to write a custom view for yourself. HTML provides you with a powerful formatting tool that enables you to create rich text documents with varying sizes of fonts and images. Safari and `UIWebView` provide a very easy way to leverage HTML and display that content inside your applications with very little effort. This makes it an extremely effective tool for documentation, or for other creative uses such as in place of complicated custom `UIViews`.

Working with Data

In This Part

Chapter 17
Storing User Defaults

Chapter 18
Implementing a Database with Core Data

Chapter 19
Connecting to the World with Networking

Chapter 20
Using the Push Notification Service

Chapter 21
Using the Game Kit API

Chapter 22
Implementing Cut, Copy, and Paste

Chapter 23
Using the Maps API

Storing User Defaults

Some of the most common types of data you'll need to store are user defaults. These usually include user preferences, minor data, and so on. Cocoa has always had an excellent user defaults system available for developers on the desktop. With the introduction of the iPhone, Apple brought the user defaults API to Cocoa Touch as well.

The class used to store user defaults is NSUserDefaults. Using this class, you can not only read and write virtually any value to the user defaults system, but you can also store "default defaults," which are the defaults if the user hasn't set a value.

When you store values to NSUserDefaults, the values are serialized and stored in your application sandbox under the directory Library/Preferences, similar to OS X for the desktop.

Acquiring the NSUserDefaults Object

NSUserDefaults is a *singleton*. This means that an application contains a single instance of NSUserDefaults that is used by the entire application, no matter where it's accessed. The class provides a static accessor to initialize and retrieve the singleton object. The accessor for this is standardUserDefaults. Listing 17.1 shows how you access the user defaults singleton.

This is all you need to do to read the user defaults singleton. There is no need to read anything or, when you're done with the object, to save anything. The NSUserDefaults object handles all of this for you, storing your data in a file specific to your application, and automatically loading that file when it is invoked for the first time. The NSUserDefaults object also caches all the values in the file in memory, and so it is perfectly acceptable to access values directly from the NSUserDefaults object when needed without being concerned about incurring the cost of reading from the disk.

In This Chapter

Storing and retrieving user preferences in NSUserDefaults

Adding and configuring a settings bundle for your iPhone app

The `NSUserDefaults` object periodically synchronizes the contents of the in-memory cache with the disk file to ensure that it is kept up-to-date. If you feel the need to force this synchronization, the `NSUserDefaults` object provides the `synchronize` method for this purpose. However, in my years of Cocoa and Cocoa Touch programming, I have rarely felt the need to call the synchronize method. More than likely, the periodic synchronization that `NSUserDefaults` does on its own is more than sufficient for your needs.

Listing 17.1

Accessing the User Defaults Singleton

```
{
    NSUserDefaults *defaults = [NSUserDefaults standardUserDefaults];

    // ... read or write defaults here

}
```

Reading and Writing Values to NSUserDefaults

Reading and writing values to `NSUserDefaults` is trivial. At its simplest level, `NSUserDefaults` behaves a bit like an `NSDictionary` in that it provides methods for setting and retrieving objects that are associated with keys using the methods `objectForKey:` and `setObject:forKey:`. In addition to this low-level method, it also provides a set of methods for storing specific kinds of values such as scalars, arrays, and dictionaries.

In order for a value to be stored in `NSUserDefaults`, it must either be a scalar value with one of these specialized setters or getters or it must be a value that can be serialized to a property list. Most of the commonly used objects in Cocoa and Cocoa Touch meet this criterion. For other objects that do not, they can be serialized by archiving them into an `NSData` and then storing the `NSData` in `NSUserDefaults`.

Table 17.1 shows the setters and getters available for storing and retrieving objects to and from `NSUserDefaults`.

Table 17.1 Methods for Getting and Setting User Defaults

Method	Purpose

`- (void)setBool:(BOOL)value forKey:(NSString *)defaultName`
 	
and
 	
`- (BOOL)boolForKey:(NSString *)defaultName`
	Sets and retrieves a Boolean scalar value.

`- (void)setDouble:(double)value forKey:(NSString *)defaultName`
 	
and
 	
`- (double)doubleForKey:(NSString *)defaultName`
	Sets and retrieves a double scalar value.

`- (void)setFloat:(float)value forKey:(NSString *)defaultName`
 	
and
 	
`- (float)floatForKey:(NSString *)defaultName`
	Sets and retrieves a float scalar value.

`- (void)setInteger:(NSInteger)value forKey:(NSString *)defaultName`
 	
and
 	
`- (NSInteger)integerForKey:(NSString *)defaultName`
	Sets and retrieves an integer scalar value.

`- (void)setObject:(id)value forKey:(NSString *)defaultName`
 	
and
 	
`- (id)objectForKey:(NSString *)defaultName`
	Sets and retrieves any object that can be encoded to a property list.

`- (NSArray *)arrayForKey:(NSString *)defaultName`	Retrieves an object as an array. To set an array, use the `setObject:forKey:` method.

continued

Table 17.1 Continued

Method	Purpose
-(NSData *)dataForKey:(NSString *)defaultName	Retrieves an object as an NSData. To set the object, use the setObject:forKey: method.
-(NSDictionary *)dictionaryForKey:(NSString *)defaultName	Retrieves an object as an NSDictionary. To set the object, use the setObject:forKey: method.
-(NSString *)stringForKey:(NSString *)defaultName	Retrieves an object as an NSString. To set the object, use the setObject:forKey: method.

NOTE
All objects retrieved from NSUserDefaults are always immutable, even if the original object you stored was mutable. If you need the objects to be mutable, then you must copy them to a mutable object.

CAUTION
Static constant strings are optimized by the compiler to reference a single object in memory. This makes it convenient to use them as constants in your code because you can simply compare the pointers themselves using the == operator. However, if you store these values in NSUserDefaults, the string that you later retrieve does not follow this convention and is not equal to the string constant that you used originally to set the value.

Listing 17.2 shows an example how you set and access values in NSUserDefaults.

Listing 17.2

Typical NSUserDefaults Usage

```
-(void)foo
{
    NSUserDefaults *defaults = [NSUserDefaults standardUserDefaults];
    // storing defaults
    [defaults setObject:@"some value" forKey:@"SOME KEY"];
    [defaults setBool:NO forKey:@"SOME OTHER KEY"];
    [defaults setInteger:10 forKey:@"A THIRD KEY"];

    NSData *someData = [NSData dataWithBytes:bytes length:128];
    [defaults setObject:someData forKey:@"DATA KEY"];
    // ...

    // retrieving defaults
    NSString *aString = [defaults stringForKey:@"SOME KEY"];
    BOOL boolValue = [defaults boolForKey:@"SOME OTHER KEY"];
```

```
    NSInteger intValue = [defaults integerForKey:@"A THIRD KEY"];
    NSData *dataValue = [defaults objectForKey:@"DATA KEY"];
    const void *bytes = [dataValue bytes];
}
```

Setting Your Default Defaults

So now you've seen how to store and retrieve data in `NSUserDefaults`, but how do you store the default values for those defaults for use before the user has actually set any of the default values? These are the values that are used when the application is first launched, before the preferences file for this application has even been created.

If you can imagine that your objects and classes can be initialized at any place in your application startup, and that you want to try to avoid having to think about whether or not your defaults have been set when you go to access them initially, then it makes sense that you should look for a place to initialize your defaults as early on in the application startup process as possible. Preferably, this would occur before any of your classes have even been initialized.

Fortunately, Objective-C has a convenient method of doing exactly this. When the runtime loads a class definition, it calls a special static class method called `initialize`. This method is called before any methods on your class are called, including your designated initializer. It is also called only once per application runtime. By leveraging the special properties of this particular method, you can place your initialization of your defaults into this special static class method and be certain that when your instance methods are ultimately called, your defaults have already been initialized.

> **NOTE**
> You can use the `initialize` method on the class that actually requires the default if you prefer, but typically, I find the best location to do `NSUserDefaults` initialization is in the initialize method of your `UIApplication` delegate. This is the first class loaded by your application and ensures that your defaults are available everywhere.

So, you have a location where you can initialize your user defaults, and that is called before any of your class methods or object methods are called. This means that your defaults are initialized for you when you need them in your classes. So the next question is, how do you initialize those defaults?

`NSUserDefaults` provides a mechanism for doing this too. The method that you use is called `registerDefaults:`, and it is called with a single parameter that is an `NSDictionary` containing the default values for your defaults. This may seem confusing, but just remember that these are the values that you receive when you ask for a value that has not been set by the user. It uses an `NSDictionary` as its parameter so that if you prefer, you can include a property list in your application bundle that you can simply read from disk at startup.

Listing 17.3 shows an implementation of the `initialize` method and setting up the defaults in code.

Listing 17.3

Typical initialize Method Implementation

```
+(void)initialize
{
    NSMutableDictionary *defs = [NSMutableDictionary dictionary];

    [defs setObject:@"default value" forKey:@"SOME KEY"];
    [defs setObject:[NSNumber numberWithBool:YES]
            forKey:@"SOME OTHER KEY"];
    [defs setObject:[NSNumber numberWithInt:20] forKey:@"A THIRD KEY"];
    [[NSUserDefaults standardUserDefaults] registerDefaults:defs];
}
```

CAUTION

The "+" before the method signature in the `initialize` method declares the method as a static class method. This is absolutely required for implementing this method. A "-" indicates that a method is an instance method. If you use "-" instead of "+" here, your `initialize` method is not called.

This example code initializes your default values right here in code. This is the way that I typically do my defaults initialization, but if you wanted to do it by using a property list included in your application bundle, you can use code such as that shown in Listing 17.4.

Listing 17.4

Initialize Method Using a Property List

```
+(void)initialize
{
        NSString *defaultsFilename = [[NSBundle mainBundle]
                            pathForResource:@"defaults"
                            ofType:@".plist"];
        NSDictionary *defs = [NSDictionary
                        dictionaryWithContentsOfFile:defaultsFilename];
            [[NSUserDefaults standardUserDefaults]
                        registerDefaults:defs];
}
```

Using the Settings App

Everything I've talked about so far with regard to defaults applies equally to both Cocoa and Cocoa Touch. There is, however, one part of the default system that is specific to Cocoa Touch in the iPhone. That is the use of settings bundles.

A settings bundle enables you to display application settings inside the Settings application on the iPhone home screen. Generally speaking, in order to maintain a consistent user experience, Apple has stated that it prefers that applications that need to store settings that are typically set once and then not changed again during the lifetime of the application should allow the user to configure those settings through the Settings application. This enables you as the developer to not have to clutter your user interface with a settings tab that is rarely used and allows the user to have a single location where she can configure the settings for all of her applications.

Think about it like this: if you have an application setting where the default value is usually appropriate for most users but some advanced users may want to have the option to change it, this is an ideal candidate for storing in the Settings application.

This is not to say that all applications should always store all of their settings in the Settings application. In fact, in practice, many users are not aware that there are settings in the Settings application at all. Instead, if you have a setting that either needs to be changed frequently or that you expect most of your users will want to customize, then you may want to provide a mechanism in the application to configure that setting.

Adding a settings bundle

To add a settings bundle to your application, simply choose New File from the File menu, and then from the Resource category, choose Settings Bundle. This adds a new bundle to your project with some property lists that you can use to configure your settings.

Inside the settings bundle, you find the `Root.plist` file, which is the one that you use to configure your settings. If you click this file, it opens the property list editor in Xcode. To view this property list as an iPhone settings bundle, choose Property List Type ➪ iPhone Settings plist from the View menu in Xcode.

NOTE
Many developers don't like the default property list editor in Xcode. Property lists are nothing but plain text, and so if you would prefer to use the text editor built in to Xcode to edit your property lists, then you can change the default association for the file type of "text.plist," in Xcode preferences under the File Types settings, to Plain Text. This will not work, however, with binary plists, so beware.

Adding settings to your settings bundle

To add settings to your settings bundle, you simply add a new property to the Preference Items array. The types of items that you can add are shown in Table 17.2.

Table 17.2 Settings Value Types

Type	Purpose
Group	Displays a group title and separates elements below it into a new group
Multi-Value	Displays a subview that allows you to select a value from a list of values
Slider	Displays a slider that enables you to configure a numeric value
Text Field	Displays a text field that enables you to configure a string
Toggle Switch	Displays an on/off switch to configure a Boolean value
Title	Displays a read-only value suitable for displaying a static read-only string
Child Pane	Enables a subpanel of preferences, which must be configured via another preferences `plist`

Each of these value types has different properties that can be set. When you choose the type of value, the possible parameters that can be set for that value are automatically populated in the item that you are configuring.

There are also some values that are shared amongst each of the value types. In particular, the identifier parameter specifies the key that is used to store the value that this setting represents in your `NSUserDefaults`. The title parameter specifies the title that is shown to the left of the configurable value in the Settings application. The default value parameter specifies the value that is displayed by default if the user has not specified the setting.

CAUTION

The Default Value parameter in settings must be set, even if you initialize the default value in code, as I described earlier. Neither default value is written to disk, and so you should be careful to make sure you are setting the defaults to the same values in both places.

Figure 17.1 shows the settings configuration for the parameters you previously accessed in code. Figure 17.2 shows the iPhone Settings application to display these values.

Figure 17.1

The configuration for settings

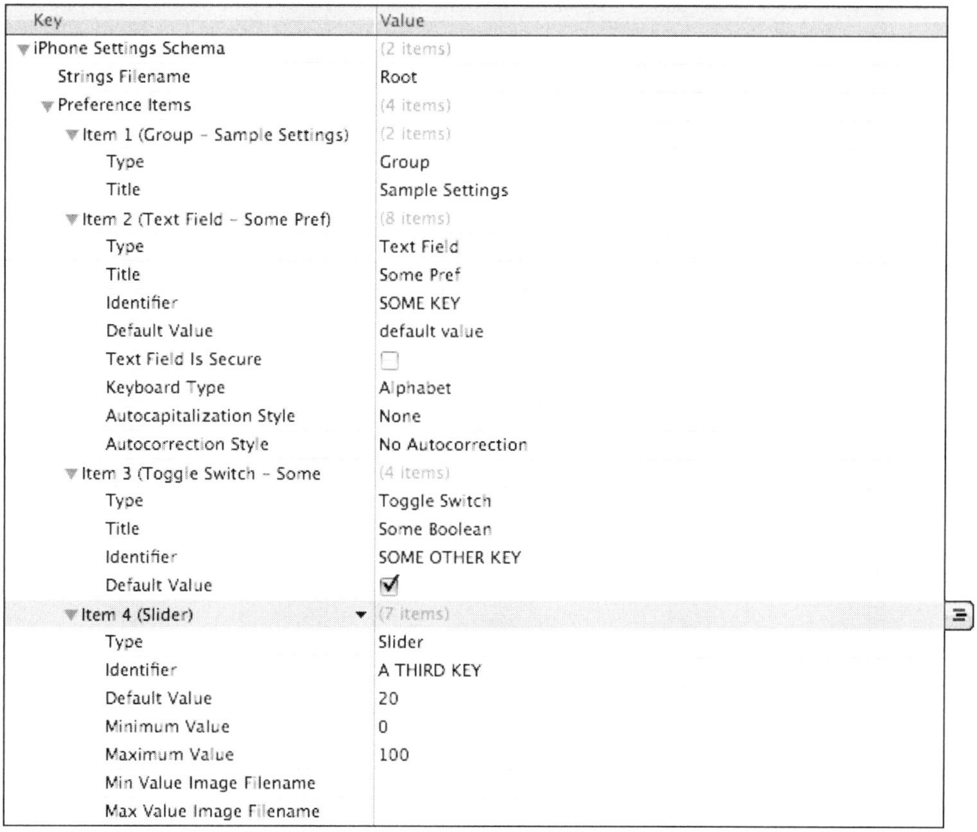

Key	Value
▼ iPhone Settings Schema	(2 items)
Strings Filename	Root
▼ Preference Items	(4 items)
▼ Item 1 (Group - Sample Settings)	(2 items)
Type	Group
Title	Sample Settings
▼ Item 2 (Text Field - Some Pref)	(8 items)
Type	Text Field
Title	Some Pref
Identifier	SOME KEY
Default Value	default value
Text Field Is Secure	☐
Keyboard Type	Alphabet
Autocapitalization Style	None
Autocorrection Style	No Autocorrection
▼ Item 3 (Toggle Switch - Some	(4 items)
Type	Toggle Switch
Title	Some Boolean
Identifier	SOME OTHER KEY
Default Value	☑
▼ Item 4 (Slider)	(7 items)
Type	Slider
Identifier	A THIRD KEY
Default Value	20
Minimum Value	0
Maximum Value	100
Min Value Image Filename	
Max Value Image Filename	

Figure 17.2
The displayed configuration screen in the iPhone Settings application

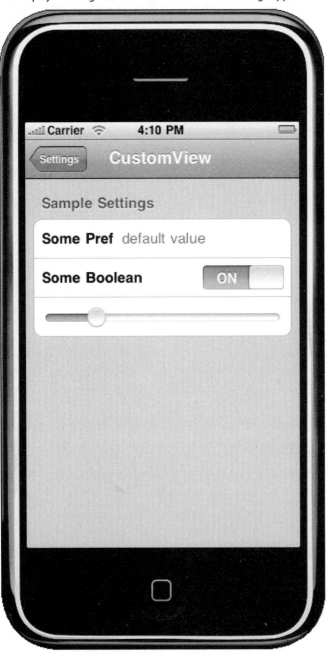

You can even add more subpanes of configuration by adding a Child Pane value type. With this value type, you can specify an additional property list to use for configuring the values of the subpane. To do this, add an additional property list to your settings bundle, and add a Child Pane value type to your root property list. The title value in the Child Pane value is the title shown on the row in the root pane that the user touches to navigate to the subpane.

Summary

Almost every application needs to store user defaults. Cocoa Touch provides an excellent mechanism to do exactly that. Interestingly, though, the `NSUserDefaults` mechanism is powerful enough that it makes it a versatile location to store more than just user preferences.

NSUserDefaults is a great API because it's simple, and it works, and that's exactly how an API should be.

Implementing a Database with Core Data

When the iPhone SDK originally became available, one of the features that developers most frequently requested was to have a framework that had been available on Mac OS X for several years. That framework was Core Data.

Core Data is a database abstraction layer developed by Apple to help make applications with a database-backed model easier to write. It provides a set of classes that enable the user to perform complex queries on a tabular data set without needing to know the details of the low-level implementation of that data set. It can be used in conjunction with a SQL database, an XML database, or a binary flat file. The format of the underlying implementation is abstracted from the developer. Additionally, it provides a set of GUI tools that allow the developer to define his schema. Finally, it provides a mechanism for doing migrations from one schema to the next, thus allowing the database and schema to grow and change as the developer's application changes.

NOTE
XML Data stores are not supported on the iPhone, only on Mac OS X.

In this chapter, you are going to take a detailed look at how to work with Core Data in conjunction with the iPhone for developing database applications and you will reimplement your Favorite Cities application, using Core Data to model your data.

 In This Chapter

Adding a database to your application using Core Data

Exploring the object modeling tool in Xcode to make your application schema

Understanding the different types of properties and relationships available in Core Data entities

Creating, reading, updating, and deleting Core Data objects

Updating the sample Favorite Cities application to use Core Data

Understanding Core Data's Building Blocks

The Core Data framework, at its most basic level, consists of four primary components.

The first component is the schema that defines the data that will be stored and manipulated in the application. This schema is known as the Managed Object Model. Through code, this component is accessed via the classes `NSManagedObjectModel` and `NSEntityDescription`, which allow access to the model as a whole, and individual object models themselves, respectively. To design the schema, Xcode provides an object modeling tool that enables direct manipulation of the Managed Object Model when you are designing your application. Figure 18.1 shows the managed object modeling tool in Xcode.

Figure 18.1

The object modeling tool in Xcode

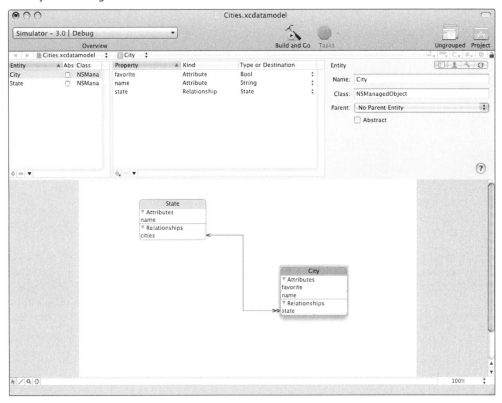

Essentially, the Managed Object Model defines the schema for entities. Entities form the blueprint for the objects stored in the database. The objects themselves are the second component of the core data framework.

Any object that is stored in and managed by Core Data is known as a Managed Object and is a subclass of `NSManagedObject`. If the definitions for the attributes of the entities are defined by `NSEntityDescription` and the `NSManagedObjectModel`, then the entities themselves are represented by `NSManagedObject`s. When working with objects stored in Core Data, you will be working with `NSManagedObject`s.

`NSManagedObject`s are stored in one of three different types of files. Those files can be a SQL database, an XML database, or a binary flat file. The SQL database supported by Core Data is limited to sqlite3. This persistent data is managed through the Persistent Store Coordinator, which manages a collection of persistent object stores. The Persistent Store Coordinator represents the third component in the Core Data framework. The different persistent object stores are what give you the ability to configure the type of actual physical storage used for storing your data. The persistent object store is accessed through the `NSPersistentStore` class. Typically, as a developer, you'll very rarely interact with the persistent store and its coordinator directly. Most Core Data work is done through the managed objects, the object model, and so on.

The final component of Core Data consists of the `NSManagedObjectContext`. The `NSManagedObjectContext` provides the in-application scratchpad upon which `NSManagedObject`s are manipulated. All operations in Core Data are done within the scope of an `NSManagedContext`.

Adding Core Data Support to Favorite Cities

To demonstrate the use of Core Data in an application, you're going to bring up your Favorite Cities application, and you are going to modify it to add Core Data to it. You could just as easily use the Core Data template in Xcode to create a new core data application and then modify it to be like the Favorite Cities application; but again, I felt that it would be more instructive to modify your existing application so that you don't come to rely on the templates as a crutch. This is also a more practical example in terms of most applications and how they are written. Most applications are fairly orthogonal in the way that they grow, meaning that you might develop the application initially storing things in a flat file, and then realize later that you want to utilize Core Data instead.

So, to do this, get the example code from Chapter 8, make a copy of it, and open it in Xcode. When you were working with this code in Chapter 8, you actually had not made it save its data to disk. Everything was being done in memory. This makes it very easy for you to transition this to Core Data, because you don't have an existing file of data that you have to convert. If, however, you were updating an existing application to Core Data, you would need to read your custom file into memory and then store it in the Core Data store the first time that you run it after the upgrade.

In this case, however, you don't need to do that. You'll just convert it straight over to Core Data and work with it as if it were a brand-new application.

Modifying the app delegate

The first thing that you need to do before you can actually begin working with the Core Data parts of the new application is to add support for setting up the `NSManagedObjectContext` and its associated persistent store.

NOTE
You must add the Core Data framework to your application. Do this by right-clicking your target, and from the General tab, under the Frameworks section, click the plus button and choose the Core Data framework.

To do this, you need to modify your application delegate. You first add several methods for setting up the object context and then you modify the `applicationDidFinishLaunching:` method and the `applicationWillTerminate:` method to load and save the data to the persistent store, respectively.

The application sandbox on the iPhone prevents you from writing data to your application bundle directories. Instead, you are expected to save data in your Documents directory. Therefore, you need to be able to find where your Documents directory is located so that you can place the persistent store there. Listing 18.1 shows the utility method that you use to get your Documents directory.

Listing 18.1

Utility Method to Retrieve the Path to the Documents Directory

```
- (NSString *)applicationDocumentsDirectory
{
    NSArray *paths = NSSearchPathForDirectoriesInDomains(NSDocumentDirectory,
                                      NSUserDomainMask, YES);
    NSString *basePath =
        ([paths count] > 0) ? [paths objectAtIndex:0] : nil;
    return basePath;
}
```

You'll need someplace to store your context, store, and model, so add these as member variables to your application delegate and make them properties. This is shown in Listing 18.2, which is the interface file for your application delegate.

Listing 18.2

New Interface for the Application Delegate

```
#import <CoreData/CoreData.h>
@interface FaveCitiesListAppDelegate : NSObject <UIApplicationDelegate>
{
    NSManagedObjectContext *context;
    NSManagedObjectModel *model;
    NSPersistentStoreCoordinator *psc;

    UIWindow *window;
    UINavigationController *navController;
}
@property (nonatomic, retain) IBOutlet UIWindow *window;
@property (retain, nonatomic) IBOutlet UINavigationController * navController;
@property (retain, nonatomic) NSManagedObjectContext * context;
@property (retain, nonatomic) NSPersistentStoreCoordinator * psc;
@property (retain, nonatomic) NSManagedObjectModel * model;
@end
```

> **NOTE**
> Remember to add the @synthesize declarations for your @properties.

You next need to modify your `applicationDidFinishLaunching:` method to add the initialization of these variables. The only thing that really requires extra work on your part is setting up the persistent store coordinator. So I'll show you that in a separate method, which I will explain momentarily. For now, Listing 18.3 shows the initialization of your required core data objects.

Listing 18.3

Implementation of the applicationDidFinishLaunching: Method

```
- (void)applicationDidFinishLaunching:(UIApplication *)application
{
    [self setModel:[NSManagedObjectModel mergedModelFromBundles:nil]];
    [self setupPersistentStore];
    context = [NSManagedObjectContext new];
    [context setPersistentStoreCoordinator:psc];

    [window addSubview:[navController view]];
    [window makeKeyAndVisible];
}
```

Essentially, everything starts by loading the object model from the bundle. You have to do this first because the object model is going to be used in the `setupPersistentStore` method. After you set up the persistent store, you set up your new `NSManagedObjectContext` and set the persistent store coordinator for that context to be the one that you have created for your application. Once you've done all of this, the Core Data housekeeping has been taken care of and you are ready to actually begin working with managed objects as defined by your model. Of course, at this point in your example application, you haven't created your object model yet. You'll do that soon, but first let's take a look at the `setupPersistentStore` method which actually shows how you choose what kind of data store you want to use for your application. Listing 18.4 shows that method.

Listing 18.4

Implementation of the setupPersistentStore Method

```
-(void)setupPersistentStore;
{
    NSString *docDir = [self applicationDocumentsDirectory];
    NSString *pathToDb = [docDir stringByAppendingPathComponent:
                          @"FaveCities.sqlite"];
    NSURL *urlForPath = [NSURL fileURLWithPath:pathToDb];
    NSError *error;
    psc = [[NSPersistentStoreCoordinator alloc]
            initWithManagedObjectModel:[self model]];
    if (![psc addPersistentStoreWithType:NSSQLiteStoreType
                          configuration:nil
                                    URL:urlForPath
                                options:nil
                                  error:&error])
    {
        // Handle error
    }
}
```

In this method, the first few lines really deal with getting the path to where your database is going to be located. For the purposes of this example, you are going to be working with a SQL database for your data store. Therefore, you are going to get the application documents directory and you are going to append to that the filename "`FaveCities.sqlite`." You will then convert that full path to a URL, which is what the `NSPersistentStoreCoordinator` requires. After all of this, the real work of this method involves actually allocating the persistent store coordinator and initializing it with the Managed Object Model, then adding a persistent store of `NSSQLiteStoreType` type associated with the file. This is done with the calls to the methods `initWithManagedObjectModel:` and `addPersistentStoreWithType:configuration:URL:options:error:`.

When adding the persistent store to the persistent store coordinator, you have to specify the type and the URL for the file that you want to use. Additionally, you can also specify an options dictionary containing key value pairs to specify, among other things, whether the store should be opened read-only and whether (in the case of an XML store in Mac OS X) the XML file should be validated against a DTD before it is read. This parameter can be nil, as it is here in this case.

The possible types of persistent stores that you can create are listed in Table 18.1.

Table 18.1 Types of Persistent Stores

Type	Description
NSSQLiteStoreType	sqlite3 database file
NSBinaryStoreType	Binary flat file
NSInMemoryStoreType	In-memory storage

The configuration parameter gives you the ability to specify an optional configuration from the Managed Object Model. In your case, you pass nil here as well, which specifies no optional configurations.

You also need to update your applicationWillTerminate: method to save your state to disk before you exit. That method is shown in Listing 18.5.

Listing 18.5

Updated applicationWillTerminate: Method

```
- (void)applicationWillTerminate:(UIApplication *)application
{
    NSError *error;
    if (context != nil)
    {
        if ([context hasChanges] && ![context save:&error])
        {
            // Handle error.
        }
    }
}
```

Once you've done all of this, you are ready to add your model to your application and modify it to contain the schema for your application.

Working with the Managed Object Model

Now that you've got your application ready to utilize Core Data, you need to add a Managed Object Model to your application bundle. Typically, these are kept in the resources folder of your application. So right-click the resources group, choose Add ⇨ New File. Then, from the file selection dialog, choose Resource, and then Data Model. You are given the option to select classes from your project to import into your data model. Because you didn't actually have any data model classes already written, you simply skip this step by choosing Finish. This brings up the model-editing window with an empty model, as shown in Figure 18.2.

Figure 18.2

You use this tool to actually design your schema.

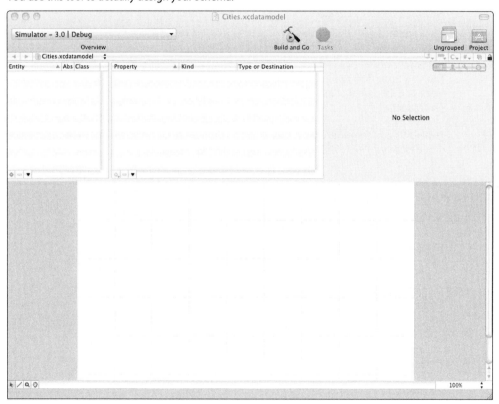

Designing your schema in Xcode

A data model consists of entities. Entities are made up of attributes and relationships. In the case of your Favorite Cities application, you will have entities consisting of states and cities. There will be a one-to-many relationship between them, meaning that any given state can contain multiple cities.

Attributes for the cities will consist of the city name and whether or not it's a favorite. Attributes on the state will consist of just the name. This, in a nutshell, is the exact same data model that you used previously in this application. The only difference is that you are going to define it using the data modeling application and it will be persisted into a database through Core Data.

Defining entities

So, to begin, let's add an entity of type "City". To do this, click the plus under the entity panel in the modeling tool. Rename the added entity to **City**. Next, click the plus in the Property panel to add an attribute to the city called "name". Do the same thing again, but this time, add an attribute called "favorite". Make the name attribute a string type. Make the favorite attribute a Boolean type and set its default value to NO. You now have your cities. Now let's add the states.

To add the states, you follow the exact same procedure. You click the plus under the Entity panel, you name it **State**, and then you add an attribute for "name". Here, however, you are going to do something a little bit different: You need to define a relationship between the states and the cities. To do this, you are going to click the plus in the Property panel, but this time, you are going to add a relationship.

> **NOTE**
> Entities can have a variety of different types of properties, but even in the cases where the properties are basic types, they are still represented as objects when you access them. This means that when you access things such as floats or integers, you are actually receiving `NSNumber` objects that you can then retrieve the actual value from.

Defining relationships among entities

When you add a relationship, you have an opportunity to select the destination of that relationship. What you need to do is define a relationship where the State has a relationship with a destination of City. You'll call this relationship "cities". You'll specify that it is a "To-Many" relationship, and you want your delete rule to be set to "Cascade". What this relationship means is that a state contains many cities. When the state is deleted, you want the cities associated with that state to also be deleted.

You also want to define the inverse relationship so that from the perspective of your city, you can move the city from one state to another. To do this, click the City entity, and add a property for the inverse relationship called "state". This time, be sure to specify the Inverse property of the relationship as "cities". From the perspective of the city, this is not a "To-Many" relationship. Furthermore, your delete rule from the perspective of the city should simply be the default, which is "Nullify".

Once you've set up your relationships and your attributes, your data model should look something like Figure 18.3.

Figure 18.3

The data model for your Favorite Cities application

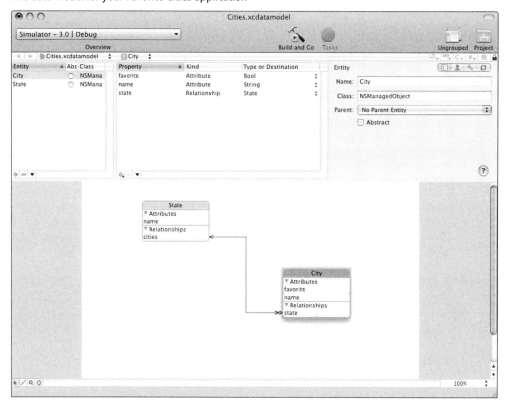

Generating classes for entities

Once you define the entities, their attributes, and their relationships, you actually have enough right here to use Core Data. You could use this model, and access your data using the low-level functions that Core Data provides. However, Core Data also provides the capability for you to automatically generate model classes that encapsulate the objects defined in your model so that, if needed, you can add additional logic.

Doing this involves a bit of Xcode manipulation. Specifically, you need to make sure that you have one of your entities selected in your data model; then you right-click one of the group folders in your project and choose New File. From the file selection dialog, you choose Cocoa

Touch Class, and then finally, an `NSManagedObject` class. This brings up a dialog that enables you to select all of the classes that you want to generate implementation files for. Go ahead and do this for this project, adding your new `NSManagedObject` subclasses to a new group in your project called Managed Objects. This should create four new files in your project consisting of the headers and implementation files for the city and state entities. Later, if you make modifications to your model, you can use the code generator in Xcode to generate the accessor definitions from the Data Model menu under Design.

> **NOTE**
> Core Data is responsible for the life cycle of all `NSManagedObject`s; therefore, there is no need for a `dealloc` method in these classes. If, however, you add custom member variables, you need to allocate and deallocate them as you would normally.

You'll also find that if you look at your data model, it has now been updated so that the city and state entities have their class properties set appropriately to the new classes that have been generated in your project.

At this point, you now have your data model built, your custom model classes created, and the application has been updated so that it can use Core Data to store and retrieve your cities and states. Now, you need to look at how you work with the managed object context to create, read, update, and delete objects in the persistent store.

Exploring entity properties

Core Data is an incredibly deep and powerful framework providing many tools to enable you to store whatever kind of data you need to, with whatever sort of relationship you might need. In addition to the plethora of data types that can be used as attributes, it also has the ability to do fetched properties, which enable a sort of loose relationship from one entity type to another. A fetched property on your cities object might represent the list of nearby cities to a given city. You won't be using fetched properties in this chapter, but to find out more about them, see the Core Data documentation.

CRUD — Creating, Reading, Updating, Deleting

Once you have Core Data, your object model, and so on integrated with your application, actually using it to access your data is reasonably straightforward. The most important thing to recognize is that at its heart, Core Data is really just an object persistence mechanism, and once you learn the API surrounding that, it's easy to reuse that knowledge throughout your application.

As with most data storage systems, operations essentially come down to four different things: creating objects, reading objects, updating objects, and deleting objects.

Creating

Creating managed objects is very simple. You use the `NSEntityDescription` class method `insertNewObjectForEntityForName:inManagedObjectContext`, passing the name of the entity type that you want to create and the managed object context within which you want to create it. For example, to create a new state, you might do something like Listing 18.6.

Listing 18.6

Creating a New State Object

```
State *state = [NSEntityDescription
                insertNewObjectForEntityForName:@"State"
                inManagedObjectContext:context];
[state setName:@"Arizona"];
NSError *error;
[context save:error];
```

You ensure that the object is written out to disk by calling save off of the managed object context. Of course, it's possible that an error might occur here, so if one does, you should probably deal with that error in some manner appropriate for your application. The save method returns a BOOL value indicating success. If it returns NO, you should check the error object for the cause of the error.

Reading

When I talk about reading objects, what I'm really talking about is using an `NSFetchRequest`. An `NSFetchRequest` enables you to query the database for objects of a particular kind. Using the `NSManagedObjectContext` instance method `executeFetchRequest:error:`, you can execute one of these queries, passing parameters to filter the results, and even sort descriptors that enable you to sort the array that is returned. The array returned contains instances of the type of entity that you have requested. So, for example, to retrieve all of the cities for a particular state, you might do something like the code shown in Listing 18.7.

Listing 18.7

Fetching Cities for a Particular State

```
-(NSArray *)citiesForState:(State *)inState;
{
    NSFetchRequest *req = [NSFetchRequest new];
    NSEntityDescription *descr =
```

```
            [NSEntityDescription entityForName:@"City"
                        inManagedObjectContext:context];
[req setEntity:descr];

NSSortDescriptor *sort = [[NSSortDescriptor alloc] initWithKey:@"name"
                                                    ascending:YES];
[req setSortDescriptors:[NSArray arrayWithObject:sort]];
[sort release];

NSPredicate *predicate =
[NSPredicate predicateWithFormat:@"state.name == \"%@\"",
                                [inState name]];

[req setPredicate:predicate];

NSError *error;
NSArray *ret = [context executeFetchRequest:req error:error];
if(ret == nil)
{
    // do something about the error
}
[req release];
return ret;
}
```

Important things to note about this particular code are that you create a new fetch request for entities of type City. You then create a sort descriptor to sort your result set by name. Finally, you create an `NSPredicate`, which enables you to specify the filter criteria for your result set. In this particular case, you are specifying that the attribute on the city called *state* must have a name that is the same as the name of the state that was passed into the method.

> **NOTE**
> `NSPredicate` is the class that enables you to construct complex queries and filter your result set. It's an extremely powerful class. See the Apple documentation for more information on how to build other kinds of queries using it.

After all of this, you finally call `executeFetchRequest:error`, which returns to you an array containing the result set that matches your criteria sorted according to your sort descriptor.

Updating

Once you have an `NSManagedObject`, updating it is as easy as using the accessors on the object to set its state and then saving it. There's really nothing else to it. Listing 18.8 shows an example.

Listing 18.8
Updating an Object

```
-(void)setFavorite:(City *)inCity;
{
    [inCity setFavorite:[NSNumber numberWithBool:YES]];
    NSError *error;
    if(![context save:error])
    {
        // handle the error
    }
}
```

Deleting

Finally, perhaps the simplest operation is deleting an object. To delete an object, you simply call the instance method `deleteObject:` on the `NSManagedObjectContext`. So, for example, to delete a city after you have fetched it, you would do something like `[context deleteObject:fetchedCity]`. This would delete the object in memory; then, to actually commit the deletion to disk, you would, of course, call `save:` on the managed context.

Bringing it together and updating your app

So, to bring all of this together and update your Favorite Cities application to use Core Data, you need to manipulate your table view controller to get its data from Core Data. Fortunately, this is really easy to do.

The first thing you need to do is to pass the `NSManagedContext` from your app delegate to the `MyViewController` class. This is done in your final version of your `applicationDidFinishLaunching:` method, as shown in Listing 18.9.

Listing 18.9
Final applicationDidFinishLaunching: Method

```
- (void)applicationDidFinishLaunching:(UIApplication *)application
{
    [self setModel:[NSManagedObjectModel mergedModelFromBundles:nil]];
    [self setupPersistentStore];
    context = [NSManagedObjectContext new];
    [context setPersistentStoreCoordinator:psc];
```

```objc
    MyViewController *viewController =
            (MyViewController *)[navController topViewController];
    [viewController setContext:context];

    [window addSubview:[navController view]];
    [window makeKeyAndVisible];
}
```

The rest of the work occurs in the `MyViewController` code, and really just consists of replacing the `NSMutableDictionary`-based model with your brand-new, shiny State and City objects. I won't go into huge amounts of detail here; I'll just show you the code. There are probably more efficient ways of doing this — a lot of the code here is a holdover from the in-memory model — and therefore, it's not really taking full advantage of some of the clever querying you can do with Core Data. Listing 18.10 shows the new `MyViewController` implementation, and Listing 18.11 shows its interface.

Listing 18.10

New MyViewController Implementation

```objc
#import "MyViewController.h"
#import "EditViewController.h"
#import "State.h"
#import "City.h"
@implementation MyViewController
@synthesize states;
@synthesize editedItem;
@synthesize context;
- (void)dealloc
{
    [self setEditedItem:nil];
    [self setStates:nil];
    [self setContext:nil];
    [super dealloc];
}
-(void)setupModel;
{
    NSError *error;
    NSFetchRequest *req = [NSFetchRequest new];
    NSEntityDescription *descr =
        [NSEntityDescription entityForName:@"State"
                    inManagedObjectContext:context];
    [req setEntity:descr];
```

continued

Listing 18.10   *(continued)*

```
    NSSortDescriptor *sort = [[NSSortDescriptor alloc] initWithKey:@"name"
                                                          ascending:YES];
    [req setSortDescriptors:[NSArray arrayWithObject:sort]];
    [sort release];

    [self setStates:[[context executeFetchRequest:req error:&error]
                    mutableCopy]];
    if([[self states] count] == 0)
    {
        // set up default data
        State *state;
        City *city;
        // New York
        state = [NSEntityDescription
                 insertNewObjectForEntityForName:@"State"
                 inManagedObjectContext:context];
        [state setName:@"New York"];
        city = [NSEntityDescription
                 insertNewObjectForEntityForName:@"City"
                 inManagedObjectContext:context];
        [city setName:@"New York"];
        [state addCitiesObject:city];
        // Chicago
        state = [NSEntityDescription
                 insertNewObjectForEntityForName:@"State"
                 inManagedObjectContext:context];
        [state setName:@"Illinois"];
        city = [NSEntityDescription
                 insertNewObjectForEntityForName:@"City"
                 inManagedObjectContext:context];
        [city setName:@"Chicago"];
        [state addCitiesObject:city];

        // San Francisco and Los Angeles
        state = [NSEntityDescription
                 insertNewObjectForEntityForName:@"State"
                 inManagedObjectContext:context];
        [state setName:@"California"];
        city = [NSEntityDescription
                 insertNewObjectForEntityForName:@"City"
                 inManagedObjectContext:context];
        [city setName:@"San Francisco"];
        [state addCitiesObject:city];
        city = [NSEntityDescription
                 insertNewObjectForEntityForName:@"City"
                 inManagedObjectContext:context];
        [city setName:@"Los Angeles"];
        [state addCitiesObject:city];
```

```objc
        // Arizona
        state = [NSEntityDescription
                  insertNewObjectForEntityForName:@"State"
                  inManagedObjectContext:context];
        [state setName:@"Arizona"];
        city = [NSEntityDescription
                  insertNewObjectForEntityForName:@"City"
                  inManagedObjectContext:context];
        [city setName:@"Phoenix"];
        [state addCitiesObject:city];
        [context save:&error];
        
        [self setStates:[[context executeFetchRequest:req error:&error]
                          mutableCopy]];
    }
}
-(IBAction)addCity:(id)sender;
{
    [self setEditedItem:nil];
    EditViewController *controller =
    [[EditViewController alloc] initWithCity:@""
                                       state:@""
                                    favorite:NO
                                    delegate:self];
    [[self navigationController] pushViewController:controller
                                           animated:YES];
    [controller release];
}
-(void)viewDidLoad;
{
    [self setupModel];
    [[self navigationItem]
     setRightBarButtonItem:[[[UIBarButtonItem alloc]
        initWithBarButtonSystemItem:UIBarButtonSystemItemAdd
                             target:self
                             action:@selector(addCity:)] autorelease]];
    [[self tableView] setDataSource:self];
    [[self tableView] setDelegate:self];
    [super viewDidLoad];
}
#pragma mark DELEGATE METHODS
- (NSIndexPath *)tableView:(UITableView *)tableView
  willSelectRowAtIndexPath:(NSIndexPath *)indexPath;
{
    return indexPath;
}
- (void)tableView:(UITableView *)tableView
didSelectRowAtIndexPath:(NSIndexPath *)indexPath;
{
```

continued

Listing 18.10   *(continued)*

```objc
    State *state = [states objectAtIndex:[indexPath section]];
    NSArray *cities = [[state cities] allObjects];
    City *city = [cities objectAtIndex:[indexPath row]];

    NSString *stateName = [state name];
    NSString *cityName = [city name];
    BOOL isFavorite = [[city favorite] boolValue];

    EditViewController *controller =
    [[EditViewController alloc] initWithCity:cityName
                                       state:stateName
                                    favorite:isFavorite
                                    delegate:self];

    [[self navigationController] pushViewController:controller
                                           animated:YES];
    [controller release];

    [self setEditedItem:indexPath];
}
-(void)editViewFinishedEditing:(EditViewController *)inEditView;
{
    NSString *stateName = [inEditView stateInfo];
    NSString *cityName = [inEditView cityInfo];
    if([self editedItem]) // we're editing an existing item
    {
        State *origState = [states objectAtIndex:[editedItem section]];

        NSString *origStateName = [origState name];
        // first we check to see if the state name changed...
        if(![origStateName isEqualToString:stateName])
        {
            // if it did, we need to move the city to the new state.
            State *newState = nil;
            for(State *state in states)
            {
                if([[state name] isEqualToString:stateName])
                    newState = state;
            }
            if(!newState)
            {
                newState = [NSEntityDescription
                        insertNewObjectForEntityForName:@"State"
                        inManagedObjectContext:context];

                [newState setName:stateName];
                [states addObject:newState];
            }
```

```objc
        City *newCity = [NSEntityDescription
                    insertNewObjectForEntityForName:@"City"
                    inManagedObjectContext:context];

        [newCity setName:cityName];
        [newCity setFavorite:
            [NSNumber numberWithBool:[inEditView favorite]]];
        [newState addCitiesObject:newCity];
        // now remove the old city and optionally
        // state if the state has no more cities.

        City *oldCity = [[[origState cities] allObjects]
                    objectAtIndex:[editedItem row]];

        [context deleteObject:oldCity];

        [origState removeCitiesObject:oldCity];
        if([[origState cities] count] == 0)
        {
            [context deleteObject:origState];
            [states removeObject:origState];
        }
    }
    else
    {
        // the state didn't change, we can just update the old one.
        NSArray *cities = [[origState cities] allObjects];
        City *city = [cities objectAtIndex:[editedItem row]];
        [city setName:cityName];
        [city setFavorite:
            [NSNumber numberWithBool:[inEditView favorite]]];
    }
}
else // it was a new add
{
    State *newState = nil;
    for(State *state in states)
    {
        if([[state name] isEqualToString:stateName])
            newState = state;
    }
    if(!newState)
    {
        newState = [NSEntityDescription
                    insertNewObjectForEntityForName:@"State"
                    inManagedObjectContext:context];
        [newState setName:stateName];
        [states addObject:newState];
    }
```

continued

Listing 18.10   *(continued)*

```objc
            City *newCity = [NSEntityDescription
                            insertNewObjectForEntityForName:@"City"
                            inManagedObjectContext:context];

            [newCity setName:cityName];
            [newCity setFavorite:
                [NSNumber numberWithBool:[inEditView favorite]]];
            [newState addCitiesObject:newCity];
        }
        [context save:nil];
        [[self tableView] reloadData];
        [self setEditedItem:nil];
}
#pragma mark DataSource Methods
- (NSInteger)numberOfSectionsInTableView:(UITableView *)tableView
{
    return [states count];
}
- (NSInteger)tableView:(UITableView *)tableView
 numberOfRowsInSection:(NSInteger)section
{
    State *state = [states objectAtIndex:section];
    return [[state cities] count];
}
- (UITableViewCell *)tableView:(UITableView *)tableView
        cellForRowAtIndexPath:(NSIndexPath *)indexPath
{
    static NSString *CellIdentifier = @"Cell";

    UITableViewCell *cell =
    [tableView dequeueReusableCellWithIdentifier:CellIdentifier];
    if (cell == nil)
    {
        cell = [[[UITableViewCell alloc]
                initWithStyle:UITableViewCellStyleDefault
                reuseIdentifier:CellIdentifier] autorelease];
    }
    State *state = [states objectAtIndex:[indexPath section]];
    NSArray *cities = [[state cities] allObjects];
    City *city = [cities objectAtIndex:[indexPath row]];

    NSString *cityName = [city name];
    [[cell textLabel] setText:cityName];

    if([[city favorite] boolValue])
        [[cell imageView] setImage:[UIImage imageNamed:@"FullStar.png"]];
    else
        [[cell imageView] setImage:[UIImage imageNamed:@"EmptyStar.png"]];
```

```
        return cell;
}
- (NSString *)tableView:(UITableView *)tableView
titleForHeaderInSection:(NSInteger)section;
{
    State *state = [states objectAtIndex:section];
    return [state name];
}
- (BOOL)tableView:(UITableView *)tableView canEditRowAtIndexPath:(NSIndexPath *)
   indexPath
{
    return YES;
}
   (void)tableView:(UITableView *)tableView
commitEditingStyle:(UITableViewCellEditingStyle)editingStyle
 forRowAtIndexPath:(NSIndexPath *)indexPath
{
    State *state = [states objectAtIndex:[indexPath section]];
    NSArray *cities = [[state cities] allObjects];
    City *city = [cities objectAtIndex:[indexPath row]];

    [context deleteObject:city];
    [state removeCitiesObject:city];
    if([[state cities] count] == 0)
    {
        [context deleteObject:state];
        [states removeObjectAtIndex:[indexPath section]];
    }

    [context save:nil];

    [[self tableView] reloadData];
}
@end
```

Listing 18.11

The MyViewController iInterface

```
#import <UIKit/UIKit.h>
#import "EditViewController.h"
#import <CoreData/CoreData.h>
@interface MyViewController : UITableViewController <UITableViewDelegate,
EditViewControllerDelegate>
{
    NSMutableArray *states;
```

continued

Listing 18.11   *(continued)*

```
    NSIndexPath *editedItem;
    
    NSManagedObjectContext *context;
}
@property (retain, nonatomic) NSManagedObjectContext * context;
@property (retain, nonatomic) NSMutableArray * states;
@property (retain, nonatomic) NSIndexPath * editedItem;
@end
```

Now if you make your view controller look like this, you should have a version of the Favorite Cities application that uses Core Data. From the perspective of using it, it should act exactly like it did before, except that if you make changes to the data set and then exit the application and return, your changes should persist.

Understanding What Core Data Makes Easier

So, you've seen that Core Data can make working with databases and object models much simpler. In particular, complex object models containing multiple entities especially benefit from being designed using the entity designer. This makes building complex applications with difficult-to-visualize relationships very easy to work with.

Additionally, virtually any application that manages tabular data can benefit from using Core Data instead of storing it in a flat file or some other proprietary database.

Understanding What Core Data Is Not Good For

It's important to remember that Core Data does have its limitations. Specifically, the types of files that it can use as a data store are relatively limited. On the iPhone, it really comes down to a binary flat file or sqlite database. The XML file format is not yet supported on the iPhone. Additionally, Core Data is not able to access external database servers such as MySQL, PostgreSQL, Oracle, and so on.

These limitations are minor, however when you consider the benefits that Core Data gives you for the specific problem set that it is trying to address. This problem set consists of any situation where you need to manage a local database of objects for your applications.

Summary

In this chapter, you've taken a look at the details of one of the most powerful new frameworks included in iPhone OS 3, Core Data. It's fortunate that Apple has provided this excellent object persistence framework. Using it makes developing certain kinds of applications much easier and reduces development time.

They say that the most bug-free code is code that's not written. By leveraging Core Data, you don't have to worry about spending your time writing complicated, potentially buggy database abstraction layers for your applications. Instead, you can focus on your specific problem domain and the problems that you need to solve.

Connecting to the World with Networking

One of the incredible things about the iPhone is that underneath it all, it really is running Mac OS X. This means that down deep, underneath everything else, lies the beating heart of UNIX. It's true, if you dig deep enough into the iPhone OS, you eventually find that there is a full UNIX file system and many of the libraries and capabilities that you've come to expect from a UNIX computer.

Normally, even as a developer, you are sufficiently isolated from UNIX that it simply doesn't matter to you that UNIX is underneath the covers. But there's one area where having UNIX available to you becomes a definite asset. That area is in the realm of networking.

UNIX was built with networking from the ground up, and so it's no surprise that its networking stack is one of the best available. As a result, the iPhone, because of its heritage, shares this excellent networking capability. The iPhone not only has built-in BSD sockets, but it also has an excellent set of Cocoa Touch classes that enable you to work with sockets with very little effort. The beauty of this design is that when you need simple things done, you have the higher-level abstractions available to you. Things such as accessing Web pages, sending e-mail, and so forth are trivial on iPhone OS. But in addition to this, when you need to go lower, when you need to do things that are different or more difficult, these are also possible through the use of Core Foundation sockets and BSD sockets.

In this chapter, you are going to see how to use some of the high-level classes for doing basic networking. You're also going to take a look at using Bonjour to discover services on the local network. You're going to take a look at how to send e-mail using the new messaging API. Finally, you're going to take a brief look at core foundation sockets to introduce them to you. By the time you're finished, you should be able to conquer just about any networking task on the iPhone.

 In This Chapter

Learning to use the Objective-C classes NSURLRequest and NSURLConnection

Sending e-mail from within your app using the MessageUI framework

Using Core Foundation networking to do low-level networking tasks

Using Bonjour to discover services on the local network

Accessing the Web

In our connected world, perhaps the most frequent type of network activity is connecting to the Web. Therefore, it's no surprise that Apple has integrated Web connectivity into all levels of Cocoa Touch. Not only can you connect to the Web using dedicated classes for doing so, such as NSURLRequest and so forth, but there are even utility methods on several of the foundation classes that encapsulate some of the most frequently needed methods for Web connectivity.

Let's take a look at how the Web can be accessed through Cocoa Touch.

Using URLs with foundational classes

The first and most basic mechanism for accessing Web content through Cocoa Touch involves utilizing factory methods on some of the foundation classes themselves, such as NSString, NSData, and so on. Each of these has an initializer called initWithContentsOfURL: encoding:error:, which enables you to initialize an instance of the class by automatically downloading the content of the file located at the given URL to be used as the contents of the object. In other words, by using this initializer, you can request that a given URL be downloaded and decoded using the given encoding, and then that the contents of that URL be used as the data for an NSString, or an NSData, and so on. This is an incredibly convenient mechanism if you only need very basic Web resource download capabilities. Listing 19.1 shows an example of downloading a URL into an NSString and then using the data.

Listing 19.1

Downloading a URL into an NSString

```
-(void)downloadAndDisplayFile:(NSString *)inUrlString;
{
    NSURL *url = [NSURL URLWithString:inUrlString];
    NSString *fileData = [NSString stringWithContentsOfURL:url];
    // ... do something with the fileData string ...
}
```

This works extremely well if your needs are very simple. However, it is not nearly as flexible as some of the other solutions that you will be looking at shortly. In particular, this call blocks, which means that your main thread is stuck waiting for this file to be downloaded. If there is some kind of network difficulty that causes the download to take an exceptionally long period of time, then your application's UI may appear to be unresponsive. This is not a very good user experience; you should always give the user the option to cancel a download if needed. A very important thing to remember with regard to writing network code on the iPhone is that the wireless network that the iPhone connects to can be, and often is, extremely unreliable and slow. Because of this, you will probably want to use one of the asynchronous mechanisms for downloading data. I am showing you this method because it's simple to use and can be used for simple tasks. Let's take a look at some of the more powerful methods now.

CAUTION
It may be tempting to utilize this method with a background thread to get around the fact that it blocks. You should avoid doing this if possible. Writing thread-safe code is a tricky business, with very little benefit on a single-processor machine such as the iPhone. You're much better off using the asynchronous mechanisms for networking instead.

Using NSURLRequest and NSURLConnection

Probably the best mechanism for downloading content from the Web in Cocoa Touch is the combination of classes `NSURLRequest` and `NSURLConnection`, which together form a powerful mechanism for asynchronously downloading content from the Web.

`NSURLRequest` encapsulates a request for a particular resource, including enabling you to configure certain options such as timeout, the base URL, and so on. Once you have an `NSURLRequest`, you then pass it to an `NSURLConnection` to actually perform the request and load the data. The `NSURLConnection` uses a delegate pattern to enable you to customize specific behavior with regard to the actual loading of the data. For example, the delegate of the `NSURLConnection` contains methods that are called as data is received, and another delegate method that's called when the request has completed.

To start the download of a resource from the Web using an `NSURLConnection`, you first create an `NSURLRequest`, and then use it to create an `NSURLConnection`, using one of its initializers, such as `initWithRequest:delegate:`. Then, you call the start method.

If you need to cancel a request while it is loading — for example, in response to a user clicking a Cancel button — you call the cancel method on the `NSURLConnection`.

Table 19.1 shows the most frequently used delegate methods for `NSURLConnection`. To utilize them, you would implement them in a class in your application, and set an instance of that class to be the delegate of your `NSURLConnection`.

Table 19.1 NSURLConnection Delegate Methods

Method	Description
`- (NSURLRequest *)connection: (NSURLConnection *)connection willSendRequest:(NSURLRequest *) request redirectResponse: (NSURLResponse *)redirectResponse`	This method is called when the request has been redirected to another URL. The delegate has the option to disallow the redirect or modify the destination by returning nil, or an alternative URL request, respectively. If the delegate chooses to allow the redirect, simply return the original request.
`- (void)connection:(NSURL Connection *)connectiondid ReceiveAuthenticationChallenge: (NSURLAuthenticationChallenge *) challenge`	This method is called when the `NSURL Connection` has received an authentication challenge as part of the request processing. The delegate must determine the appropriate action to take, such as providing appropriate credentials, or canceling the download. The sender property of the challenge parameter is an instance of `NSURLAuthentication ChallengeSender`, which implements methods that allow you to send credentials back to the challenger if needed.
`- (void)connection:(NSURL Connection *)connectiondid CancelAuthenticationChallenge: (NSURLAuthenticationChallenge *) challenge`	This method is called when an authentication challenge is canceled as a result of an error.
`- (void)connection:(NSURL Connection *)connection didReceiveResponse: (NSURLResponse *)response`	This method is called when the request has completed successfully, but before the data for the request has been downloaded. After receiving this message, you begin receiving `didReceiveData:` messages containing the data from the resource.
`- (void)connection:(NSURL Connection *)connection didReceiveData:(NSData *) data`	This method is called as data is downloaded from the resource. It may be called multiple times, and you should plan on appending the data received to a buffer until you receive the message `connectionDid FinishLoading:` or `connection: didFailWithError:`.
`- (void)connectionDidFinishLoading: (NSURLConnection *)connection`	This method is called when the resource has finished downloading. It is at this point that you can take whatever data you have received and assume that it is complete. After receiving this message, no further messages are received.
`- (void)connection:(NSURL Connection *)connection didFailWithError:(NSError *) error`	This method is called if the connection fails for any reason at any time during the loading of the data. This signals that the download has failed due to the error provided. After receiving this message, no further messages are received.

Now that you've seen the basics of how to work with an `NSURLRequest`, `NSURLConnection`, and so on, let's see some example code that actually implements a download.

Let's imagine an example application where you want to download an image from the Web and display that image in a `UIImageView`. In this case, the interface of the controller might look something like Listing 19.2.

Listing 19.2

Interface for the Image View Downloader View Controller

```
@interface ImageDownloaderViewController : UIViewController
{
    IBOutlet UIImageView *imageView;
    IBOutlet UITextField *urlField;

    NSURLConnection *connection;
    UIActionSheet *cancelSheet;
    NSMutableData *data;
}
@property (retain, nonatomic) NSURLConnection * connection;
@property (retain, nonatomic) UIActionSheet * cancelSheet;
@property (retain, nonatomic) NSMutableData * data;
-(IBAction)loadImage:(id)sender;
@end
```

In this case, you've added `IBOutlets` for an image view and the URL field. What happens is that the user types the URL of an image into the URL field and then clicks a Load button. This causes the app to download the image and then display the image in the image view. The application can also display a Cancel dialog so that the user has the option to cancel the download at any time.

To start the download off, you need to implement the `loadImage:` method. This is shown in Listing 19.3.

Listing 19.3

Implementation of the loadImage: Method

```
-(IBAction)loadImage:(id)sender;
{
    NSURL *url = [NSURL URLWithString:[urlField text]];
    NSURLRequest *request = [NSURLRequest requestWithURL:url];
```

continued

Listing 19.3 *(continued)*

```
    [self setConnection:[NSURLConnection connectionWithRequest:request
                                                     delegate:self]];

    cancelSheet = [[UIActionSheet alloc] initWithTitle:@"Downloading..."
                                              delegate:self
                                     cancelButtonTitle:@"Cancel"
                                destructiveButtonTitle:nil
                                     otherButtonTitles:nil];
    [cancelSheet showInView:self.view.window];
    [self setData:[NSMutableData data]];
    [connection start];
}
```

So you're taking the text from the URL field, and you are creating an `NSURL` with that. You then use that to make an `NSURLRequest`, which you then use to create your `NSURLConnection`. You then create a `UIActionSheet` to display a Cancel button, which you display over the top of your window. Finally, you start the connection and thus start the download.

With the request sent, you now need to implement the `connection:didReceiveData:`, `connectionDidFinishLoading:`, and `connection:didFailWithError:` methods to appropriately handle the data that you are now going to be receiving.

Listing 19.4 shows these three methods.

Listing 19.4

Implementation of the Delegate Methods

```
- (void)connection:(NSURLConnection *)connection
    didReceiveData:(NSData *)inData
{
    [data appendData:inData];
}
- (void)connection:(NSURLConnection *)connection
  didFailWithError:(NSError *)error
{
    [cancelSheet dismissWithClickedButtonIndex:0 animated:YES];
    [self setCancelSheet:nil];
    [self setConnection:nil];
    [self setData:nil];

    UIAlertView *alert =
    [[UIAlertView alloc] initWithTitle:@"Download Error"
                               message:[error localizedDescription]
```

```
                            delegate:nil
                   cancelButtonTitle:@"OK"
                   otherButtonTitles:nil];
    [alert show];
    [alert release];
}
- (void)connectionDidFinishLoading:(NSURLConnection *)connection
{
    [cancelSheet dismissWithClickedButtonIndex:0 animated:YES];
    [self setCancelSheet:nil];
    [self setConnection:nil];
    UIImage *image = [UIImage imageWithData:data];
    [imageView setImage:image];
    [self setData:nil];
}
```

This code is very straightforward and handles, in order, receiving data, handling errors, and finishing the connection and displaying the image. Looking at each one of these methods in detail, the implementation of `connection:didReceiveData:` simply appends the data received to your data member variable.

If an error occurs while trying to load the data or making the request, the method `connection:didFailWithError:` is called. In it, you dismiss your cancel sheet, free any of your member variables that you will no longer be using, and then display an alert to the user to tell him about the error.

If everything goes successfully, you receive the message `connectionDidFinishLoading:` to tell you that the download has completed. In this method, you dismiss the cancel sheet, free the cancel sheet and the connection, and then initialize a new image using your data. Finally, you set the image view's image property to your new image.

In short, this pattern can be reused for virtually any data that you need to download from the Web. Furthermore, it does not require launching an additional thread or doing anything outside of a normal delegate pattern.

Next, let's take a look at a new feature with iPhone OS 3, the Message UI framework.

Sending E-mail from within Your App with the Message UI Framework

While it has always been possible to send e-mail from within an iPhone application using a `mailto:` URL, doing so has been a jarring experience for the user, as using a `mailto:` URL kicks the user out of the currently running application and launches the Mail application. This, needless to say, does not result in an ideal user experience. Furthermore, the amount of customizability in terms of not just composing the e-mail itself, but also in terms of being able to take action after the e-mail has been sent, has been very limiting.

Because of this, with iPhone OS 3, Apple has introduced the Message UI framework, which enables you to present to the user a fully customizable interface inside your application for the user to compose and send an e-mail. It provides a class called `MFMailComposeViewController`, and the associated delegate protocol for `MFMailComposeViewController`, `MFMailComposeViewControllerDelegate`.

The `MFMailComposeViewController` is a view controller just like any other, and can be presented to the user either through a navigation controller or modally using the method `presentModalViewController:animated:`. Before attempting to display an `MFMailComposeViewController`, however, you should first check to see if the device has been configured to send mail. You do this by calling the class method `canSendMail` on the `MFMailComposeViewController` class. If this method returns NO, then you should not attempt to send mail. If, however, this method returns YES, then you can proceed.

The `MFMailComposeViewController` has several methods on it that enable you to set various attributes of the e-mail that will be sent. They are listed in Table 19.2.

Table 19.2 Mail Customization Methods on MFMailComposeViewController

Method	Description
`-(void)setSubject:(NSString*)subject`	Sets the subject of the e-mail message.
`-(void)setToRecipients:(NSArray*)toRecipients`	Sets the recipients of the e-mail message.
`-(void)setCcRecipients:(NSArray*)ccRecipients`	Sets the carbon copy recipients of the e-mail message.
`-(void)setBccRecipients:(NSArray*)bccRecipients`	Sets the blind carbon copy recipients of the e-mail message.
`-(void)setMessageBody:(NSString*)body isHTML:(BOOL)isHTML`	Sets the message body. If the `isHTML` parameter is set to YES, then the message body is interpreted as HTML when displayed.
`-(void)addAttachmentData:(NSData*)attachment mimeType:(NSString*)mimeType fileName:(NSString*)filename`	Adds attachments to the e-mail of the given types with the given filenames.

NOTE
If you want to display images in the e-mail, you must attach the images as attachments as well.

These methods should be called before the view is displayed to the user. Do not call these methods after displaying the view.

So, for example, to display an e-mail composition view to send an e-mail with attached images to someone, you might do something like Listing 19.5.

Listing 19.5

Sending E-mail within an Application

```
-(void)sendFeedbackMailAboutImage:(UIImage *)inImage;
{
    if(![MFMailComposeViewController canSendMail])
    {
        UIAlertView *alert =
        [[UIAlertView alloc] initWithTitle:@"Mail Config Error"
                                  message:@"Mail not configured."
                                 delegate:nil
                        cancelButtonTitle:@"OK"
                        otherButtonTitles:nil];
        [alert show];
        [alert release];
        return;
    }
    MFMailComposeViewController *controller =
                          [MFMailComposeViewController new];
    [controller setSubject:@"Pics you'd like..."];
    [controller setToRecipients:@"pictures@yourdomain.com"];
    [controller setMessageBody:
     @"Thought you'd like the attached images." isHTML:NO];

    NSData *data = UIImagePNGRepresentation(inImage);
    [controller addAttachmentData:data
                         mimeType:@"image/png"
                         fileName:@"Image.png"];
    [controller setMailComposeDelegate:self];

    [parentController presentModalViewController:controller
                                        animated:YES];
    [controller release];
}
```

Essentially, first you check to see if this device can send e-mail at all. If it can't, then you display an error to the user explaining the situation. If it can, then you create a new `MFMailCompose ViewController`, and set the attributes of the e-mail that you would like to send. In this case, you are attaching an image, so you need to convert the passed-in `UIImageView` object to a representation that can be sent via e-mail — in this case, a PNG. You then attach it to the e-mail using a mime type of image/PNG. After that, you set yourself as the mail compose delegate and then present the controller modally as a child of whatever your parent controller is. Because the controller has now been added to the hierarchy, you can release it here.

NOTE

In this code, we just display an error message if the device is not configured for mail. You can also compose a mailto URL and use it to launch the Mail app on the iPhone. If it's not configured, the user will be prompted to configure mail then.

After the e-mail has been composed and sent, your delegate method `mailComposeController:didFinishWithResult:error:` is called with the result of whether or not the application was able to send the e-mail successfully. For the purposes of this example, you simply check to see if the result was an error so that you can display the error to the user. This is shown in Listing 19.6.

Listing 19.6

Implementation of the Delegate Method

```
- (void)mailComposeController:(MFMailComposeViewController*)controller
          didFinishWithResult:(MFMailComposeResult)result
                        error:(NSError*)error
{
    [controller dismissModalViewControllerAnimated:YES];
    if(result == MFMailComposeResultFailed)
    {
        // an error occurred.
        UIAlertView *alert =
        [[UIAlertView alloc] initWithTitle:@"Download Error"
                                   message:[error localizedDescription]
                                  delegate:nil
                         cancelButtonTitle:@"OK"
                         otherButtonTitles:nil];
        [alert show];
        [alert release];
    }
}
```

The result codes that can possibly occur here are listed in Table 19.3.

Table 19.3 Possible Mail Result Codes

Code	Meaning
`MFMailComposeResultCancelled`	The user canceled sending the e-mail.
`MFMailComposeResultSaved`	The user saved the e-mail to the Drafts folder.
`MFMailComposeResultSent`	The e-mail was saved to the Sent folder and will be sent the next time the e-mail client connects to the mail server.
`MFMailComposeResultFailed`	An error occurred during mail composition. The mail was not sent or queued.

Using Core Foundation Sockets

If working with the higher-level networking classes isn't enough for you, iPhone OS provides a set of low-level network wrappers that act as a thin veneer over the lowest-level socket API, BSD sockets. Though I use the phrase *thin veneer* here, don't be confused into thinking that these wrappers are not powerful; they are quite powerful, and they provide functionality that makes writing networking code much easier. This API is called Core Foundation Network, or `CFNetwork`.

BSD sockets are certainly ubiquitous in terms of their implementation across platforms. However, they were developed over 20 years ago and have not changed much in the intervening years. As a result, using them directly can be a bit of a chore. Typically, developers use BSD sockets and threads to do asynchronous networking, as many of the BSD calls block. The `CFNetwork` API provides a mechanism for you to work with BSD sockets without having to resort to threads. It provides a mechanism that enables you to schedule socket events through the standard run loop mechanism provided by the runtime. This means that you don't have to spawn a thread to check to see if data is waiting on your sockets or if it's okay to send data on your sockets. You simply register a callback on the run loop for when either of those events occurs, and then do whatever processing you need to do in those callback functions.

`CFNetwork` is not only limited to low-level capabilities, however; it also provides several wrappers for popular protocols including HTTP, FTP, and SSL. In fact, the higher-level Objective-C classes actually utilize the low-level `CFNetwork` API to implement their functionality.

For the purposes of these examples, you are going to mainly focus on the low-level API encapsulated primarily in `CFSocket` and `CFStream`. This is because if you learn to use the lower-level API, using the higher-level API is relatively simple to pick up.

NOTE
In this section, you are working with a Core Foundation API again, which means that it obeys the Core Foundation retain-and-release rules. To review them, see Chapter 10.

The two most basic APIs that you will work with here are the `CFSocket` API and `CFStream` API. These provide the interface for creating sockets of all kinds, both TCP and UDP, and an API for treating those sockets as streams, so that you can simply write data into them or read data from them. The third API that you will look at in this section is the `CFHost` API, which provides an API for doing DNS resolution.

NOTE
Technically, `CFStream` is not strictly part of `CFNetwork` and can be used with any communications channel, not just sockets. For the purposes of this example, though, you'll only be using it for sockets.

Exploring CFSocket

The `CFSocket` API is the lowest-level API in `CFNetwork`, and it is the basis for the rest of the framework. It provides a mechanism to create a socket either from scratch or by using an existing BSD socket, and for adding that socket to the application's run loop.

To create a `CFSocket`, you call one of the `CFSocket` create methods. The method `CFSocket Create` has a method signature of `CFSocketRef CFSocketCreate (CFAllocatorRef allocator, SInt32 protocolFamily, SInt32 socketType, SInt32 protocol, CFOptionFlags callBackTypes, CFSocketCallBack callout, const CFSocket Context *context)`. It takes several parameters, and looks imposing, but it's really very straightforward. The parameters to the method are shown in Table 19.4.

Table 19.4 Parameters to the CFSocketCreate Method

Parameter	Purpose
`CFAllocatorRef allocator`	This parameter enables you to specify a custom memory allocator for allocating the memory for the socket object that will be returned. If you pass NULL for this parameter, it uses the default allocator.
`SInt32 protocolFamily`	This parameter specifies the protocol family of the socket. The default value, if you pass 0, is PF_INET.
`SInt32 socketType`	This parameter specifies the socket type to create. The default value, if `protocolFamily` is specified to be PF_INET, is SOCK_STREAM. Possible other values might be SOCK_DGRAM for a UDP socket.

Parameter	Purpose
`SInt32 protocol`	This parameter specifies the protocol for the socket. Possible values include IPPROTO_TCP or IPPROTO_UDP. If `protocolFamily` is PF_INET, and `socketType` is SOCK_STREAM, this defaults to IPPROTO_TCP, and if `protocolFamily` is PF_INET and `socketType` is SOCK_DGRAM, then this defaults to IPPROTO_UDP.
`CFOptionFlags callBackTypes`	This parameter specifies a bitwise-OR of expected events that might trigger a call to callout.
`CFSocketCallBack callout`	This is a function to call when one of the expected events specified in `callBackTypes` occurs.
`const CFSocketContext *context`	This is a user-specified contextual data object; it can be NULL.

Interestingly, most of these parameters allow you to pass null or zero to receive their default values. This means that if you simply accept the defaults for all of the parameters that you can, you get a TCP stream socket, which is probably the most common-use case. You only really need to specify the parameters differently from the defaults if what you want is a UDP socket or something a little bit more exotic, such as a UNIX domain socket.

The callback types that you can specify are listed in Table 19.5.

Table 19.5 Possible Callback Types

Value	Description
`kCFSocketNoCallBack`	No callbacks are received for any events.
`kCFSocketReadCallBack`	You receive a callback when data is available to be read or when a new connection is ready to be accepted. This callback is called when the data is ready to be read but has not yet been read. Your code needs to read the data or accept the connection when this callback is received.
`kCFSocketAcceptCallBack`	This is used with listening sockets; you receive this callback when a connection has been requested. The new connection is automatically accepted, the callback is called, and the data argument is a handle for the new child socket.
`kCFSocketDataCallBack`	When data is received on the socket, it is automatically read in chunks in the background and your callback function is called. The data argument is a `CFData` object containing the data that was read.
`kCFSocketConnectCallBack`	When attempting to connect a socket to a port, this callback is called when the connection has completed either successfully or not. If the connection fails, the data pointer is an error code. If the connection succeeds, the data pointer is null.
`kCFSocketWriteCallBack`	When writing large amounts of data to a socket, this callback is called when the socket is ready for more data.

CAUTION
You do not actually receive your callbacks until you add the socket to a `runloop` using `CFRunLoopAddSource`.

As you can see, right out of the gate, you have the ability to create a socket that has an event-driven run loop associated with it. You can receive callbacks for reading of any important events associated with the socket. In order to actually receive those callbacks, you need to create a `CFRunLoopSourceRef` object using the function `CFSocketCreateRunLoopSource`, and then add it to the main runloop using the function `CFRunLoopAddSource`. In order to actually connect to a port with that socket, you use the function `CFSocketConnectToAddress`.

Let's take a look now at an example code snippet that does all this. Listing 19.7 shows the creation of a `CFSocket`.

Listing 19.7

Creating a Socket and Connecting to a Port

```
-(void)openSocketToHost:(CFDataRef)inHostInfo;
{
    CFSocketRef socket =
        CFSocketCreate(NULL, PF_INET, SOCK_STREAM, IPPROTO_TCP,
                       (kCFSocketConnectCallBack|kCFSocketDataCallBack),
                       (CFSocketCallBack)&myCallback, NULL);

    CFRunLoopSourceRef source =
        CFSocketCreateRunLoopSource(NULL, socket, 0);
    CFRunLoopAddSource(CFRunLoopGetCurrent(), source,
                       kCFRunLoopCommonModes);

    // we've created the socket, it's running in the runloop...
    // now let's connect it!

     CFSocketConnectToAddress(sock, inHostInfo, -1);

    // now free up our resources. - they are in the runloop so it's safe
    // to release them here

    CFRelease(source);
    CFRelease(socket);
}
```

In this code, you've already resolved the host information, so all you need to do is create the socket and connect it! The `myCallback` callback method handles transitioning to sending data after the connection is made. It's shown in Listing 19.8.

Listing 19.8

myCallback Callback Function

```
void myCallback(CFSocketRef inSocketRef,
                CFSocketCallBackType inCallbackType,
                CFDataRef inAddress,
                const void *inData,
                void *inUserInfo)
{
    if(inCallbackType == kCFSocketConnectCallBack)
    {
        // we connected... start sending data to the remote host.
        // ...
    }
    else if(inCallbackType == kCFSocketDataCallBack)
    {
        CFDataRef data = (CFDataRef)inData;
        // the data read is in inData do something with it.
        // ...
    }
    // so on...
}
```

As you can see, the callback is called both when the connection is completed, as well as when the other events you've requested are triggered. For example, it will also be called when you receive new data.

Of course, this code assumes that you already resolved the hostname. To do this, you use the `CFHost` API, which you will review next.

Getting host info with CFHost

`CFNetwork` also provides you with an excellent API for doing DNS resolution asynchronously. That API is the `CFHost` API.

Using the `CFHost` API, you can resolve DNS hostnames and determine reachability to IP addresses very easily. Listing 19.9 shows the `CFHost` API being used to schedule resolution of a hostname.

Listing 19.9
Resolving a Hostname

```
-(void)resolveHost:(NSString *)inHostName
{
    CFHostRef hostRef = CFHostCreateWithName(NULL, inHostName);


    CFHostClientContext ctx = { 0, NULL, NULL, NULL, NULL };
    CFHostSetClient(hostRef,
                    (CFHostClientCallBack)&resolutionCallback, &ctx);

    CFHostScheduleWithRunLoop(hostRef, CFRunLoopGetCurrent(),
                              kCFRunLoopCommonModes);

    CFStreamError error = { 0, 0 };
    CFHostStartInfoResolution(hostRef, kCFHostAddresses, &error);

    // now our callback will be called when the resolution completes.

}
```

Again, you need to pass a callback, in this case `resolutionCallback`, by using the `CFHostSetClient` function. The implementation of this method is shown in Listing 19.10.

NOTE
This code shows asynchronous resolution of hostnames. To resolve the hostname synchronously, you simply don't schedule it on a run loop before starting the resolution.

Listing 19.10
CFHost Callback Function

```
void resolutionCallback (CFHostRef inHostInfo,
                         CFHostInfoType inType,
                         const CFStreamError *inError,
                         void *inUserInfo)
{
    if(inError->error = noErr)
    {
        // pull out an address from our hostinfo, and set the port...
        // note: hosts can have multiple addresses
```

```
            NSArray *addresses = CFHostGetAddressing(inHostInfo, NULL);
            CFDataRef address = [addresses objectAtIndex:0]; // just grab 1st
            struct sockaddr *addr =
                    (struct sockaddr *)CFDataGetBytePtr(address);

            struct sockaddr_in finalSocketInfo;
            // copy the address from the retrieved structure, and
            // add in our port information
            memset(&finalSocketInfo, 0, sizeof(finalSocketInfo));
            finalSocketInfo.sin_addr = addr->sin_addr;
            finalSocketInfo.sin_family = AF_INET;
            finalSocketInfo.sin_port = htons([myObject getPort]);

            CFDataRef host = CFDataCreate(NULL,
                                          (unsigned char *)&finalSocketInfo,
                                          sizeof(finalSocketInfo));
            [myObject openSocketToHost:host];
            CFRelease(host);
        }
        else
        {
            // handle the error
        }
    }
}
```

In this case, you're simply checking to see if an error occurred, and if not, you're opening the socket using the method shown in the previous section. The majority of this code is actually making a copy of the `sockaddr` structure you got back from `CFHost`. This is because the `CFHostRef` does not contain the port information for the connection, so you need to add it. The pointer you get from a `CFDataRef` is read-only, so you need to copy the data to another object you can modify. After this, you use the `openSocketToHost:` method using your new `sockaddr_in` structure wrapped in a new `CFDataRef`.

This all might look a bit complicated, but it's really not.

Using CFStreams

The final API you're going to look at in the `CFNetwork` API is `CFStream`. An important item to remember with regard to `CFStream` is that it is the lowest-level API you can use that will trigger the iPhone OS to connect to the cellular or Wi-Fi network. Anything lower than `CFStream` (for example, `CFSocket` or BSD sockets) will *not* cause the cellular network to start. Therefore, it's important that you at least use `CFStream` or something higher level to start your network, even if you ultimately use some other mechanism for actual communication.

Like the `CFHost` API, `CFStream` can be scheduled on a run loop with a client callback function to be called when events occur. Listing 19.11 shows the creation of a write stream.

Listing 19.11
Creating a Write Stream

```
-(void)startStreamToHost:(NSString *)inHost andPort:(int)inPort;
{
    CFStreamClientContext ctx = {0, NULL, NULL, NULL, NULL};
    CFStreamCreatePairWithSocketToHost(kCFAllocatorDefault, inHost, inPort,
                                       &rStreamRef, &wStreamRef);

    // tell the stream to close the socket when we release it.
    CFWriteStreamSetProperty(wStreamRef,
                             kCFStreamPropertyShouldCloseNativeSocket,
                             kCFBooleanTrue);

    // set up the client
    CFWriteStreamSetClient(wStreamRef, (kCFStreamEventOpenCompleted |
                                        kCFStreamEventErrorOccurred |
                                        kCFStreamEventEndEncountered),
                           streamCallback, &ctx);

    // schedule the stream on the runloop
    CFWriteStreamScheduleWithRunLoop(wStreamRef, CFRunLoopGetCurrent(),
                                     kCFRunLoopCommonModes);

    // annnd connect!
    if (!CFWriteStreamOpen(wStreamRef))
    {
        // handle any error here...
    }
}
```

And again, you have a callback for this code. In this case, it's called `streamCallback`, and it's shown in Listing 19.12.

Listing 19.12
Write Stream Callback

```
void streamCallback(CFWriteStreamRef inStream, CFStreamEventType inType,
                    void *inUserInfo)
{
    if(inType == kCFStreamEventCanAcceptBytes ||
       type == kCFStreamEventOpenCompleted)
    {
```

```
            int bytesWritten = CFWriteStreamWrite(inStream, someData,
                                         strlen(someData));
            if(bytesWritten != strlen(someData))
            {
                someData += bytesWritten;
            }

        }
        else if(inType == kCFStreamEventEndEncountered)
        {
            CFWriteStreamUnscheduleFromRunLoop(inStream, CFRunLoopGetCurrent(),
                                         kCFRunLoopCommonModes);
            CFWriteStreamClose(inStream);
            CFRelease(inStream);
        }
        else if (inType == kCFStreamEventErrorOccurred)
        {
            CFErrorRef error = CFWriteStreamCopyError(inStream);
            // handle the error...

        }
}
```

In this code, when you initially connect, you start sending your data from your buffer. If the socket could not send all the data at once, it returns the amount it was able to send. You then adjust your buffer pointer and wait for the `kCFStreamEventCanAcceptBytes` event, which tells you that you can write more data to the socket again.

If you were listening for data using a read stream, you might have a callback such as the one shown in Listing 19.13.

Listing 19.13

Read Stream Callback

```
void readCallback(CFReadStreamRef inStream, CFStreamEventType inType,
                  void *inUserInfo)
{
    if(inType == kCFStreamEventHasBytesAvailable)
    {
        UInt8 readBuffer[MAX_MSG_SIZE];
        CFIndex bytesRead =
             CFReadStreamRead(inStream, readBuffer, MAX_MSG_SIZE);
        if (bytesRead > 0)
        {
```

continued

Listing 19.13 *(continued)*

```
            // do something with the data in readBuffer here
        }
    }
    else if(inType == kCFStreamEventEndEncountered)
    {
        CFReadStreamUnscheduleFromRunLoop(inStream, CFRunLoopGetCurrent(),
                                         kCFRunLoopCommonModes);
        CFReadStreamClose(inStream);
        CFRelease(inStream);
    }
    else if (type == kCFStreamEventEndEncountered)
    {
        CFErrorRef error = CFWriteStreamCopyError(inStream);
        // handle the error...

    }
}
```

As you can see, working with `CFStreams` is very straightforward. Usually, you only need to work with these APIs if you need to implement some kind of proprietary protocol or something of that nature.

Exploring Bonjour

You've seen how to retrieve data from the Internet and from other hosts on your network using both the high-level Objective-C classes and the low-level core foundation API. But how do you discover services available on your local network? The answer is Bonjour.

Bonjour is the name for the API that Apple provides for interfacing with its multicast DNS resolution service. Multicast DNS is an Internet standard that allows for resolution of hostnames, services, and even IP addresses, with no preconfiguration necessary. It's used in a variety of applications on Mac OS X, such as iTunes for sharing, iPhoto, and even file sharing. And it is also available on the iPhone with Cocoa Touch.

Bonjour allows you to advertise services that you provide, as well as to discover services that are provided. This capability is limited to only your local network, but within that context, it works very effectively.

An important distinction should be made here. Bonjour does not in any way handle transport-level network activity. This means that it does not actually handle sending application data from one host to another. Bonjour is strictly for the purposes of service publishing and discovery.

Browsing for Bonjour services

To discover local services on your network, you use the `NSNetServiceBrowser` class. It enables you to search for services of particular types, and for services in particular domains. It provides a delegate API that enables you to work with it asynchronously. The types of services that you can search for can be anything from HTTP to FTP to your own custom protocol.

To begin using an `NSNetServiceBrowser`, you simply create a new one using its `init` method, configure its delegate object, and then tell it to begin searching for services of an appropriate type by using the method `searchForServicesOfType:inDomain:`.

When telling the `NSNetServiceBrowser` to begin searching for services of a particular type, the service type parameter is specified as a string, which follows a specific convention. The string must specify both the service name and the transport layer name with an underscore character preceding both, and a period character following both. So, for example, to specify that you want to browse for HTTP services, you would specify a service type of "`_http._tcp.`". Similarly, to specify that you want to browse for FTP services, you might specify a service type of "`_ftp._tcp.`".

The domain argument can be specified to be "local" for the local network domain, or you can specify an explicit domain name. Typically, an empty string is passed for this parameter, which indicates the default registration domain.

Listing 19.14 shows an example of initializing and `NSNetServiceBrowser` and setting it to browse for HTTP services.

Listing 19.14
Creating a New NSNetServiceBrowser

```
-(void)browseForHttpServers;
{
    browser = [NSNetServiceBrowser new];
    [browser setDelegate:self];
    [browser searchForServicesOfType:@"_http._tcp" inDomain:@""];
}
```

The method `searchForServicesOftype:inDomain:` returns immediately; the `NSNetServiceBrowser` automatically adds itself to the current run loop, and begins its search asynchronously in the background. When a service matching your requirements is found, the delegate method `netServiceBrowser:didFindService:moreComing:` is called, enabling you to take action with the service that was found. An implementation of this is shown in Listing 19.15.

Listing 19.15

Receiving Services on the Delegate

```
- (void)netServiceBrowser:(NSNetServiceBrowser *)netServiceBrowser
        didFindService:(NSNetService *)netService
          moreComing:(BOOL)moreServicesComing
{
    [serviceList addObject:netService];
    if(!moreServicesComing)
    {
        // finish updating the UI and
        // allow the user to select a service...
    }
}
```

This delegate method is called for each net service that is found on the network. In each case, the `moreComing` parameter signals whether or not there are more services being discovered. When this flag returns NO, you should then take the opportunity to update your UI and allow the user to select one of the services. The `NSNetService` object that you receive here contains all the information needed to connect to the host and service. This data includes things such as the IP address, the port number, and so on. It also includes any human-friendly name that might be associated with that particular service, for example, "Joe's Web Server." A listing of the more commonly used properties of `NSNetService` is shown in Table 19.6.

Table 19.6 Commonly Used Properties on NSNetService

Property Name	Purpose
addresses	An array of `NSData` objects containing socket addresses for the service. Each element of the array contains a `sockaddr` structure, appropriate for use in conjunction with the `CFNetwork` APIs to connect to the service.
hostName	The hostname of the computer providing the service.
name	The name of the service.
port	The port the service is running on.

The `addresses` property requires additional resolution to be performed using the `NSNetService`. To do this resolution, you must set yourself as the delegate for the `NSNetService` and then call the method `resolveWithTimeout:`, which causes the `NSNetService` to query the multicast DNS for the appropriate information. When the resolution process is complete, the delegate method `netServiceDidResolveAddress:` is called, informing you that you can now use the `NSNetService` data to connect to the host. If an error occurs during this resolution process, or if the host could not be resolved, the method `netService:didNotResolve:` is called.

When a service is no longer available, the delegate method `netServiceBrowser:didRemoveService:moreComing:` is called. Just like when the services are being discovered, this method is called for each service as it is removed. Additionally, the `moreComing` parameter again signals whether there are multiple services that are going to be removed. You should again continue reading the services until this parameter returns NO, and then update your UI.

Using NSNetServices

Finally, to publish your services to Bonjour, you use the `NSNetService` class again.

Essentially, to advertise that you are providing a service on a particular port at a particular host, you instantiate an instance of `NSNetService`, initializing it with the type, name, and port of your service. You then call the `publish` method. This causes your service to then be advertised to any `NSNetServiceBrowsers` that are looking for services of the type that you are providing.

Listing 19.16 shows an example of advertising a Web service using the `NSNetService` class.

Listing 19.16

Advertising a Web Service

```
-(void)advertiseWebService;
{
    // set up your web server here...
    // ...
    // advertise the web server
    NSNetService *svc = [[NSNetService alloc]
                        initWithDomain:@"local."
                        type:@"_http._tcp"
                        name:@"Joe's Web Service"];
    [svc setDelegate:self];
    [svc publish];
}
```

Again, just like before, the `publish` method returns immediately and events associated with publishing are handled by the delegate. In this case, probably the most important event that you want to handle on the delegate is the event signifying that your service was published successfully. When this occurs, the delegate is sent the message `netServiceDidPublish:`. Your application should take whatever appropriate action is necessary in response to this event. Additionally, you should probably also implement the delegate method `netService:didNotPublish:`, which is called in the event that your service could not be published, for example, if there is some kind of network error.

The signatures for these delegate methods are shown in Table 19.7.

Tables 19.7 NSNetService Delegate Methods

Method	Purpose
`- (void)netServiceDidPublish: (NSNetService *)sender`	This method is called when the net service is published.
`- (void)netService: (NSNetService *) sender didNotPublish: (NSDictionary *) errorDict`	This method is called in the event of an error while publishing.

Summary

In this chapter, I covered a lot of ground with regard to accessing the Internet and local hosts. You've seen how you can use the high-level Objective-C classes such as `NSURLRequest` and `NSURLConnection` to access data from the Web. Additionally, you've seen how to use the new Message UI framework to send e-mail from within your application. If these high-level APIs aren't powerful enough for you, you've also seen how you can drop down into the core foundation socket API to do more low-level socket coding when needed. Finally, you've seen how you can use Bonjour to both publish and discover services on the local network.

Using the knowledge that you've gained in this chapter, you can implement any number of different types of network services in your application. It's important to remember that the iPhone is more than just a smart phone. In many ways, it is a handheld computer, capable of communicating on the Internet as a first-class citizen.

Using the Push Notification Service

As you know, on the iPhone, resources such as memory and CPU cycles are extremely scarce. As a result, Apple has chosen to restrict applications that can run in the background to only official applications written by Apple developers. This means that your applications are unable to run in the background. This also means that if you want to produce an application that maintains a persistent connection to a server even when your application is not running, you will be unable to do so.

Apple recognizes that this is a significant barrier for certain classes of applications. An example might be instant messaging applications, where you might send a message and then wait a period of time before receiving a response to that message. It would be inconvenient to the user to have to stay in the instant messaging application all the time in order to receive new messages.

To at least partially solve this problem, Apple has introduced the Apple Push Notification Service, or APNS.

The APNS enables application developers to send notifications to their users' iPhones or iPod touches directly from their own server-side applications. The iPhone OS displays these notifications when they are received in a way similar to the way it displays SMS messages. If an application has registered to receive notifications, when notifications for that application arrive, the iPhone gives the user the option to view the notification. If the user chooses to view the notification, the application is launched and given the notification so that it can take action. Figure 20.1 shows what one of these notifications might look like.

In This Chapter

Using the Apple Push Notification Service

Understanding the intricacies of the push protocol

Sending and receiving notifications

Implementing a push provider in Ruby

Figure 20.1

A push notification

In this chapter, you're going to see exactly how to use the Apple Push Notification Service to add push notifications to your applications. You will look first at the architecture of the push notification service and the message formats that are used in sending notifications to your applications. Finally, I'll show you some example code of how to make a push notification server, and you'll build an example application that will receive notifications from your example notification server.

Understanding the Push Notification Service Architecture

Before I delve too deeply into actual implementation of a push notification server and client, I need to review the Apple Push Notification Service architecture and define some terms. It's important that you understand how the architecture functions so that you can work with it effectively.

Understanding the push notification communications

Because of the nature of server-side technologies, the interface for working with the APNS on the server-side is somewhat low-level. It utilizes a binary message format that is streamed over a socket connection to servers that Apple maintains. Notifications are streamed to Apple's servers, and then Apple sends the notification from its servers to the designated device running your application. The iPhone OS on the device maintains a persistent connection to Apple's notification servers. When the notification is received, the iPhone OS decodes the notification, determines if an application has registered for this notification, and if it has, displays the notification.

If no application has registered to receive the notification, the iPhone OS notifies Apple's servers of this fact and does not display the notification. Developers are expected to occasionally query another set of servers, called the "Feedback" servers, to determine if notifications being sent to devices are being rejected in this manner. If so, the developer's application server is expected to discontinue sending notifications to that particular device. This sort of situation might occur when a user has installed an application, selected to be notified of certain events, and then subsequently uninstalled the application.

Looking at the overall architecture

Figure 20.2 shows the APNS from a high-level point of view.

Essentially, an application is installed on an iPhone OS device. This application communicates with a developer's application server using whatever protocol or mechanism the developer chooses.

The developer's application server communicates with the APNS servers using the protocol specified by Apple. The iPhone OS device that is registered for the notifications also opens a socket connection to the APNS servers.

When a push notification needs to be sent, the developer's application server sends the message to the APNS servers, which then send the message to the iPhone OS device specified in the message.

NOTE
Apple provides two sets of servers: one for production and another for sandboxed development during testing. Whether you are using a development versus a distribution certificate determines which server you connect to within your application, but you want to ensure that your application server is also connecting to the correct APNS server when testing. The production server is located at `gateway.push.apple.com`, port 2195. The development server is located at `gateway.sandbox.push.apple.com`, port 2195.

Figure 20.2

The APNS architecture

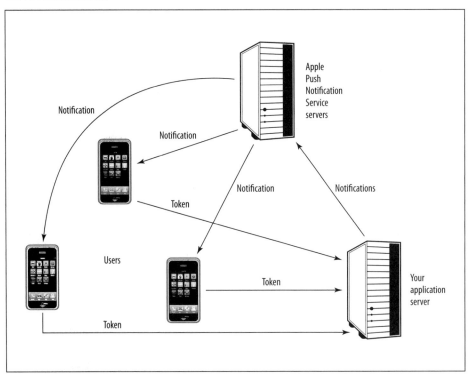

If no application on the device is currently registered to receive the specified notification, the device sends a message to the feedback servers to indicate that the message could not be delivered. The developer's application server is expected to query this server occasionally to determine if messages it is sending are not being delivered. If this happens, it should stop attempting to deliver them to the devices that are rejecting them.

NOTE
The APNS servers cache only the last sent notification for the purposes of resending if it's not received due to network difficulties. This means that if you send two messages to a device that is currently off or disconnected, only the last one is cached and sent to the device when it reconnects.

To register for notifications, the application uses the method `registerForRemote NotificationTypes:` to register for the specified notification types. The types the application can ask to register for are shown in Table 20.1.

Table 20.1 Possible Notification Types

Value	Description
UIRemoteNotificationTypeNone	The application ignores all notifications.
UIRemoteNotificationTypeBadge	The application receives notifications that badge the application icon.
UIRemoteNotificationTypeSound	The application receives notifications that play a sound.
UIRemoteNotificationTypeAlert	The application receives notifications that display an alert message.

These different notification types can be combined so that you can indicate that you receive any combination of notification, including all of them.

Once the application has registered to receive notifications, its application delegate receives the message `didRegisterForRemoteNotificationsWithDeviceToken:`. This method takes as a parameter an `NSData` object containing the device token for this device. This token must be sent to the developer's application server, which then uses it to communicate with the APNS servers when it needs to send the notification to that particular device.

The device token received here is unique to that particular device and is what enables the developer's application server to send the notification through the push notification service to that device. Applications are expected to register and send their device token to the developer's application server every time the application is launched.

Examining the push notification message formats

A push notification consists of a JSON dictionary object with another dictionary contained within it associated with the dictionary key "aps". This internal dictionary contains a set of properties that specify the components of the notification itself. This JSON object is encoded as a string, and put into a binary message, which is then sent to the APNS servers.

CAUTION
The maximum allowed size on a payload is 256 bytes.

The JSON dictionary components consist of an alert message, a number for displaying as the badge on the application icon, and a sound to play.

NOTE
The sound included in notifications must be contained within the application bundle of the application that handles the notification.

Exploring the payload

The alert component of the notification can be either a string or a dictionary containing several specific subelements. If the alert is simply a string, the text given is displayed as the message text for the alert window that is displayed on the device. With it, two buttons are also displayed, Close and View. If the user touches View, then the application that has registered for that notification is launched, and given the notification.

If the alert component is a dictionary, it may contain several subelements. These include `body`, `action-loc-key`, `loc-key`, and `loc-args`.

The `body` key specifies the alert message when using this message format.

The `action-loc-key`, if specified, specifies the text to display in place of the View button on the display. If its value is null, the system displays the alert with only a single button that says OK. This button dismisses the alert when touched. There is not a button that launches your application.

The `loc-key`, if specified, provides a key to an alert message string in your localization strings file for the current localization. This key is used to get a localized message string to display as the text of the alert message. This value can be formatted using printf-style format strings in conjunction with the `loc-args` key.

Finally, the `loc-args` key provides an array of strings for use in conjunction with the print format specifiers used in the `loc-key` property.

NOTE
The JSON dictionary must adhere to RFC 4627.

A typical payload message, when fully constructed, should look something like this: `{"aps":{"alert":"<messageText>", "badge":1}}`.

Wrapping the payload in the APNS message

Once you have encoded your payload as specified here, you need to deal with actually sending it to the APNS servers. The interface for doing this is a plain TCP socket. You'll see more about how to connect to the socket and send your message shortly, but I want to address the message format that's used to send the message here while I am discussing the payload.

The actual message that your application server sends to the APNS servers consists of a binary format, as shown in Figure 20.3.

Figure 20.3

Push message format

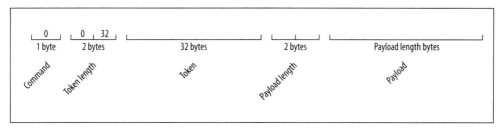

This format consists of the first byte being a command byte of 0. The next two bytes indicate the length of the following field, which is the device token. After the token length is a 32-byte value giving the device token. Following the device token are two bytes to indicate the length of the payload. Finally, the remaining length of the message, based on your payload length, consists of your JSON-encoded payload dictionaries.

All lengths specified in the format are specified to be big endian.

Understanding the feedback message

The feedback service uses a similar binary format to communicate to you the device tokens of the devices that are not receiving your notifications. Figure 20.4 shows the format of that message.

Figure 20.4

Format of a feedback message

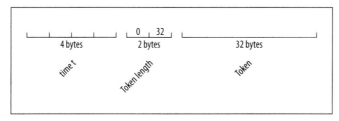

Essentially, this message consists of three components. The first component, "`time_t`", indicates the timestamp at which the push notification service determined that the application no longer exists on the device and thus is no longer registered to receive the notifications. The second element is the length of the device token. And the third element is the device token itself.

The APNS servers, as well as the feedback servers, deal with these messages as asynchronous streams. This means that when sending your messages to the APNS servers, you simply stream the messages into the socket. You do not receive any acknowledgment messages while streaming. Similarly, when receiving data from the feedback server, all you need to do is open a socket and you begin receiving the messages that are designated for your application. You do not need to acknowledge them. The push notification service servers determine your application's identity based on the certificates that you use to connect to it.

Understanding push notification security

The APNS secures its data using two different and complementary mechanisms.

The first security mechanism is used to secure the connection point from the application developer's application server to the APNS servers and from the APNS servers to the devices. This is called *Connection Trust*.

The second security mechanism is called *Token Trust* and ensures that messages are delivered and routed properly from the developer's application server to the developer's application on the iPhone OS device.

Between these two mechanisms, the push notification service can ensure that only authorized devices are connecting to its push notification network and that messages are being properly delivered to the appropriate applications.

Understanding connection security

The Connection Trust provided between the APNS servers and the devices is provided by Apple and requires no action from a developer. However, the Connection Trust between the APNS server and the developer's application server does require work on the part of the developer.

Figure 20.5 shows the Connection Trust mechanism from a high-level point of view.

When the developer's application server connects to the APNS servers, it is required to use TLS. By using TLS, you can specify the use of particular certificates in the connection. The certificates that you use on your application server are provided from Apple for use with the push notification service. By using the certificates provided by Apple, Apple is able to authenticate and verify the connection from your application server to theirs.

NOTE
TLS is short for Transport Layer Security and is a mechanism by which TCP connections encrypt their data using secure certificates.

Figure 20.5

Connection Trust

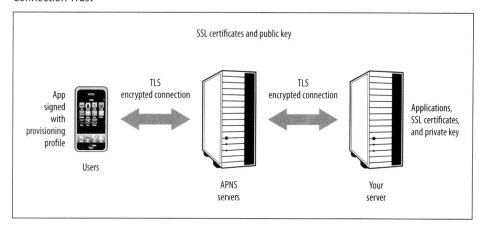

Understanding token security

The idea behind Token Trust is that when an application registers to receive notifications, you want to be able to ensure that the message that you send is received by your application and your application only. You do not want your message to be able to be intercepted by another application on the device.

Figure 20.6 shows a high-level view of the Token Trust mechanism.

To do this, the iPhone OS application needs to communicate with the APNS servers to retrieve a unique token that can then be passed through your application server back to the APNS servers again for when the notifications need to be delivered.

The token that the application receives from the APNS servers is opaque to both the application and the developer's application server. Its purpose is strictly intended to provide for a security mechanism between the APNS servers and the recipient of the message, to provide a trust between those two.

CAUTION

Just because the APNS communication is secured using these mechanisms, you should never use it to send personal information such as credit card or Social Security card numbers because you can't predict who might be reading the message on the other end.

Figure 20.6

The Token Trust mechanism

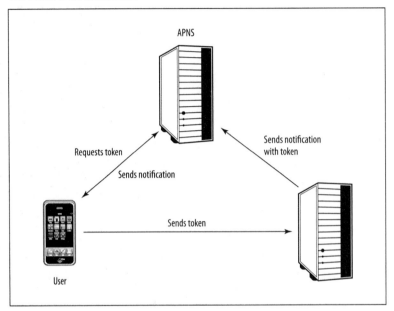

Acquiring Push Notification Certificates

Before you can begin using the APNS, you have to acquire certificates for use with your application server in communicating with the APNS servers. This is done through the iPhone Developer Center on the Apple Web site via the iPhone Developer Program Portal.

I'm going to go through the present (as of early June 2009) process that's required for getting a certificate. Apple can and does change the iPhone developer program portal on a regular basis, and this process may change, so be sure to check the latest instructions on the Apple Web site for updated information on this process before beginning.

In order to acquire a certificate, the first thing you need to do is create an application ID for the applications that are going to use your notifications. In order to load an iPhone application onto a provisioned iPhone, you have to have an application ID for that application. Typically, these application IDs are created with wildcards so that you can use them for all of your applications. However, in order to use an application ID in conjunction with the APNS, it must be a specific, non-wildcard ID. To create an application ID, go to the iPhone Developer Program Portal, and select App IDs from the panel on the left side. Then choose Add ID. This brings you to the screen shown in Figure 20.7.

Figure 20.7

The Create App ID screen

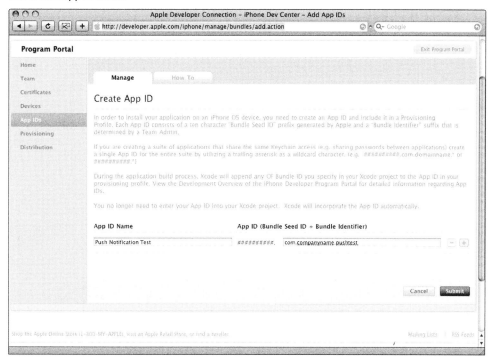

Once you have created the application ID, you are returned to the application ID listing screen, which shows your new application ID. Now you have to configure the application ID to be used with a push notification service. To do this, click the "Configure" link. This brings you to the screen shown in Figure 20.8.

Here, you have the option of configuring your development and your production SSL certificates.

> **NOTE**
> I previously mentioned that there is both a sandbox development server available for testing as well as a production server used for actual deployed applications. Which server you want to push your traffic to depends on which certificate you use with your application server.

Figure 20.8

Configuring an application ID for push notification

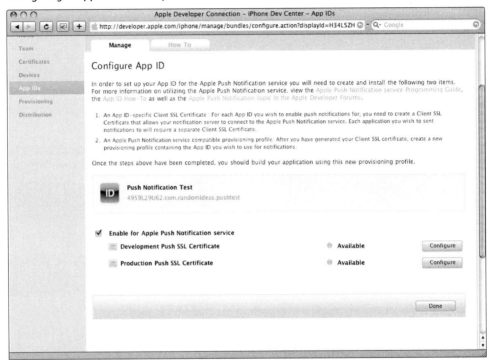

The portal next instructs you to open the Keychain Access application and to generate a certificate signing request. Follow the instructions until it has generated your APNS SSL certificate. You should then download the certificate and copy it and your private key from your keychain to your application server or development machine where you will be generating your push notifications.

Lastly, in order to use your push notification certificate with your iPhone client application, you need to generate a new provisioning profile that you will use when signing your application. Visit the provisioning link in the left panel of the program portal in order to create this provisioning profile. Make sure when you are creating this new provisioning profile that you select the application ID that you used with the APNS SSL certificate generation. This ensures that when your application connects to the APNS servers, they correctly authenticate your application.

Developing Your Server-Side Push Notification Service

There are so many different technologies available for creating an application server for push notifications using SSL that it's impossible for me to give you instructions on how to configure them all. You're going to have to look at the implementation of the specific technology that you choose to use.

That said, for the purposes of this example, I'm going to demonstrate implementation of a push notification supplier built using Ruby.

I'm not going to be able to go into a lot of the intricacies of how the application is built or teach you how to read Ruby. You're just going to have to take my word for it that it works. However, I will include the full source code in the example, and, of course, the source code is available on the book's Web site.

The important thing to recognize is that this part of the push notification service is entirely implementation-dependent when you choose to configure it. The example that I'm giving you here is one way to do it.

Implementing a Ruby push notification supplier

To implement this push notification supplier in Ruby, you're going to use Ruby's built-in support for OpenSSL, connect up and send the message, and then disconnect. There are certainly other ways to do this. In particular, you really shouldn't connect, send, and then disconnect for every single message; in fact, if you do this, Apple will probably assume that you are attempting to attack their servers and will revoke your certificate. The APNS model is based on the idea that you can connect, stay connected, and just stream in your messages as needed. There are also ways to do it in other languages, such as Perl, Python, or C.

So, without further ado, let's take a look at the code. Listing 20.1 shows the code for a push notification supplier in Ruby.

Listing 20.1

A Push Notification Supplier Implemented in Ruby

```
#!/usr/bin/ruby
# ./notify.rb '<device token>' '<message text>'
require 'socket'
require 'openssl'
# just checking our arguments...
```

continued

Listing 20.1   *(continued)*

```ruby
if(ARGV.length < 2 or ARGV[0].length < 32)
  puts "Usage: ./notify.rb '<device token>' '<message>'"
  exit(-255)
end
# strip out the spaces in the token
# the token is copied from an NSLog in our client app
tokenText = ARGV[0].delete(' ')
messageText = ARGV[1]
# pack the token to convert the ascii representation back to binary
tokenData = [tokenText].pack('H*')
# construct the payload
payload = "{\"aps\":{\"alert\":\"#{messageText}\", \"badge\":1}}"
# construct the packet
packet = [0, 0, 32, tokenData, 0, payload.length, payload].pack("ccca*cca*")
# read our certificate and set up our SSL context
cert = File.read("devcerts.pem")
openSSLContext = OpenSSL::SSL::SSLContext.new
openSSLContext.cert = OpenSSL::X509::Certificate.new(cert)
openSSLContext.key = OpenSSL::PKey::RSA.new(cert)
# Connect to port 2195 on the server.
sock = TCPSocket.new('gateway.sandbox.push.apple.com', 2195)
# do our SSL handshaking
sslSocket = OpenSSL::SSL::SSLSocket.new(sock, openSSLContext)
sslSocket.connect
# write our packet to the stream
sslSocket.write(packet)
# cleanup
sslSocket.close
sock.close
```

Let's take a look at this code step-by-step to see exactly what it is that you need to do to send a push notification.

First, you take your message and your device ID from the command line parameters passed to this script, and you use them to build the packet that you are going to send to the APNS servers. This is shown in the extract in Listing 20.2.

Listing 20.2

Constructing Your Packet

```ruby
# strip out the spaces in the token
# the token is copied from an NSLog in our client app
tokenText = ARGV[0].delete(' ')
```

```
messageText = ARGV[1]
# pack the token to convert the ascii representation back to binary
tokenData = [tokenText].pack('H*')
# construct the payload
payload = "{\"aps\":{\"alert\":\"#{messageText}\", \"badge\":1}}"
# construct the packet
packet = [0, 0, 32, tokenData, 0, payload.length, payload].pack("ccca*cca*")
```

Essentially, you take your `commandline` argument for the message and use it to build the JSON serialized object that needs to go into the payload. You place the message into a JSON dictionary with a key of "alert." You are also adding another key of "badge" to set the badge number to 1. Finally, you take this dictionary object and put it into another dictionary associated with the key "aps".

Next, you have some code related to reading your certificate from disk. This is shown in Listing 20.3.

Listing 20.3

Reading the Certificates

```
# read our certificate and set up our SSL context
cert = File.read("devcerts.pem")
openSSLContext = OpenSSL::SSL::SSLContext.new
openSSLContext.cert = OpenSSL::X509::Certificate.new(cert)
openSSLContext.key = OpenSSL::PKey::RSA.new(cert)
```

The important part here is that when you receive your certificate from the iPhone Developer Portal, it is in a .p12 format. Most OpenSSL libraries prefer to work with .pem format certificates. Therefore, you have to convert them from the .p12 format to the .pem format. To do this, you run the terminal commands:

```
openssl pkcs12 -in Certificates.p12 -out devcerts.pem -nodes
    -clcerts
```

With this file extracted, you can then use it to create your OpenSSL context. On the context, you set the `cert` and `key` properties accordingly by extracting them from your certificate.

Next, you open the connection to the APNS server. Once you have the connection open, you initiate a new SSLSocket using the TCP socket. You then call `connect` on your SSL socket, which causes SSL to do its handshaking. This is shown in Listing 20.4.

Listing 20.4

Connecting to the Server and Sending Your Packet

```
# Connect to port 2195 on the server.
sock = TCPSocket.new('gateway.sandbox.push.apple.com', 2195)
# do our SSL handshaking
sslSocket = OpenSSL::SSL::SSLSocket.new(sock, openSSLContext)
sslSocket.connect
# write our packet to the stream
sslSocket.write(packet)
```

Finally, you send the packet to the server.

Pushing notifications

So now that you have an implementation of a notification supplier, to use it, you simply run the command:

```
./notify.rb '<token>' '<message>'
```

The token here is received from your client application. For the purposes of testing and experimentation, you could make the application simply log its token to the console. You could then copy and paste that into your command line. In your actual implementation, however, you will need to work out some mechanism to send the token to your server using sockets and automating this process.

Once the message is received on the device, it will be displayed on the screen.

Checking delivery using feedback

If a user removes your application from her device, obviously it makes no sense to continue sending the notifications. Notifications received for applications that are no longer installed are not displayed. The device also sends a notice back to the APNS server to indicate that the message was unable to be delivered. The APNS server then subsequently adds that device to a list that it keeps on the APNS feedback servers. You should occasionally connect to the feedback servers and download the list of tokens that have not received the notifications you've sent. You should then cease sending messages to those tokens.

Integrating Push Notifications with Your iPhone Client

Building an application that registers for push notifications is incredibly simple. You first need to register for the notifications, and then when you get your token, you send it to your application server. Finally, when notifications are received, your application is launched with the notification information as a parameter. If a notification occurs while your application is running, your application receives a message that you can also use to take action based on the notification.

NOTE
If a notification is received for your application while your application is running, then no message displays. You are expected to take whatever action is necessary for it, including displaying a message yourself if appropriate.

Registering for notifications

To register for notifications, you simply call the method `registerForRemoteNotificationTypes:` on your `UIApplication` singleton. This is shown in Listing 20.5.

Listing 20.5

Registering for Remote Notification Types

```
- (void)applicationDidFinishLaunching:(UIApplication *)application {
    [[UIApplication sharedApplication] registerForRemoteNotificationTypes:
        (UIRemoteNotificationTypeAlert|UIRemoteNotifcationTypeSound|
        UIRemoteNotificationTypeBadge)];
    // Override point for customization after app launch
    [window addSubview:viewController.view];
    [window makeKeyAndVisible];
}
```

After the application has contacted the APNS and registered for the notifications, your application delegate receives a call to `didRegisterForRemoteNotificationsWithDeviceToken:`. You should use this opportunity to send your device token to your push supplier. In the example shown in Listing 20.6, you use the fictional method `sendTokenToServers:` for this purpose.

Listing 20.6

Sending the Device Token to the Server

```
- (void)application:(UIApplication *)application didRegisterForRemoteNotification
  sWithDeviceToken:(NSData *)deviceToken
{
    [self sendTokenToServers:deviceToken];
}
```

Receiving notifications in your application

Finally, when you actually receive a notification, there are two paths by which it can enter your application.

The first occurs when your application is not running, but the user chooses the View button on the notification. In order to intercept this notification, you need to have implemented the method `application:didFinishLaunchingWithOptions:`, which is called in place of `applicationDidFinishLaunching:` in this circumstance. Listing 20.7 shows an implementation of this method.

Listing 20.7

Implementation of the application:didFinishLaunchingWithOptions: Method

```
- (BOOL)application:(UIApplication *)application didFinishLaunchingWithOptions:
  (NSDictionary *)launchOptions
{
    NSDictionary *thePayload = [launchOptions objectForKey:@"aps"];
    NSDictionary *alertMsg = [thePayload objectForKey:@"alert"];
    NSNumber *badgeCount = [thePayload objectForKey:@"badge"];
    // .. do something with this ..
    [self applicationDidFinishLaunching:application];
}
```

Notice that the launch options dictionary contains a key for "aps". The object associated with this key contains the actual payload from your notification. You can then access the elements of that dictionary and take action accordingly.

The second way that a notification can enter your application is if your application is already running. In this case, your application delegate receives a call to `application:didReceiveRemoteNotification:`. Just like when you receive the `application:didFinishLaunchingWithOptions:` message, this method also receives a user info dictionary that contains the notification. Listing 20.8 shows an implementation of this method.

Listing 20.8

Handling Remote Notifications While Your Application Is Running

```
- (void)application:(UIApplication *)application didReceiveRemoteNotification:(NS
  Dictionary *)userInfo
{
    NSDictionary *thePayload = [userInfo objectForKey:@"aps"];
    NSDictionary *alertMsg = [thePayload objectForKey:@"alert"];
    NSNumber *badgeCount = [thePayload objectForKey:@"badge"];
    // .. do something with this ..
}
```

You should take this opportunity, for example, to display a message to the user or update your badge count.

Summary

In this chapter, you have examined in detail the Apple Push Notification Service and how to work with it. You've taken an in-depth look at the architecture and message formats needed to communicate with the APNS servers, you've looked at how APNS handles security, and you've seen how you acquire the certificates necessary for communicating with the APNS servers. You've also seen a full example of a push notification supplier that can be deployed on Linux or Mac servers. Finally, you've seen how you handle remote notifications in your own iPhone OS applications.

Push notification is one of the most powerful new features in iPhone OS 3. By mastering this technology, you can enable your applications with new features to delight your users.

Using the Game Kit API

As part of iPhone OS 3, Apple introduced the Game Kit API. Game Kit is a new framework that aims to solve two main problems facing application and game developers.

The first problem is the ability to create ad hoc peer-to-peer networks among devices within close proximity to each other. To satisfy this need, Game Kit provides a sophisticated, yet simple-to-use mechanism for doing exactly this over the Bluetooth networking capability built into iPhone and the second-generation iPod Touch.

The second problem is the ability to add in-game voice to applications. This is a common requirement among applications that enable multi-user play, and including this capability is generally beyond the abilities or resources of most developers. To solve this, Game Kit provides an easy-to-use in-game voice capability that leverages Apple's prior experience in adding voice to iChat, and brings that power to the realm of developers. Developers don't need to know about codecs, or protocols; they just use the API, and everything is handled for them.

In this chapter, you'll examine these two frameworks and see how you can leverage them in your games and applications. In particular, it should be pointed out that just because it's called Game Kit doesn't mean that the features in this API are only applicable to games. The peer-to-peer networking capability, in particular, could be leveraged by applications to exchange data amongst multiple users of a particular application when they are in close proximity.

In This Chapter

Adding peer-to-peer ad hoc networking to your iPhone applications

Adding voice chat to your iPhone applications

Providing Peer-to-Peer Connectivity

The first feature of Game Kit that I'm going to discuss is its ability to provide an ad hoc peer-to-peer network to devices within close proximity to each other. This functionality is provided primarily by the `GKPeerPickerController` and `GKSession` classes.

Before using the Game Kit API, you must first add the Game Kit framework to your project. Do this by right-clicking your target, choosing Get Info, and then choosing the General tab. Click the + button at the bottom of the window under Linked Libraries, and then choose the Game Kit framework. You must also include the `GameKit.h` file in your source file.

Finding peers

The `GKPeerPickerController` class provides the mechanism by which you find other peers on your local Bluetooth ad hoc network. Using it, you can retrieve a list of all of the other devices in the vicinity that are currently available for networking with your application. It also allows the user to find peers over the Internet, but when you choose to use the Internet instead of Bluetooth, you need to dismiss the `GKPeerPickerController` and present your own user interface to complete the connection. `GKPeerPickerController` does not handle actually making peer connections over the Internet.

To allocate and display a `GKPeerPickerController`, you simply need to initialize it, configure the allowed network types, and then set your delegate. Finally, to actually display it on-screen, you call the `show` method. Listing 21.1 shows a typical implementation of setting up and displaying a `GKPeerPickerController`.

Listing 21.1

Initialization and Display of the GKPeerPickerController

```
#import <GameKit/GameKit.h>
        ...
-(void)showPeerPicker;
{
    controller = [[GKPeerPickerController alloc] init];
    [controller setDelegate:self];
    [controller show];
}
```

NOTE
Your delegate must conform to the `GKPeerPickerControllerDelegate` protocol to be able to be a delegate for the `GKPeerPickerController`.

When you call `show`, your application displays a dialog showing that it is looking for other Game Kit–enabled devices nearby.

As you can see, displaying the peer picker controller is reasonably trivial. The actual heavy lifting is going to be handled by your `GKPeerPickerControllerDelegate`, which is going to take action when a peer has been selected. In this particular code, I've set your delegate to self, and so this class needs to implement the appropriate delegate messages to handle the events that it is interested in.

The most important of these delegate methods is, of course, the method `peerPickerController:didConnectPeer:toSession:`, which is called when the user has selected a peer. The parameter passed to this delegate method for the session is, in fact, a `GKSession` instance that has been initialized so that it can be used to communicate with the selected peer. If you are using the `GKSession` class for your peer-to-peer communication, then at this point, you are ready to actually begin sending data to and from your peer.

When two peers instantiate and show GKPeerPickerControllers, they are both shown a list of all the devices discovered on the network. If one of the peers chooses one of the devices to connect to, the peer displays a connection accept dialog to allow them to accept the connection request. Once accepted, the peer delegate receives a call to the `peerPickerController:didConnectPeer:toSession:` method.

Listing 21.2 shows an implementation of this method.

Listing 21.2

Typical Implementation of the peerPickerController:didConnectPeer:toSession: Method

```
- (void)peerPickerController:(GKPeerPickerController *)picker
           didConnectPeer:(NSString *)peerID
                toSession:(GKSession *)session
{
    [self setPeerSession:session];
    [controller dismiss];
    [controller release];
    // begin sending data to the peer.
    { ... }
}
```

This method is also called on the *other* peer when the connection is completed, and also for assuming ownership of the newly created `GKSession`.

This may seem confusing, so I'm going to repeat it here for clarity. The point here is that you are connecting two peers, and they both use the `peerPickerController:didConnectPeer:toSession:` method to complete the connection.

Imagine two iPhones, A and B. They both initialize and show instances of `GKPeerPickerController`. Both devices then display lists of devices. The list on A shows iPhone B, and vice versa. If the user of iPhone A then chooses iPhone B, iPhone B displays a connection confirmation dialog box. If the user of iPhone B chooses to accept, then the delegate method `peerPickerController:didConnectPeer:toSession:` is then called on both devices.

If the user cancels the peer selection dialog, then the method `peerPickerController
DidCancel:` is called on the `GKPeerPickerControllerDelegate`. This is shown in
Listing 21.3.

Listing 21.3
Typical Implementation of the peerPickerControllerDidCancel: Method

```
- (void)peerPickerControllerDidCancel:(GKPeerPickerController *)picker
{
    [controller release];
}
```

In some cases, you may want to provide a special custom `GKSession` instance for your sessions. The method `peerPickerController:sessionForConnectionType:` is provided on the delegate for exactly this purpose. In it, you can allocate a session of whatever custom subclass of `GKSession` you want and return it. Listing 21.4 demonstrates this functionality.

Listing 21.4
Allocating a Custom GKSession

```
- (GKSession *)peerPickerController:(GKPeerPickerController *)picker
         sessionForConnectionType:(GKPeerPickerConnectionType)type
{
    GKSession *session = [[MyCustomSession alloc] init];
    [session setDelegate:self];
    return [session autorelease];
}
```

In cases where you are giving the user the opportunity to choose an Internet connection type instead of a Bluetooth connection type, the method `peerPickerController:didSelect
ConnectionType:` is called on your `GKPeerPickerControllerDelegate`, giving you the opportunity to dismiss the peer picker and present your own dialog for making the connection over the Internet.

The GKPeerPickerController is used only for forming peer-to-peer connections. If your application requires client-server connectivity, you need to use GKSession directly, which is what I discuss next.

Working with sessions

The class used to provide actual network connectivity between peers is the `GKSession` class. This class provides an easy-to-use mechanism for sending arbitrary data to one or multiple peers over an ad hoc Bluetooth network. Additionally, the `GKSession` class allows you to configure client-server connections instead of strict peer-to-peer connections when your application requires it. In this case, you bypass the `GKPeerPickerController` entirely and simply configure your connections using the `GKSession` class directly.

To allocate a `GKSession` object, you call the initializer `initWithSessionID:displayName:sessionMode:`, which takes as parameters a session ID, a display name, and the session mode.

Your `GKSession` instance can act as client, server, or peer, depending on the `sessionMode` configured on initialization. As a server, you have clients connect to you, but the application does not search for servers. As a client, your application searches for servers, but does not advertise itself as a server. As a peer, your application both searches for servers and advertises itself as a server.

When needing to create a server instance of a `GKSession`, you simply allocate and initialize the `GKSession` object with a `sessionMode` parameter of `GKSessionModeServer`. To create a client instance of GKSession, pass a parameter of `GKSessionModeClient`. To create a `GKSession` as a peer, pass a parameter of `GKSessionModePeer`.

Servers become visible on the local network by setting the `available` property of the `GKSession` object to `YES`.

The `GKSessionDelegate` provides methods that can be implemented to handle reception of data from other peers. As usual, the heavy lifting is handled by the delegate of the `GKSession` class.

Each session on a network is uniquely identified by its sessionID. This can be provided by the application, or if you choose to pass nil, it creates one for you by using the application bundle ID. In this way, the session for users of your application is unique.

When working with GKSession directly, you are expected to present whatever user interface you feel is required to allow the user the ability to choose the server they wish to connect to. The `displayName` property is used to provide a human-readable name for your UI when displaying this UI.

Once you have created the `GKSession`, you set its `isAvailable` property to `YES` in order to begin searching for servers and advertising as a server. If you are handling the connection yourself (that is, you are not using a `GKPeerPickerController`), then you can later call `connectToPeer:withTimeout:` to actually initiate a connection to a server, for example, in response to the user selecting a server to connect to.

NOTE

If you received your `GKSession` instance from the `GKPeerPickerController`, then you needn't allocate it. It's already initialized and ready to go.

Once your `GKSession` is initialized, when a peer requests a connection to you, the delegate method `session:didReceiveConnectionRequestFromPeer:` is called. Here, you have the option of accepting the connection or denying it, using the methods `acceptConnectionFromPeer:error:` or `denyConnectionFromPeer:`, respectively.

CAUTION

In the case where you have created a custom `GKSession` object for a `GKPeerPickerController`, you should ignore the message to `session:didReceiveConnectionRequestFromPeer:` while the peer picker is visible, as it will handle the message itself.

A typical implementation of `session:didReceiveConnectionRequestFromPeer:` is shown in Listing 21.5.

Listing 21.5

Implementation of session:didReceiveConnectionRequestFromPeer:

```
- (void)session:(GKSession *)session
didReceiveConnectionRequestFromPeer:(NSString *)peerID
{
    NSError *error = nil;
    if(![allowedPeers containsObject:peerId])
    {
            [session denyConnectionFromPeer:peerID];
            return;
     }
    if([session acceptConnectionFromPeer:peerID error:error])
    {
        [self setConnectedPeer:peerID];
        [self setConnectedSession:session];
    }
    else
    {
        // handle the error
    }
}
```

The method `session:peer:didChangeState:` is provided to allow you to handle changes in state of peers relative to the session. For example, if clients connect or disconnect, you receive messages indicating their change in state. This method is also used to notify your application when servers are detected as available. Therefore, it is in this method that you will populate your server selection UI on your client application.

NOTE
You access a human-readable display name for peers by using the method `displayNameForPeer:` on a `GKSession` object.

Listing 21.6 shows an example of how to do this.

Listing 21.6

Implementation of the session:peer:didChangeState: Method

```
- (void)session:(GKSession *)session peer:(NSString *)peerID
 didChangeState:(GKPeerConnectionState)state
{
    switch (state)
    {
        case GKPeerStateConnected:
            [connectedClientIds addObject:peerID];
            [session setDataReceiveHandler:self withContext:nil];
            // display the connected user information...
            break;
        case: GKPeerStateDisconnected:
            [connectedClientIds removeObject:peerID];
            break;
        case: GKPeerStateAvailable:
            [self addServer:[session displayNameForPeer]];
            break;
        case: GKPeerStateUnavailable:
            [self delServer:[session displayNameForPeer]];
            break;
        // ...
    }
}
```

NOTE
When using a `GKPeerPickerController`, most of the state changes are handled for you. That is to say, the UI provides the capability to the user to choose which peers to connect to and so on. Therefore, you don't need to implement state changes such as `GKPeerStateAvailable` and `GKPeerStateUnavailable` in this method. However, if you choose to utilize your own UI for browsing the list of available servers, you should handle those state changes in this method.

Once your connection is established, you can send data to your peers using the methods `sendDataToAllPeers:withDataMode:error:` and `sendData:toPeers:withDataMode:error:`. In the case of the former method, you are sending your data to all peers, and in the case of the latter method you are sending them only to a subset. In both cases, however, the data that is sent is an `NSData` instance containing an arbitrary set of bytes that you define. Apple recommends that you keep this chunk of data relatively small, under 1KB in length. Larger data sets may result in excessive latency.

The mode parameter on these methods enables you to specify whether you prefer reliable or unreliable transmission of the data. When working with games, unreliable transmission modes are often preferable because it is more important for the application to receive updated information than it is to receive accurate but out-of-date information. Unreliable transmission mode does not ensure that every packet you send is received by the peer. Reliable transmission mode will resend packets if they fail to arrive. It is up to you to determine what type of transmission your application requires.

To receive data, you implement the method `receiveData:fromPeer:inSession:context:` on an object that you have previously set to be your data handler via the method `setDataReceiveHandler:withContext:`, which is available on `GKSession`. So, in other words, you designate an object that will be your data receiver, and you set it as your "data receive handler" using the method `setDataReceiveHandler:withContext:` on `GKSession`. You then implement the method `receiveData:fromPeer:inSession:context:` on this object, and use the data in whatever way you want when it is called.

Providing In-Game Voice

The Game Kit in-game voice feature enables devices to send real-time voice communication from one device to another over an Internet link or utilizing the Game Kit Bluetooth network by allowing you to transmit voice data over a network connection provided by your application. This network connection can be anything from a normal socket connection to an instance of `GKSession` that you have allocated for this purpose.

If you utilize a `GKSession` instance, you can, of course, use its built-in mechanisms for finding peers. However, if you choose to use the Game Kit voice chat feature over the Internet, then you need to provide a mechanism that enables a peer directory service and that enables peers to translate a directory entry to a network address for establishing the network connection. This directory service may require a server-side service to be implemented by you.

The important thing to understand is that when initiating a voice chat, you're required to provide a participant ID and a mechanism for sending data to the participant. Where you get this participant ID, and the network connection, is up to your implementation. If, for example, you utilize the `GKSession` as your low-level peer-to-peer network, then you could use the peer ID as the participant ID and use the method `sendData:toPeers:withDataMode:error:` to send the data.

Before beginning to use the Game Kit voice chat features, you must first initialize the audio system to allow for recording and playback of audio. You do this by getting the `AVAudioSession` singleton and setting its category to `AVAudioSessionCategoryPlayAndRecord`. Listing 21.7 demonstrates how you do this.

Listing 21.7

Initializing the Audio Session

```
-(void)setupAudio;
{
    NSError *error = nil;
    AVAudioSession *audioSession = [AVAudioSession sharedInstance];
    if(![audioSession setCategory:AVAudioSessionCategoryPlayAndRecord
                                                  error:error])
            [self handleError:error];
    if(![audioSession setActive:YES error:error])
            [self handleError:error];
}
```

The `GKVoiceChatService` utilizes a delegate-like pattern in the form of a client property. The client is responsible for implementing the methods necessary to actually transport the data between the peers and must implement the `GKVoiceChatClient` protocol. There are two important methods from the `GKVoiceChatClient` protocol that must be implemented on the client in order for the `GKVoiceChatService` to do its work.

The first is the `voiceChatService:didReceiveInvitationFromParticipantID: callID:` method, which is called on the client when the service receives an invitation from a peer to begin a chat. The default implementation of this method allows all clients to connect and initiate a chat. If you want to give the user the opportunity to accept or decline an invitation to chat, then you should implement this method accordingly. When you implement this method, in order to indicate that the user has accepted the request to chat, call the method `acceptCallID:error:` on the service. Alternatively, to decline the chat request, call the method `denyCallID:`.

The second and more important method to implement on your client is the method `voice ChatService:sendData:toParticipantID:`, which must be implemented on the client in order to send data between the peers. If you do not implement this method, no data is sent and voice chat does not work.

Essentially, when the chat service needs to send data to the peer, it calls this method with an `NSData` object containing the payload to be sent to the remote host. This can be as a result of an invitation, voice data, or other things. The intent is not for you to have to do anything with the data other than pass it on to your transport mechanism. For the most part, you can consider the `NSData` parameter to be opaque. Your job here is simply to pass it on to the peer.

Listing 21.8 shows an example implementation of this method utilizing a proprietary communication class, which theoretically would use a socket connection to send data between the peers.

Listing 21.8

Example Implementation of voiceChatService:sendData:toParticipantID:

```
- (void)voiceChatService:(GKVoiceChatService *)voiceChatService
            sendData:(NSData *)data
     toParticipantID:(NSString *)participantID
{
    [commObject sendData:data toHost:[hosts ipForPeerId:participantID]];
}
```

> **NOTE**
> The GKVoiceChatService is a singleton and can be accessed via the defaultVoiceChatService static class method.

On the receiving end, your communication object must implement a mechanism to receive the data and similarly forward it on to the GKVoiceChatService via the method receivedData:fromParticipantID:. For example, with my fictional communication class, I might do something similar to Listing 21.9.

Listing 21.9

Example Implementation of a Fictional Receive Method on the Peer

```
-(void)receiveData:(NSData *)data fromHost:(NSString *)inHost
{
    NSString *peerId = [peers idForHost:inHost];
    [[GKVoiceChatService defaultVoiceChatService] receivedData:data
                                          fromParticipantID:peerId];
}
```

> **CAUTION**
> Keep in mind that the communication mechanism shown here is fictional and expected to be provided by you. The key element is that you provide the low-level transport networking capability, and the GKVoiceChatService provides the application-level data to send on it. This transport mechanism could, for example, leverage the GKSession API as its transport mechanism if you only need local chat, or it could use something more traditional over UDP for use over the Internet. The key issue is that it just needs to move bytes.

Having implemented all of this, you can instantiate your client, and configure it as the client for the GKVoiceChatService. You can then initiate a voice chat with a peer by calling the method startVoiceChatWithParticipantID:error: on the GKVoiceChat Service singleton after having established a network connection with a peer via your proprietary mechanism. Listing 21.10 shows an implementation using my fictional communication class.

Listing 21.10

Initiating Chat

```
-(void)setupChat;
{
    GKVoiceChatService *svc = [GKVoiceChatService defaultVoiceChatService];
    MyClient *client = [[MyClient alloc] initConnectionToHost:hostName];
    [svc setClient:client];

    NSString *peerId = [peers idForHost:hostName];
    [svc startVoiceChatWithParticipantID:peerId error:nil];
}
```

When the method startVoiceChatWithParticipantID:error: is called, the GKVoice ChatService bundles its configuration information and its voice chat initiation data into an NSData and calls the method voiceChatService:sendData:toParticipantID: on your client. Your client then forwards that through its networking code over the network to the remote client, which then receives the data and forwards that NSData to its instance of GK VoiceChatService via the method receivedData:fromParticipantID:. This then establishes the chat connection by negotiating the type of codec to use and so forth. Both clients then send and receive voice data packets via the methods voiceChatService:send Data:toParticipantID: and receivedData:fromParticipantID: as the users talk.

Finally, when you're ready to discontinue voice chat and disconnect from your peer, you can call the method stopVoiceChatWithParticipantID: on the GKVoiceChatService. This subsequently calls the method voiceChatService:didStopWithParticipant ID:error: on your client class, giving you the opportunity to disconnect from the peer.

The GKVoiceChatService also provides methods to enable you to monitor and adjust audio properties while using voice chat. In particular, it allows you to do things such as mute the microphone, modify remote participant volume, and monitor metering levels both on input and output. To find out more, see the GKVoiceChatService documentation.

Summary

In this chapter, I've introduced you to the Game Kit API, which is new with iPhone OS 3. Even though it's called Game Kit, it can be used to provide both peer-to-peer networking and voice communication for more than just games. The possibilities are virtually limitless. You could use it in applications to transfer contact information among people standing side by side, or even to play head-to-head games.

The voice chat API gives you the opportunity to easily add voice chat to your games without having to know any of the underlying details of how to efficiently handle voice communication. Apple has already done the work for you and has a great deal of experience in solving this problem. By leveraging this API, it opens a new class of applications to everyday developers.

Implementing Cut, Copy, and Paste

One of the most eagerly anticipated new features in iPhone OS 3 is also seemingly one of the most fundamental. That feature is cut, copy, and paste. Prior to iPhone OS 3, there was no systemwide cut, copy, and paste feature provided as part of the operating system. With the introduction of iPhone OS 3, you not only have this feature, but also application-specific and even vendor-specific copy-and-paste capabilities.

The systemwide pasteboard enables you to copy and paste any type of content, from text to images, and share it among any applications on the system that support the data type that you've placed into it.

To share data amongst a specific subset of applications, an individual application, or even a suite of applications, you can also create a named pasteboard. Applications that know the name of the pasteboard can access the data on that pasteboard. This makes it possible for applications to share data amongst themselves, but not necessarily with the rest of the system.

Pasteboards in iPhone OS 3 can also be persistent, meaning that their contents are retained even if the application is terminated.

In This Chapter

Understanding copy and paste with default UI components

Working with the UIPasteboard class

Adding copy and paste to a custom UIView

Copying and Pasting with Standard Controls

When it comes to working with the standard `UITextView`, `UIWebView`, and `UITextField` objects, there's really nothing you need to do to implement copy and paste; they are already enabled on these standard components. Data selected on these components, when cut or copied, is put onto the general system pasteboard and is available to any other applications on the system. Similarly, data that is available on the general system pasteboard is also available for use in any of these controls via the paste mechanism.

Again, this is all done for you; you don't have to do anything to get this support out-of-the-box.

The only time that you really need to do anything special with regard to cut, copy, and paste is when you have custom views or other special cases where you specifically need to put data onto a pasteboard to send it to other applications. This is where you need to interact with the Main class that implements the copy-and-paste mechanism in iPhone OS 3. That class is the `UIPasteboard` class, and it is that class that I will spend the remainder of this chapter discussing.

Understanding Pasteboard Types

There are really two different kinds of pasteboards that an application can interact with. The first is the general, systemwide pasteboard. This pasteboard can be accessed by all applications on the system and is the default pasteboard that is used by `UITextView`, `UITextField`, and so on.

The second kind of pasteboard is called a named pasteboard. These pasteboards are created by your application and given a name. Other applications that know the name of your pasteboard can also access that pasteboard and retrieve your data. In this way, you can share data amongst a select set of applications if you need to.

The `UIPasteboard` class provides the mechanism by which applications interact at a low level with both the general system pasteboard and specific named pasteboards that they themselves create.

To access the general system pasteboard, you call the method `generalPasteboard` on the `UIPasteboard` class. This returns the general pasteboard singleton.

To access a named pasteboard, you call the method `pasteboardWithName:create:` on the `UIPasteboard` class. If you would like the `UIPasteboard` class to create a unique name for you, then you can pass nil as the first parameter. The second parameter indicates whether or not you want the `UIPasteboard` class to create the pasteboard for you if it does not exist.

NOTE
When calling `pasteboardWithName:create:`, Apple recommends using reverse DNS notation when creating a name for your pasteboard to avoid unintentional naming collisions, for example, `com.MyCompany.exampleapp`.

Which kind of pasteboard you choose depends on your application needs.

Interacting with UIPasteboard

Once you have the pasteboard, you can put any kind of data that you want on it. To place data on the pasteboard, you call the method `setData:forPasteboardType:` or `setValue:forPasteboardType:`. This enables you to place raw data onto the pasteboard in the case of `setData:forPasteboardType:`; in the case of the method `setValue:forPasteboardType:`, you're placing data that can be serialized into a standard property list.

Examples of data like this are `NSString`, `NSArray`, `NSDate`, and `NSNumber`. For both of these methods, the pasteboard type parameter should specify a UTI value for the type of data that is being placed on the pasteboard. In this way, the pasteboard acts a bit like a dictionary, allowing you to associate different data representations with different types that can subsequently be accessed by other applications according to the pasteboard type. An application can, for example, first ask for a rich text representation, and then, if that's not available, ask for a plain text version. In this way, you can provide your data in different formats to increase your interoperability opportunities with other applications.

The `UIPasteboard` class also provides a set of specific properties for adding particular kinds of data to the pasteboard or getting particular kinds of data from the pasteboard. These properties are shown in Table 22.1.

Table 22.1 Properties for Setting and Getting Data of Standard Types from the UIPasteboard

Property Name	Purpose
`string`	Accesses the first pasteboard item as a string
`strings`	Accesses all pasteboard items as an array of strings
`image`	Accesses the first pasteboard item as a `UIImage`
`images`	Accesses all pasteboard items as an array of `UIImages`
`URL`	Accesses the first pasteboard item as an `NSURL`
`URLs`	Accesses all pasteboard items as an array of `NSURLs`
`color`	Accesses the first pasteboard item as a `UIColor`
`colors`	Accesses all pasteboard items as an array of `UIColors`

CAUTION
When accessing the standard type properties, if the type requested is not available on the pasteboard, the property returns nil.

NOTE

It is possible to use a non-UTI value for the pasteboard type parameter when the data is of a proprietary type. Again, in that case, Apple recommends reverse DNS notation for this value, for example, `com.mycompany.myproduct.mytype`.

When calling these methods, if data already exists for the pasteboard type provided, that data is overwritten. Multiple elements of a particular type can be stored on the pasteboard using the methods `addItems:` or by setting the items property on the `UIPasteboard`.

Let's take a look now at how you might implement copying custom data from your application to a `UIPasteboard`.

Implementing Cut, Copy, and Paste on a Custom View

First off, let's imagine that you have an application that displays an image and you want to take that image and put it onto the general system pasteboard.

Imagine then that you have a UIView on your UI that you want to use to initiate a copy from. Perhaps you want to copy the data after the user has held their finger on the image for a period of time and then lifted it, just like the standard controls work.

In this case, you can use the `touchesEnded:withEvent:` method on this imaginary custom `UIView` to display the copy-and-paste menu; then, when the user selects an option, it calls either the `copy:` or `paste:` method on your custom `UIView` to actually do the work with the `UIPasteboard` object.

Implementing touchesEnded:withEvent: to display the menu

Listing 22.1 shows the implementation of the `touchesEnded:withEvent:` method. In it, you're determining that the user held their finger on the control for a while, and then asking the `UIMenuController` to display the copy menu.

Listing 22.1

Implementation of touchesEnded:withEvent: on the Custom View

```
- (void)touchesEnded:(NSSet *)touches withEvent:(UIEvent *)event
{
    UITouch *touch = [touches anyObject];
    if(([touch timestamp] - [firstTouch timestamp]) > 3)
    {
```

```
        UIMenuController *menu = [UIMenuController sharedMenuController];
        [menu setTargetRect:[self frame] inView:self];
        [menu setMenuVisible:YES animated:YES];
    }
}
```

> **NOTE**
> The `UIMenuController setTargetRect:inView:` method sets a rectangle within the `UIView` that is requesting the menu. The rectangle is used so that the menu displays in an appropriate location according to the data being copied. The menu pops up just above the specified rectangle.

This code is using the new class `UIMenuController`, which is provided in iPhone OS 3 to display the cut, copy, and paste standard menu when doing pasteboard operations. Your custom view class is assumed to be implementing the `UIResponderStandardEditActions` protocol as needed to make the menu appear.

Implementing the copy: method

The `UIMenuController` class interrogates the first responder to see if it implements the methods defined in the `UIResponderStandardEditActions` protocol. In this case, you want to implement the `copy:` method from that protocol so that you can copy the data from your custom view.

Listing 22.2 shows the implementation of the protocol `copy:` method that you need to display the copy menu option. This is where the actual work in interacting with the `UIPasteboard` begins.

Listing 22.2

Implementation of copy: from the UIResponderStandardEditActions Protocol

```
-(void)copy:(id)sender
{
    UIImage *theImage = [self imageRepresentation];
    UIPasteboard *pb = [UIPasteboard generalPasteboard];
    [pb setImage:theImage];
}
```

Notice that you are getting your custom view data as `UIImage` and then placing that data on the pasteboard using the image property.

Implementing the paste: method

Similarly, if you want to display a paste menu item that would allow you to paste in an image from the pasteboard, you can implement a `paste:` method as shown in Listing 22.3.

Listing 22.3

Implementation of paste: from the UIResponderStandardEditActions Protocol

```
-(void)paste:(id)sender
{
    MyView *imageView = (MyView *)sender;
    UIPasteboard *pb = [UIPasteboard generalPasteboard];
    UIImage *theImage = [pb image];
     if(theImage)
          [self setImage:theImage];
}
```

Here, you are asking the general pasteboard if the current item on the pasteboard is available as a `UIImage` representation. If it is, then you are setting your image property to whatever image you have received.

Understanding the interactions

To summarize what's happening here: Your custom `UIView` is implementing `touchesEnded:withEvent:`. In that method, you check to see if the user has had their finger down on the view for more than three seconds. If they have, then you ask the `UIMenuController` to display the standard cut-and-paste menu for your view. Your view is currently the first responder and implements the methods `copy:` and `paste:` from the `UIResponderStandardEditActions` protocol, so that when the menu displays it shows the menu options for copy and paste. When the user selects the copy or paste menu items, the associated method is called on your custom view, in this case, `copy:` or `paste:`, respectively.

When the `copy:` or `paste:` methods are called, your data is copied either from your custom view to the pasteboard or from the pasteboard to your custom view, respectively.

Summary

In this chapter, I have introduced you to one of the fundamental new features of iPhone OS 3, systemwide copy and paste. Additionally, I have introduced you to the `UIMenuController` class, which displays the menu that's used for copy-and-paste operations.

Thankfully, if you use the standard controls, most of the work is already done for you. However, if you have custom data, or custom views, you may need to interact with the `UIPasteboard` class directly to implement your own copy-and-paste operations. With the knowledge you have gained in this chapter, you should be able to do that.

Using the Maps API

A highly anticipated feature in iPhone OS 3 from a developer standpoint is the addition of the Map Kit API.

The Map Kit API gives developers the ability to finally embed maps within their application. It comes with everything needed to display a map in your application, including maps from Google maps as well as the ability to annotate the maps with notes for points of interest. You can even reverse geocode map coordinates, which allows you to take a set of coordinates and acquire a street address from it. This is an incredibly powerful API and one that will surely be used extensively by applications written for iPhone OS 3.

Showing an Embedded Map with MKMapView

The core of the Map Kit API centers on the MKMapView class, which encapsulates, in its entirety, a view that contains a map. The map itself is centered on a given coordinate, and is scaled by specifying a region span value. In addition to the ability to set these properties programmatically, the view also provides the capability for the user to position and zoom the map by using a combination of touch and pinch gestures. This capability can be enabled or disabled, depending on whether you want to give the user the ability to do this.

An MKMapView can also be annotated to display notes about locations that are displayed on the map. The annotation itself is an instance of MKAnnotation, and the view associated with the annotation is an instance of MKAnnotationView. Using these classes, you can gain quite a bit of customization when annotating your maps.

Creating an MKMapView

To add an MKMapView to your project, you can simply open one of your view nibs in Interface Builder, and drag-and-drop a map view object from the palette onto your view. This gives you an instance of an MKMapView in your nib. You can then connect this up to IBOutlets, for configuration purposes. Figure 23.1 shows a nib containing a map view.

In This Chapter

Working with MKMapView to display a map in your applications

Narrowing the map to particular regions

Annotating maps

Understanding performance with regard to map annotations

Using reverse geocoding to convert coordinates into addresses

Figure 23.1
Interface Builder with an instance of MKMapView

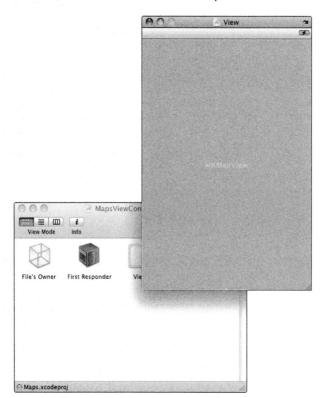
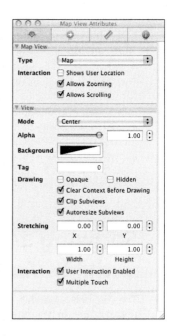

NOTE
In order to use the `MapKit` framework, you must remember to include it in your project. To do this, simply add the `MapKit` framework as a linked framework to your target.

Once you have your `MKMapView` connected to an `IBOutlet` on your view controller, you can configure its default properties in your `viewDidLoad` method. Listing 23.1 shows an implementation of `viewDidLoad` configuring some default properties on an MKMapView that's connected to an `IBOutlet` called `mapView`.

Listing 23.1

viewDidLoad Method Setting Some Properties on an MKMapView

```
- (void)viewDidLoad
{
```

```
[mapView setDelegate:self];
[mapView setShowsUserLocation:YES];
[super viewDidLoad];
}
```

Specifying the map region

The default map that an `MKMapView` displays if you change none of the configuration is an entire world map. For the purposes of this example, you'd like it to be just a bit more specific, so you're going to tell the map to display your current location, and when it adds your current location to the map, you're going to update your span and center to zoom in on that location. To do that, implement the `MKMapViewDelegate` method `mapView:didAddAnnotationViews:`, which is called when annotations are added to the map. The current user location is, in fact, an annotation, and so you will simply use this method as a trigger to then determine what the user's location is, using the `MKMapView` method `userLocation`. You'll then use that to update your map view's region. Listing 23.2 shows the implementation of this method.

Listing 23.2

Centering the View on Your Current Location

```
- (void)mapView:(MKMapView *)inMapView
didAddAnnotationViews:(NSArray *)views;
{
    CLLocation *userLoc = [[mapView userLocation] location];
    CLLocationCoordinate2D userCoords = [userLoc coordinate];
    MKCoordinateRegion region =
        MKCoordinateRegionMakeWithDistance(userCoords, 500000, 500000);
    [mapView setRegion:region animated:YES];
}
```

Interestingly, this code uses the function `MKCoordinateRegionMakeWithDistance` to actually create the region that you pass to the map view. There are several utility functions like this one that can be used for creating some of the structures used in the Map Kit API. In this particular case, you pass it a center location coordinate and longitude and latitude distances (in meters) that correspond to the span of the map that you want to display. In this example, if you run it in the simulator, your central location should come back as the Apple campus. This code sets the distance for the region to 500 km, or 500,000 m. This shows the whole Bay Area of California with your location shown as a blue dot in the middle of the screen. You can see this in Figure 23.2.

Figure 23.2
Displaying your current location as a region

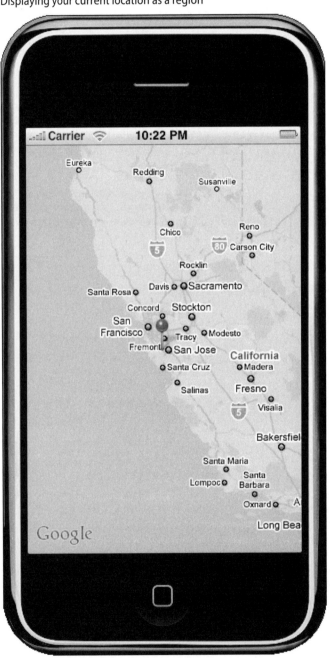

The `MKMapViewDelegate` protocol defines several methods that are useful to application developers, such as this one. Probably the most useful methods that it defines, however, are the ones related to when the map location is updated. The methods that are called during this time are `mapView:regionWillChangeAnimated:` and `mapView:regionDidChangeAnimated:`; they are called just before and after the map center location has been changed, respectively. When the user scrolls the map view with his finger, your delegate receives numerous calls to this method. Therefore, it's important that your implementation of these methods avoid any kind of complex calculation or redrawing of views.

Another useful `MKMapViewDelegate` call is `mapView:annotationView:calloutAccessoryControlTapped:`, which is called whenever any of the annotation views that are currently displayed on your map are tapped. Using it, you can trigger actions to occur based on when the user has selected a particular annotation. You can then do things such as bring in an additional view that shows detailed information about that particular annotation.

Annotating Maps

Annotating an `MKMapView` is a very simple procedure, but it involves implementation from two different perspectives.

The first is implementation of the model data for your annotations. The Map Kit framework provides a protocol that defines the interface that an annotation object must implement. This protocol is called the `MKAnnotation` protocol. Your model object should implement this protocol and, at a minimum, must implement the coordinate method. The protocol also defines a `title` and `subtitle` method that you may want to implement to provide a title and subtitle label, respectively, but these are optional.

The second part of the implementation is that your `MKMapViewDelegate` must implement the method `mapView:viewForAnnotation:`, which returns the view associated with a given annotation. When you add your annotations to the view, this method is called for each annotation. It is expected that this method will return an instance of `MKAnnotationView` customized with the pertinent details for the annotation in question. This method should be implemented similar to the `UITableView` method `cellForRowAtIndexPath:` in that, rather than instantiating a new `MKAnnotationView` for each annotation, you should instead utilize the `MKMapView` method `dequeueReusableAnnotationViewWithIdentifier:`, which returns an unused annotation view if one is available. Only if this method returns nil should you actually instantiate an instance of `MKAnnotationView`. Listing 23.3 shows an implementation of this method.

Listing 23.3

Getting the Views for Annotations

```
- (MKAnnotationView *)mapView:(MKMapView *)mapView viewForAnnotation:(id
  <MKAnnotation>)annotation
{
    MKPinAnnotationView *view = [mapView
            dequeueReusableAnnotationViewWithIdentifier:@"PIN_ANNOTATION"];
    if(!view)
    {
        view = [[MKPinAnnotationView alloc]
                    initWithAnnotation:annotation
                        reuseIdentifier:@"PIN_ANNOTATION"];
    }
    [view setCanShowCallout:YES];
    [view setPinColor:MKPinAnnotationColorPurple];
    [view setAnimatesDrop:YES];

    return [view autorelease];
}
```

NOTE
A callout is the view that is displayed when the user touches the annotation on the map. It shows the title and subtitle text from the annotation and can be customized using the left and right callout accessor view properties.

In this particular case, you are actually working with `MKPinAnnotationViews`, which display a pin for the annotation on the map view. The view that you return can be anything that you specify, including a custom view. It simply has to inherit from `MKAnnotationView`. Additionally, `MKAnnotationView` itself has quite a few methods available that enable detailed customization of the views that are displayed. For example, you have access to the image that's displayed when it's on the map. You also have access to the views that are used on the callouts. So, before you instantiate your own custom `MKAnnotationView`, take a close look at the default `MKAnnotationView` to see what kind of customization you can do with it.

NOTE
Because of the way iPhone uses the delegate method to request the annotations it actually needs to display, it is safe for you to add all your annotations for your current map display at once. The MKMapView will only request the annotations that are on-screen.

Thinking about performance with annotations

An important thing to remember when working with annotations is that when the user zooms in to a particular region, you may want to add more detail and more annotations to that region. Additionally, when the user zooms out, you may want to reduce the amount of detail that the user sees. To do this, you should implement the `MKMapViewDelegate` method `mapView:regionDidChangeAnimated:`, and from it, you should examine the span of the map after each region change and use the `MKMapView` methods `addAnnotations:` and `removeAnnotations:` as appropriate.

The point here is that if you imagine that your map view is annotating city names, for example, as your map is zoomed out, you probably only want to display annotations for the larger cities. Therefore, you need to remove annotations for cities below a particular population. Similarly, when the user zooms in, you probably want to display more detail, and thus cities with smaller populations. You should therefore add back in the smaller cities annotations at this point.

Converting Coordinates

The final aspect of the Map Kit framework that can be useful for your applications is the Map Kit reverse geocoder, which is provided via the class `MKReverseGeocoder`. The process of reverse geocoding enables your application to take a set of coordinates and find out real-world information about those coordinates. For example, you can find out what country a particular set of coordinates is in, or even go all the way down to finding out the street address for a particular location. This has tremendous potential in iPhone software because it enables you to take information about someone's location and turn it into something that is human-readable, and not just coordinates.

The process of reverse geocoding involves contacting a service on the Internet to do the reverse geocoding for you. Therefore, its implementation is asynchronous. That is to say, you instantiate an `MKReverseGeocoder`, you ask it for a reverse geocode lookup on the set of coordinates, and then you tell it to begin the lookup. When the lookup is complete, the reverse geocoder contacts your `MKReverseGeocoderDelegate` to inform you that the lookup has completed either successfully or unsuccessfully.

Listing 23.4 shows how to instantiate an `MKReverseGeocoder` and display some human-readable text for the annotation for the user's current location. In this case, you're updating your annotation from your previous example.

Listing 23.4
Using the MKReverseGeocoder

```
- (void)mapView:(MKMapView *)inMapView
didAddAnnotationViews:(NSArray *)views;
{
    CLLocation *userLoc = [[mapView userLocation] location];
    CLLocationCoordinate2D userCoords = [userLoc coordinate];
    MKCoordinateRegion region =
         MKCoordinateRegionMakeWithDistance(userCoords, 500000, 500000);
    [mapView setRegion:region animated:YES];

    geocoder = [[MKReverseGeocoder alloc] initWithCoordinate:userCoords];
    [geocoder setDelegate:self];
    [geocoder start];
}
```

Once you have the geocoder running, you then implement the delegate method to do something with the data when you receive it. This is shown in Listing 23.5.

Listing 23.5
Receiving the Reverse Geocoded Data

```
- (void)reverseGeocoder:(MKReverseGeocoder *)geocoder
       didFindPlacemark:(MKPlacemark *)placemark
{
    MKUserLocation *userLoc = [mapView userLocation];
    [userLoc setTitle:[placemark locality]];
    NSString *address = [NSString stringWithFormat:@"%@ %@",
                         [placemark subThoroughfare],
                         [placemark thoroughfare]];
    [userLoc setSubtitle:address];
}
```

In this case, you're simply taking the user location annotation and updating the title and subtitle to show the city name and street address, respectively.

The object that represents a geocoded location is encapsulated in the `MKPlacemark` class. It has a variety of accessors for information about the location. These accessors are shown in Table 23.1.

Table 23.1 MKPlacemark Properties

Property	Purpose
`addressDictionary`	This is a dictionary containing the address book keys and values defined for the location.
`thoroughfare`	This specifies the street-level information for the location. This does not contain the actual address; it is simply the street name area.
`subThoroughfare`	This specifies additional information at the street level of the location. For example, this might contain the actual street number for the location.
`locality`	This specifies the city name associated with the location.
`subLocality`	This specifies a sublocation within the city for a particular location, such as a neighborhood name or landmark.
`administrativeArea`	This specifies the state associated with the location.
`subAdministrativeArea`	This specifies a government-defined region within an administrative area such as a county, parish, and so on.
`postalCode`	This specifies the postal code associated with the location.
`country`	This specifies the country associated with the location.
`countryCode`	This specifies the country abbreviation associated with the location.

Summary

It's amazing to think that today we have handheld devices such as the iPhone that contain the powerful geolocation-sensing capabilities that they do. It's incredible to think about the kinds of applications that can be developed when you combine the capabilities of the GPS with the ability to draw maps, and reverse geocode those coordinates to human-readable information such as addresses and landmarks. It's fortunate that Apple has made the API for dealing with these capabilities incredibly easy to use. This chapter introduced you to the `MKMapView` and its associated APIs. Using them, you can develop all kinds of new and exciting applications that harness the power of GPS.

Working with Media

In This Part

Chapter 24
Exploring the Supported Media Types

Chapter 25
Playing Audio

Chapter 26
Accessing the iPod Library

Chapter 27
Recording Audio

Chapter 28
Playing Video in Your Application

Exploring the Supported Media Types

At its heart, the iPhone is a sophisticated media device. It is, at the end of the day, an iPod as well as a phone. It can play a variety of audio and video formats and contains ample storage for this media. In the upcoming chapters, I'm going to discuss how you play audio and video files, but before that I'll discuss the types of formats the iPhone is capable of playing and the limitations involved.

Supported Audio Formats

The first iPod shipped in 2001 and had a 5GB hard disk, which enabled it to store 1,000 songs. Today, over 175 million iPods have shipped, with the largest among them storing 160GB of data. The iPhone is more than just an iPod, but it can thank its predecessors for its amazing support for audio formats.

Compressed audio

When it comes to compressed audio, the iPhone OS utilizes a hardware codec for playback. This means that it is only capable of playing one such sound at a time. This makes these formats suitable for "long play" types of applications, where your audio is playing for a lengthy period of time, perhaps the duration of your application runtime. It does not, however, make it possible to mix sounds together. To do this, you must use an audio format that is played using a software codec.

The audio formats that are played with a hardware codec are AAC, ALAC (Apple Lossless), and MP3. You can only play one of these audio files at a time. That said, it is possible to play one of the formats that require the software codec concurrently with an audio file requiring a hardware codec.

In This Chapter

Audio and video formats that the iPhone OS supports

Limitations in audio playback of compressed versus uncompressed audio

Uncompressed audio

You'd never be able to make much of a game if you couldn't play more than one audio file at a time. iPhone OS does provide the ability to play multiple sounds at a time and even mix them together at different volumes. To do this, iPhone OS uses a software codec, which utilizes the CPU of the iPhone to play the audio, as opposed to the hardware decoder.

The formats that use the software codec are Linear PCM, IMA4, iLBC, μ-law, and a-law.

Using sounds of this type, you can play many of them simultaneously, as well as while a compressed audio file is playing. Unfortunately, their file sizes tend to be much larger, which means that they quickly inflate the size of your application bundle. As a result, you should probably limit their use to short sounds.

Supported Video Formats

As with audio, the iPhone is also an amazingly capable video-playing device. The video format supported is H.264 MPEG-4. You can bundle these files in your application, and play them.

When a video is played back, the video playback takes over the entire screen. You can choose to display playback controls as part of the video playback. If you choose not to display playback controls, the video plays to completion without the ability to be interrupted or paused. This is useful in cases where you want to display a cut scene in a game.

Summary

In this chapter I've introduced you to some of the multimedia capabilities of iPhone OS and shown you the types of video and audio files that the iPhone OS, and thus your applications, can play. By leveraging the iPhone's powerful audio and video capabilities, you can bring an incredibly rich user experience to your games and applications. You should never underestimate the value of using high-quality ambient sounds and dramatic video to enhance the immersive qualities of your product.

Playing Audio

The iPhone isn't just a phone. It's a media player. Thanks to this capability, it sports some of the best audio hardware in the business, and fortunately that hardware is ripe to be manipulated using code.

In the bad old days of iPhone OS 2.x, playing audio was a black art, limited to code that used the foundational framework, Core Audio. In today's world of iPhone OS 3, however, playing and recording audio has become infinitely easier, and you now have access to a set of shiny, new Objective-C classes that make playing audio a snap.

Using the AV Foundation Framework

The AV Foundation Framework consists of three primary classes — `AVAudioPlayer`, `AVAudioRecorder`, and `AVAudioSession` — which correspond to playing, recording, and configuration, respectively. Each of them also has a delegate protocol that's used to respond to events related to each of the classes.

NOTE
To use the AV Foundation Framework in your application, be sure to add it as a framework to your target's linked frameworks. The framework name is `AVFoundation`. Remember also to import the correct header files in code you want to use it in. The header file is AVFoundation/AVFoundation.h.

Setting your audio configuration with AVAudioSession

The `AVAudioSession` class provides the mechanism that enables you to configure the behavior of your audio with regard to things such as the mute switch and whether the audio causes iPod playback to pause. Additionally, its delegate enables your application to take action when the audio is interrupted for phone calls and the like.

In This Chapter

Working with the AV Foundation Framework

Using the AVAudioSession class to set the audio category

Understanding how audio categories affect audio playback and the mute switch

Using AVAudioPlayer to play audio

Handling interruption of audio playback

Playing audio with OpenAL

`AVAudioSession` is a singleton and therefore is accessed using the singleton accessor method `sharedInstance`. Once you have the `AVAudioSession` shared instance, you can use it to configure your sample rate, your IO buffer sizes, and perhaps the most important configuration item, the audio session category.

The audio session category defines the behavior of your audio by categorizing it as a particular type.

A common, but misguided question from first-time iPhone developers when they begin working with audio is, "How do I monitor for the mute switch so that I can stop my audio?" The answer is that you don't. You handle the mute switch by setting your audio session category to one that handles it for you.

You can set the audio session category by calling the method on the `AVAudioSession` singleton of `setCategory:error:`. The possible categories, and how they work, are shown in Table 25.1.

Table 25.1 Audio Session Categories

Category	Purpose
`AVAudioSession CategoryAmbient`	This category is intended for long-duration background sounds and audio that should be mixed with iPod audio. Audio from other applications is mixed with the audio from your application when played. This category silences audio when the mute switch is enabled or the phone is locked.
`AVAudioSession CategorySoloAmbient`	Like the previous category, this category is intended for long-duration background sounds. However, this category causes the audio from other applications to be silenced when your application's audio is playing. The audio is not mixed. This is the default category if you do not set one. This category silences audio when the mute switch is enabled or the phone is locked.
`AVAudioSession CategoryPlayback`	This category is for audio that is intended to play regardless of whether or not the mute switch is set. It also continues to play when the screen is locked. Typically, this would be used in an application that features audio playback as part of its central functionality.
`AVAudioSession CategoryRecord`	This category is intended strictly for recording audio. It silences all audio playback.
`AVAudioSession CategoryPlayAndRecord`	This category is intended for playback and recording of audio. It does not silence audio playback while recording.

Listing 25.1 shows how an application can utilize setting the category. Typically this is something that you would do in your application delegate when your application launches.

Listing 25.1

Setting the Audio Category

```
#import <AVFoundation/AVFoundation.h>
// ...
- (void)applicationDidFinishLaunching:(UIApplication *)application
{
    AVAudioSession *session = [AVAudioSession sharedInstance];
    [session setDelegate:self];
    NSError *error = nil;
    if(![session setCategory:AVAudioSessionCategorySoloAmbient
                    error:&error])
    {
        // handle error
    }
    if(![session setActive:YES error:&error])
    {
        // handle error
    }

    // ...
}
```

> **NOTE**
> Your audio session must be set to active in order for your configuration to take effect. Do this using the `setActive:error:` method.

Another useful feature of the `AVAudioSession` object is the ability to determine if a device has an input method available. For example, the iPod Touch, if you do not plug in an external microphone, has no capability for recording audio. It is useful to your application to be able to determine this before presenting the user with the option to record audio. To do this, you call the method `inputIsAvailable`, which returns true if an audio input device is present.

Finally, you can reduce your system usage by matching the audio sample rate of the hardware to the audio sample rate of the files that you are playing. This can help conserve battery life by preventing the audio hardware from having to sample up or down the audio before playing. To do this, you utilize the methods `currentHardwareSampleRate` and `setPreferredHardwareSampleRate:error:` to get the current hardware sample rate and set the hardware sample rate you prefer, respectively.

NOTE
The hardware may not be able to support the sample rate you request. You should therefore check to verify that the sample rate you requested was available.

Using an AVAudioSessionDelegate

The AVAudioSessionDelegate protocol allows the object that you have configured as a delegate to respond to specific events related to the AVAudioSession. Specifically, it enables you to be notified when your application audio has been interrupted, for example, during a phone call, or when certain audio parameters have been changed. For example, you can be notified if the sample rate has been changed or if the availability of an audio input device has changed. Thus, if you can imagine an application that records audio (which requires audio input), that application might want to be notified if the user unplugs her microphone.

The kinds of methods defined on the AVAudioSessionDelegate are shown in Table 25.2.

Table 25.2 AVAudioSessionDelegate Methods

Method	Purpose
-(void)beginInterruption	This method is called when your audio is interrupted, such as during a phone call.
-(void)endInterruption	This method is called when your audio is resumed after being interrupted, such as when a phone call ends.
-(void)categoryChanged:(NSString*)category	This method is received when the AVAudioSession category is changed.
-(void)inputIsAvailableChanged:(BOOL)isInputAvailable	This method is received when the available input device has changed.
-(void)currentHardwareSampleRateChanged:(double)sampleRate	This method is received when the hardware sample rate has changed.
-(void)currentHardwareInputNumberOfChannelsChanged:(NSInteger)numberOfChannels	This method is received when the hardware input number of channels has changed.
-(void)currentHardwareOutputNumberOfChannelsChanged:(NSInteger)numberOfChannels	This method is received when the hardware output number of channels has changed.

Playing audio with AVAudioPlayer

So you've set your audio category, you've configured your sample rate, and now you're ready to actually play some audio. To do this, you use the class `AVAudioPlayer`. The `AVAudioPlayer` class is provided for the purposes of playing any audio that does not require stereo positioning, precise synchronization with UI elements, or audio captured from a network stream. For these purposes, you should probably use something else, such as `OpenAL`. In other words, the `AVAudioPlayer` class is good for lightweight audio such as that used in the most basic applications. You could probably use the `AVAudioPlayer` class even in some game applications, but `OpenAL` is more suited to things such as high-performance immersive games, where the audio positioning helps to enhance the game experience and must be changed dynamically and programmatically. The `AVAudioPlayer` class is suitable for most other applications.

To create an `AVAudioPlayer` instance, you simply call the constructors `initWithContentsOfURL:error:` or `initWithData:error:`, which enable you to initialize the object using either a file or data already existing in memory. If you want to play multiple files simultaneously and have their audio mixed together, you simply instantiate multiple instances of the `AVAudioPlayer` class and use them individually.

Once the `AVAudioPlayer` instance is created, to play the audio you simply call the play method. Similarly, to pause or stop the audio, you call the pause or stop methods. You can also configure whether you want the audio to loop a certain number of times by setting the `numberOfLoops` property. To cause the audio to loop indefinitely, set the `numberOfLoops` property to any negative number. The default value of this property, which is zero, causes the audio to play once and then stop.

In some cases, you may want to minimize the amount of delay between calling the play method and when the audio actually begins to play. To do this, you can call the `prepareToPlay` method on your `AVAudioPlayer` object. This tells the `AVAudioPlayer` object to preload its caches and buffers so that it is ready to play when the play method is called. When the sound stops playing, either by using the stop method or by allowing the audio to complete, the buffers are once again empty. So if you want the audio to play again with a minimal amount of delay, you should call `prepareToPlay` again before calling the play method.

Listing 25.2 shows how you load and play an audio file from within your application bundle. In this case, you're going to play the audio file continually until you tell it to stop. Therefore, you're going to set the `numberOfLoops` property to a negative value.

Listing 25.2

Playing Audio

```
-(void)playAudioFile:(NSString *)inSoundFileName;
{
    NSString *fileName =
        [[NSBundle mainBundle] pathForResource:inSoundFileName
                                        ofType:@".mp3"];
    NSURL *fileUrl = [NSURL fileURLWithPath:fileName];
    NSError *error = nil;
    AVAudioPlayer *player =
        [[AVAudioPlayer alloc] initWithContentsOfURL:fileUrl error:&error];
    if(!player)
    {
        // handle error
    }
    else
    {
        [player setNumberOfLoops:-1];
        [player setDelegate:self];
        [player play];
    }
}
```

In this code, you initially get the actual path to the filename out of your main bundle. You then convert that path into a URL. Finally, you initialize your `AVAudioPlayer` with that URL, set the number of loops to –1, set the delegate to `self`, and call the play method on the player. At this point, the audio begins playing.

During playback, you can find out certain information about the audio file, such as your current location in the file by time using the `currentTime` property, as well as the total duration of the file using the `duration` property.

Perhaps one of the most intriguing items that has been added to this API is the ability to get metering information about the audio while it's playing. To do this, you set the property `meteringEnabled` to YES; you then occasionally call the methods `averagePowerFor Channel:` and `peakPowerForChannel:` on the player, which returns the power levels in decibels for the channel specified. Before requesting this information, it's important to remember that you have to tell the player to update the meters. You do this using the method `updateMeters`.

CAUTION
You can play multiple uncompressed audio files simultaneously on the iPhone, but you can only play one compressed audio file at a time. This means that if you want to mix your audio at runtime, you need to use uncompressed audio files. This includes even audio from the iPod application, regardless of the category you have set.

Using an AVAudioPlayerDelegate

The `AVAudioPlayer` also has a delegate protocol called `AVAudioPlayerDelegate`. The most useful method that your delegate will likely receive is the `audioPlayerDidFinish Playing:successfully:` method, which tells your delegate that the audio player completed playing the requested sound successfully. If there is an error during audio decoding, the delegate method `audioPlayerDecodeErrorDidOccur:error:` is also called. The error parameter that is passed to this method indicates the error that occurred during decoding.

Again, just like the `AVAudioSession` class, the `AVAudioPlayer` can respond to interruption events by implementing the methods `audioPlayerBeginInterruption:` and `audioPlayerEndInterruption:`. When the `AVAudioPlayer` receives an interruption, the audio is automatically paused, and your audio session is deactivated. When the interruption is ended, the `AVAudioSession` is automatically reactivated, but you must resume play by calling the play message again on your `AVAudioPlayer`. When you call play, playback should resume where it left off.

Playing Audio with OpenAL

The AV Foundation Framework is excellent for most lightweight audio requirements. However, when it comes to developing a highly immersive game application, you may find that you want a bit more control over the way that the audio is played back. For example, you might want to have the audio for an explosion appear to come from a particular side of the screen, or you may want to have an environmental sound move when the user changes his perspective. To support this, the iPhone includes full support for the open standard API OpenAL.

OpenAL, like its close cousin OpenGL, is a platform-independent API. Where OpenGL handles three-dimensional graphics, OpenAL is designed to enable three-dimensional audio and effects. To support this, OpenAL provides an environment mapping system that includes both a listener object and sound sources, all of which can be positioned within an environment. Thus, rather than individually mixing the sound channels in order to imitate stereo sound that is appropriately positioned for your application, you tell the OpenAL engines that you want a particular sound to play at a particular volume at a particular location inside the environment and that your listener is located at a particular location inside the environment. The OpenAL engine handles the mixing of the sound so that it appears that the sound is coming from the location you specified.

Also like OpenGL, a truly comprehensive coverage of this API is beyond the scope of this book. If you really want to fully understand OpenAL, the OpenAL Web site, located at www.OpenAL.org, provides you with more than enough information. That said, however, I will give you a brief overview.

Essentially, to work with OpenAL, you have to go through several steps. First, you have to open the system's default output device. To do this, you call the method `alcOpenDevice` passing a parameter of NULL. Next, you have to create an OpenAL context; to do this, you call the method `alcCreateContext`, passing as parameters the device that you got back from the prior call and zero. Whenever you work with OpenAL, you need to have a context, and that context needs to be set as the default context for your particular session. To set the current context, you use the function `alcMakeCurrentContext`. Finally, you create your sources, and you configure your listener's location. To create an audio source, you use the function `alGenSources`. To configure your listener location, you use the function `alListenerfv`, which configures the location of the listener using an array of floats. To configure the location of your source, you similarly use the function `alSourcefv`.

Once your environment, source, and listener are all configured, you can trigger individual sources to begin playing by using the `alSourcePlay` function. Similarly, to stop the source, you use the `alSourceStop` function.

Summary

In this chapter, you've seen how to play basic audio in your applications, and you've also seen the AV Foundation Framework, the new API that iPhone OS 3 gives you for playing audio. Additionally, you've seen the basics of how you use OpenAL to play audio, and where you'd use OpenAL instead of the AV Foundation Framework.

Accessing the iPod Library

A new feature that was added to iPhone OS 3 was the ability to access the user's iPod media library. This is enabled through the Media Player framework. Using the Media Player framework, you can perform a variety of tasks with the user's iPod media library. For example, you can search for items, access the user's playlists, and play audio items from the library. The Media Player library includes support for all audio types that the iPod can play. This capability has several potential applications. For example, imagine a car-driving game that enables you to tune the radio to a radio station that plays media from your iPod library. Or imagine a skiing game that enables you to program a virtual iPod to play your media while you ski down the slopes. The possibilities are endless.

In this chapter, you'll see how to leverage the Media Player framework to access and play elements from the user's iPod library. When you're finished, you should be able to use this capability in your own applications.

Working with the Media Player Framework

The Media Player framework provides a variety of classes for working with the iPod library. At its core, songs are represented by the MPMediaItem class. When working with songs, the MPMediaItem class will create the objects that are returned from the iPod library and that you use to tell the iPod what to play.

To query for songs, you use the MPMediaPredicate and its subclass MPMediaPropertyPredicate, in conjunction with the MPMediaQuery class, to construct queries that return lists of MPMediaItems. Alternatively, you can present the user with a view controller that they can use to select the songs directly from their library. The class that you use to do this is the MPMediaPickerController. It encapsulates an entire user interface that enables the user to select media from their iPod library.

 In This Chapter

Browsing the iPod Library using the MPMediaPicker Controller

Searching for a song by artist, album, or playlist

Controlling the built-in iPod application from your application

Playing audio from the user's iPod library within your application

Once you have a selection of `MPMediaItems`, to play them you utilize the `MPMusicPlayerController` class. It has the ability to both control the iPod functionality of the device and play music inside your application. I'll talk more about this distinction shortly.

Accessing the Media Library

I'll talk about the specifics of how you actually play the media shortly. Before you can play the media, however, you need the user to be able to select media. There are really two ways of doing this: The first is to present a view controller that allows the user to browse their media library and select individual items. The second is to construct queries for items and query the library directly. I will talk about each of these options in turn.

Using the MPMediaPickerController

The `MPMediaPickerController` is the most "hands-off" method for selecting items from the user's iPod library. It is an entire user interface, accessed through the `UIViewController` standard model. It inherits from `UIViewController`, which means that you can use it anywhere that you would a `UIViewController`. For example, you can push it onto a navigation controller or, as is recommended by the documentation, you can present it modally, using the method `presentModalViewController:animated:` on its parent view controller.

Figure 26.1 shows the `MPMediaPickerController` browsing some songs. In this case, the picker is allowing multiple selection and has a prompt of "Choose Some Tunes."

As you can see, the user can select by artist, podcast, individual songs, or from their playlists.

Initializing the `MPMediaPickerController` is as simple as calling its standard initializers: `init`, which returns all the media types in the library, or `initWithMediaTypes:`, which returns specific media types. The media types that are supported are shown in Table 26.1.

Table 26.1 Media Types Supported by the MPMediaPickerController

Type	Description
`MPMediaTypeMusic`	Music only
`MPMediaTypePodcast`	Podcasts only
`MPMediaTypeAudioBook`	Audio books only
`MPMediaTypeAnyAudio`	Any type of audio item
`MPMediaTypeAny`	Any type of item

Figure 26.1

Displaying the MPMediaPickerController

NOTE
In order to use the Media Player framework, you must add it to your project as a linked framework.

The interface of the `MPMediaPickerController` allows the user to select items, podcasts, music tracks, playlists, and so on that are then provided to your application when the user has finished his selection. Your application is notified that the user has completed his selection through the `MPMediaPickerControllerDelegate`. When the user has finished his selection, the delegate method mediaPicker:didPickMediaItems: is called, giving the application the ability to store and play the selected items.

CAUTION
The Media Player framework only works on a device with an iPod library. Because the simulator does not have a library, the Media Player does not work on it.

Listing 26.1 shows an application initializing and presenting an `MPMediaPickerController`.

Listing 26.1

Presenting an MPMediaPickerController

```
-(IBAction)browsePushed:(id)sender
{
    MPMediaPickerController *controller = [MPMediaPickerController new];
    [controller setPrompt:@"Choose Some Tunes"];
    [controller setAllowsPickingMultipleItems:YES];
    [controller setDelegate:self];
    [self presentModalViewController:controller animated:YES];
    [controller release];
}
```

Its corresponding delegate method implementation is shown in Listing 26.2.

Listing 26.2

Implementation of the mediaPicker:didPickMediaItems: Method

```
- (void)mediaPicker:(MPMediaPickerController *)mediaPicker
  didPickMediaItems:(MPMediaItemCollection *)mediaItemCollection;
{
```

```
    [mediaPicker dismissModalViewControllerAnimated:YES];
    MPMusicPlayerController *player =
            [MPMusicPlayerController iPodMusicPlayer];
    [player setQueueWithItemCollection:mediaItemCollection];
    [player play];
}
```

I'll talk soon about the details of using the `MPMusicPlayerController`, but for now all you need to know is that this particular code is going to play your musical selection using the built-in iPod application. So even if you exit the application, the music continues to play. Also note that you need to dismiss your `MPMediaPickerController` here as shown in the first line of this method.

As you saw in Figure 26.1, this `MPMediaPickerController` allows selection of multiple items and has a prompt that says "Choose Some Tunes."

If the user cancels a selection, there's also a delegate method specifically for that event. That method is `mediaPickerDidCancel:`.

Searching for media

Using the `MPMediaPickerController` is fine if your application uses standard controls. But what if you're making an immersive game, one that perhaps would not be well suited to displaying a standard control? The second means for selecting media from the iPod library is designed for applications that want to present their own interface for selecting the songs and simply want a low-level interface for querying the library and retrieving the elements. It allows you to build complex queries that you can then use to query the iPod library and allows you to retrieve information about the library that you can then use to play media or to allow the user to select the media to be played.

The classes used for interacting with the iPod library at this level revolve around the `MPMediaPredicate`, its subclass `MPMediaPropertyPredicate`, and the `MPMediaQuery` class. I will review each of these in turn.

Understanding media property predicates

I dealt with predicates briefly in Chapter 18. When working with the iPod library, generally speaking, you work with the `MPMediaPropertyPredicate` class, which is a concrete subclass of `MPMediaPredicate`. The `MPMediaPropertyPredicate` class allows you to define a filter that determines the subset of the user's iPod library that you retrieve. The items that match the filter are returned in the query. Items that do not match the filter are not. Predicates can be combined so that multiple levels of filtering can be performed. Once you have defined the set of predicates that outline the result set that you want to retrieve, you query the results using the `MPMediaQuery` class.

An `MPMediaPropertyPredicate` can be constructed using one of its two factory methods: `predicateWithValue:forProperty:`, which does a simple "is equal" comparison, and `predicateWithValue:forProperty:comparisonType:`, which allows you to define the type of comparison performed. In each case, you pass a value and a property name. The value is the value that you want to match, and the property name is the name of the property that you want to search. Possible property names are shown in Table 26.2.

Table 26.2 Searchable Property Names for MPMediaPropertyPredicate

Property	Description
`MPMediaItemPropertyPersistentID`	Each item in the library has a unique ID. This property can be used to select a specific item using that ID.
`MPMediaItemPropertyMediaType`	This property allows filtering for particular types of media, such as podcasts, audio books, and so on.
`MPMediaItemPropertyTitle`	The item's title.
`MPMediaItemPropertyAlbumTitle`	The item's album title.
`MPMediaItemPropertyArtist`	The item's artist.
`MPMediaItemPropertyAlbumArtist`	The item's album artist.
`MPMediaItemPropertyGenre`	The genre of the item.
`MPMediaItemPropertyComposer`	The composer of the item.
`MPMediaItemPropertyPlaybackDuration`	The duration of playback of the item.
`MPMediaItemPropertyAlbumTrackNumber`	The track number of the item from the album.
`MPMediaItemPropertyAlbumTrackCount`	The track count of the album the item is from.
`MPMediaItemPropertyDiscNumber`	The disc number of the item.
`MPMediaItemPropertyDiscCount`	The count of discs from the item.
`MPMediaItemPropertyArtwork`	The item's artwork.
`MPMediaItemPropertyLyrics`	The lyrics for the item.
`MPMediaItemPropertyIsCompilation`	A Boolean value indicating if the item is part of a compilation.

So, for example, to search for all songs by the artist Daft Punk that are not part of a compilation, you could build an `MPMediaPropertyPredicate` using the code shown in Listing 26.3.

Listing 26.3

Searching for All Songs by Daft Punk That Are Not Part of a Compilation

```
-(NSArray *)browseForDaftPunk;
{
    MPMediaPropertyPredicate *artist =
    [MPMediaPropertyPredicate predicateWithValue:@"Daft Punk"
                                 forProperty:MPMediaItemPropertyArtist];

    MPMediaPropertyPredicate *notCompliation =
    [MPMediaPropertyPredicate
                predicateWithValue:[NSNumber numberWithBool:NO]
                       forProperty:MPMediaItemPropertyIsCompilation];

    NSSet *predicates = [NSSet setWithObjects:artist, notCompliation, nil];
    MPMediaQuery *query =
    [[MPMediaQuery alloc] initWithFilterPredicates:predicates];
    [query setGroupingType:MPMediaGroupingAlbum];

    return [query items];
}
```

So in this code, you are creating two predicates: one for the artist name and one for whether or not it's in a compilation. You then combine those two predicates into a set. I will talk more about the `MPMediaQuery` shortly, but here you use the `MPMediaQuery` to create a query using the set of predicates. The `items` property contains the individual items from the iPod library that matched the predicates. The types of objects in this array are `MPMediaItem` objects. You can use them to access metadata about the item, such as the song name, artist, and artwork, or you can use them to play the item.

If you want the list of items to be grouped, for example, by album name, you can use the property `groupingType` to set the grouping type. Subsequently, the collections property contains the items as a list of collections, grouped by your selected group type. The types of objects returned from this property are `MPMediaItemCollection` objects. This class is actually a collection of `MPMediaItems`, but it has the useful property of being able to find a representative item for the collection, which contains data that is shared among the items. This is more efficient than iterating over the collection yourself and building such a representative item.

CAUTION
If the query can't find anything matching your predicate, it times out and logs an error message to the console. This can be confusing, but don't read too much into it. The point is that you got nothing back, and so therefore you have a null result instead of an array.

As you can see, you can build reasonably complex queries by combining different predicates that search for different properties and different values.

Now let's take a closer look at the `MPMediaQuery` class, which is the class that actually does the querying.

Building queries

Now you've seen how to use the `MPMediaPropertyPredicate` class to build predicates that you can then use with the `MPMediaQuery` class to query items from the user's iPod library. I also briefly discussed how the `items` property contains the `MPMediaItems` that matched that query, and how the `collections` property contains collections of `MPMediaItems` that are grouped according to a grouping type if you specify one.

The `MPMediaQuery` class also contains a collection of factory methods that have canned queries already available. Using them, you can quickly and easily define queries that group by artist, playlist, or genre.

Table 26.3 shows the `MPMediaQuery` factory methods that are available.

Table 26.3 MPMediaQuery Factory Methods

Method	Purpose
`albumsQuery`	Returns items from the iPod library grouped by album name
`artistsQuery`	Returns items from the iPod library grouped by artist name
`songsQuery`	Returns items from the iPod library grouped by song name
`playlistsQuery`	Returns items from the iPod library grouped by playlist name
`podcastsQuery`	Returns items from the iPod library grouped by podcast name
`audiobooksQuery`	Returns items from the iPod library grouped by audio book name
`compilationsQuery`	Returns items from the iPod library grouped by compilation name
`composersQuery`	Returns items from the iPod library grouped by composer
`genresQuery`	Returns items from the iPod library grouped by genre

Each of these methods returns an `MPMediaQuery` object that groups its results accordingly. If you want to add additional filtering to one of these query objects, you can simply use the instance method `addFilterPredicate:` to the returned object. Then, when you call the associated `items` or `collections` methods, you get the items, filtered as you have requested, in the selected collections.

Working with Player Controllers

Earlier, you briefly saw how to work with a player controller to play the media items in the user's iPod library. Now I want to talk about that in more detail.

The `MPMusicPlayerController` essentially operates in two different modes. The first is as an application music player. In this mode, the music plays as if it were playing within your application. It's not affected by any settings in the built-in iPod application, nor does it cause the iPod application to take any actions.

The second mode is as an iPod music player. In this mode, when you request that an item be played, it actually contacts the built-in iPod application and uses it to do the playback. This means that if your application exits, the music continues to play because it's actually playing through the iPod. Additionally, if you then go over to the iPod application, the Now Playing window displays the item that you have selected to play.

These two modes have important distinctions, because they drastically impact the behavior of the playback.

Once you have allocated and initialized an instance of `MPMusicPlayerController`, you can determine what's played by it by setting its queue. The `MPMusicPlayerController` queue can be set either by using a query, and `MPMediaQuery` instance, or by using a collection of `MPMediaItems` contained in an `MPMediaItemCollection` instance. Then, to start the items playing, you simply call play on it.

Playback can further be controlled by use of pause (to pause playback) or stop (to stop playback), or you can seek forward or backward using the `beginSeekingForward`, `beginSeekingBackward`, and `endSeeking` messages. You can also skip within the collection by using the `skipToNextItem`, `skipToBeginning`, and `skipToPreviousItem` messages.

You can control whether the playback repeats or shuffles by setting those properties on the player accordingly.

Finally, the player generates notifications when playback events occur. This is controlled by sending the message `beginGeneratingPlaybackNotifications` and then monitoring for the associated notifications via the `NSNotificationCenter`. The notifications that are generated are shown in Table 26.4.

Table 26.4 MPMusicPlayerController Notifications

Notification	Purpose
`MPMusicPlayerControllerPlaybackStateDidChangeNotification`	Received when the playback state changes either programmatically or through a user action
`MPMusicPlayerControllerNowPlayingItemDidChangeNotification`	Received when the currently playing item changes
`MPMusicPlayerControllerVolumeDidChangeNotification`	Received when the volume for the music player changes

Accessing Media Artwork

The last item that I want to talk about with regard to the media framework is how you access the artwork associated with a particular media item. The properties associated with the `MPMediaItem` are accessible using the standard method `valueForProperty`. The types of property available to be queried are the same ones from Table 26.2.

This means that to access the artwork for a given media item, you would use code similar to that shown in Listing 26.4.

Listing 26.4

Accessing the Artwork from an MPMediaItem

```
-(void)setImageForItem:(MPMediaItem *)item;
{
    MPMediaItemArtwork *artwork =
    [item valueForProperty:MPMediaItemPropertyArtwork];

    [myImageView setImage:[artwork imageWithSize:
                           [myImageView frame].size]];
}
```

As you can see, the `MPMediaItemArtwork` object has methods that enable you to access the image at a particular size. This means that if you want to access the artwork and display it in your application while the music is playing, this is how you would do it. In this particular case, I have an image view that is being initialized to the artwork from the item.

Summary

In this chapter, you've seen how you can access the iPod media library both to control the iPod player when playing media and to access the media from the user's iPod library to play it within your own application. Not only does this enable makers of various applications to simply play media, but it also opens up new classes of applications that perhaps could make recommendations based on the user's iPod library contents. Using the knowledge that you've gained here, you should be able to easily work with the user's iPod library.

Recording Audio

You've seen how to play audio in your application using the AV Foundation Framework as well as using OpenAL. You've also seen how to access your iPod library from within your application so that you can play the media from the user's music collection.

In this chapter, you're going to see how you record audio in your application. Recording audio is about as simple as playing audio. It uses the same AV Foundation Framework as you used to play audio. In this case, however, you're going to use the `AVAudioRecorder` class to do your work.

Setting up Your AVAudioSession

Just like when playing audio, you have to initialize an `AVAudioSession` appropriately before you can begin recording audio. There are two session categories associated with recording audio. The first category is `AVAudioSessionCategoryRecord`, which allows you to just record audio. This category mutes all playback while recording. The second category, `AVAudioSessionCategoryPlayAndRecord`, allows playback to continue while you're recording. This is more typically suited to voice-over IP applications and the like. Typically, for simply recording audio, you may prefer to use the `AVAudioSessionCategoryRecord` category.

In This Chapter

Setting the appropriate audio category for recording audio

Recording audio with AVAudioRecorder

Understanding the settings associated with audio recording

NOTE
Remember, to use the AV Foundation Framework, you must add it as a linked framework to your project.

To configure the audio category, you use the instance method `setCategory:error:`, which sets the audio category for you. You also need to set your audio session as active, using the instance method `setActive:error:`. Listing 27.1 shows how to set up the audio session for recording.

Listing 27.1

Setting up the Audio Session for rRecording

```
- (void)applicationDidFinishLaunching:(UIApplication *)application
{
    if(![[AVAudioSession sharedInstance] inputIsAvailable])
    {
        // display some error or something...
    }
    NSError *error = nil;
    if(![[AVAudioSession sharedInstance]
          setCategory:AVAudioSessionCategoryRecord error:&error])
    {
        // display error
    }
    if(![[AVAudioSession sharedInstance] setActive:YES error:&error])
    {
        // display error
    }
    [[AVAudioSession sharedInstance] setDelegate:self];
    // Override point for customization after app launch
    [window addSubview:viewController.view];
    [window makeKeyAndVisible];
}
```

NOTE
The `inputIsAvailable` property tells you if an audio input device is present on the machine. If it's not, you should probably alert the user. Keep in mind, however, that the user can, for example, plug in a microphone after you have checked. To detect when a user does this, be sure to implement the `AVAudioSessionDelegate` method `inputIsAvailableChanged`, which is called when this occurs.

Once you've done this, you are ready to use the `AVAudioRecorder` for recording.

Allocating an AVAudioRecorder

Initializing an `AVAudioRecorder` is very straightforward. The audio recorder needs to know where to store the data that it's recording. It also needs to know certain parameters such as what sample rate to use and what file format to use. It infers the recording format from the extension of the file that you tell it to record to. Other settings, however, must be provided via a dictionary that you pass to the initializer.

So, to allocate and initialize an instance of `AVAudioRecorder`, you call the initializer `initWithURL:settings:error:`, which takes as arguments a URL designating the place to store the recorded data, and a settings dictionary, containing the properties for the audio recording itself. The settings dictionary contains keys and values for a variety of properties relating to the audio recording. The available keys and their purposes are shown in Table 27.1.

Table 27.1 Keys Used for the AVAudioRecorder Settings Dictionary

Key	Purpose
`AVFormatIDKey`	This key specifies the audio data format identifier.
`AVSampleRateKey`	This key specifies the sample rate in hertz as a floating-point value wrapped in an `NSNumber`.
`AVNumberOfChannelsKey`	This key specifies the number of channels as an integer value wrapped in an `NSNumber`.
`AVLinearPCMBitDepthKey`	This key specifies the bit depth for a linear PCM audio format recording. This value is an integer wrapped in an `NSNumber`. Possible values are 8, 16, 24, or 32. This is only relevant for the linear PCM audio format.
`AVLinearPCMIsBigEndianKey`	This key specifies whether the audio format is big endian or little endian. It is only relevant for the linear PCM audio format.
`AVLinearPCMIsFloatKey`	This key specifies whether the audio format is floating point. It is only relevant for the linear PCM audio format.
`AVEncoderAudioQualityKey`	This key specifies the recording audio quality. It can be one of `AVAudioQualityMin`, `AVAudioQualityLow`, `AVAudioQualityMedium`, `AVAudioQualityHigh`, or `AVAudioQualityMax`.
`AVEncoderBitRateKey`	This key specifies the bit rate of the recording in hertz.
`AVEncoderBitDepthHintKey`	This key specifies the bit depth hint of the recording. Possible values range from 8 to 32.
`AVSampleRateConverterAudioQualityKey`	This key specifies the sample rate conversion quality. It can be one of `AVAudioQualityMin`, `AVAudioQualityLow`, `AVAudioQualityMedium`, `AVAudioQualityHigh`, or `AVAudioQualityMax`.

Once you've initialized an `AVAudioRecorder`, to begin recording, all you need to do is call the method record. Like the `AVAudioPlayer`, `AVAudioRecorder` provides an instance method of `prepareToRecord`. This method creates the file and prepares buffers for recording so that there is minimal delay when you actually want to begin recording. If the file that you have requested it to record to already exists, it automatically overwrites that file. `AVAudio Recorder` also provides the method `recordForDuration`, which causes the `AVAudio Recorder` to record only for the time period requested. To pause and stop recording, there are methods specifically for pausing and stopping. Finally, if you decide that you want to remove the audio file that you have recorded, the method `deleteRecording` deletes the file from disk.

NOTE
To delete a recording, the audio recorder must be stopped.

Like the `AVAudioPlayer`, the `AVAudioRecorder` also has the ability to monitor audio meters for the input. Just like before, you call the `updateMeters` method to update the meter values, and then you access the `peakPowerForChannel:` and `averagePowerForChannel:` methods on the `AVAudioRecorder` to access the meters.

`AVAudioRecorder` also has a delegate that, like the delegate for the `AVAudioPlayer`, informs your application when recording has finished, or is interrupted.

Let's take a look now at an example application that records audio.

Creating a Voice Recorder

Firstly, you need to set up a UI for your voice recorder. For the purposes of this example, it's going to be very simple. Nothing fancy, no special graphics; it's just going to have record, play, and stop buttons. Figure 27.1 shows the finished application with the UI in place. Use Interface Builder to create this interface, being sure to connect up the `UIProgressViews` to appropriate `IBOutlets`, as well as connecting the buttons to appropriate `IBActions`. In my case, I created actions called `playPushed:`, `stopPushed:`, and `recordPushed:`.

Most of the action that's going to occur in this particular application takes place in these `IBActions`. However, just to have a bit of additional fanciness, I added two `UIProgress Views`, which display the metering levels when recording and playing. The top one displays the average, and the bottom one displays the peak. I'm going to show you the sample code for this first to get it out of the way. Listing 27.2 shows the function that updates the level meters. I call this function every quarter-second using an `NSTimer`. Listing 27.2 also shows the `view DidLoad` method where I have set up the `NSTimer`.

Figure 27.1

The finished recorder

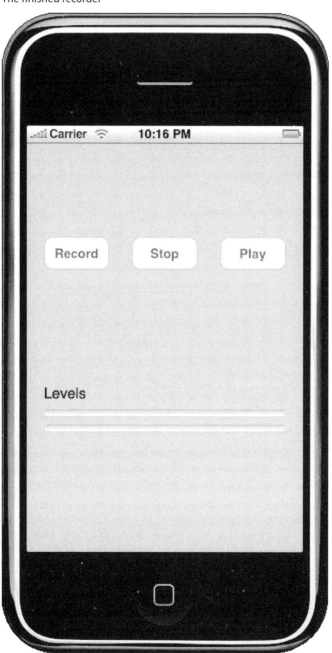

Listing 27.2

Setting up Level Metering

```
-(void)updateMeters;
{
    float peak = 0;
    float avg = 0;
    if([recorder isRecording])
    {
        [recorder updateMeters];
        peak = 100+[recorder peakPowerForChannel:0];
        avg = 100+[recorder averagePowerForChannel:0];
    }
    else if([player isPlaying])
    {
        [player updateMeters];
        peak = 100+[player peakPowerForChannel:0];
        avg = 100+[player averagePowerForChannel:0];
    }
    [peakLevel setProgress:peak/100];
    [avgLevel setProgress:avg/100];
}
- (void)viewDidLoad
{
    [NSTimer scheduledTimerWithTimeInterval:.25 target:self
                                    selector:@selector(updateMeters)
                                    userInfo:nil repeats:YES];
    [super viewDidLoad];
}
```

NOTE

The metering value is a negative floating-point number indicating decibels. 0db indicates the maximum volume. For the purposes of this demo, I'm normalizing this value to a percentage for the progress views by simply adding 100 to it. You may want to do something a bit more fancy. Most applications start off metering around -80db.

Once you have the metering set up, the actual recording and playing are very simple. They take place in the IBActions. Listing 27.3 shows the code for the recordPushed: action. Note that this is where you are actually initializing your recorder and setting all of your settings. In your application, it may make more sense to set up your recorder ahead of time, so that you're not allocating buffers and so forth when you're ready to start recording.

Listing 27.3

Implementation of the recordPushed: Method

```
-(IBAction)recordPushed:(id)sender;
{
    if([recorder isRecording])
        return;

    if([player isPlaying])
        [player stop];
    NSError *error = nil;
    [[AVAudioSession sharedInstance]
            setCategory:AVAudioSessionCategoryRecord error:&error];
    [[AVAudioSession sharedInstance] setActive:YES error:&error];

    NSMutableDictionary *settings = [NSMutableDictionary dictionary];
    [settings setValue:[NSNumber numberWithInt:kAudioFormatLinearPCM]
            forKey:AVFormatIDKey];
    [settings setValue:[NSNumber numberWithFloat:44100.0]
            forKey:AVSampleRateKey];
    [settings setValue:[NSNumber numberWithInt:1]
            forKey:AVNumberOfChannelsKey];
    [settings setValue:[NSNumber numberWithInt:16]
            forKey:AVLinearPCMBitDepthKey];
    [settings setValue:[NSNumber numberWithBool:NO]
            forKey:AVLinearPCMIsBigEndianKey];
    [settings setValue:[NSNumber numberWithBool:NO]
            forKey:AVLinearPCMIsFloatKey];

    NSString *filePath =
    [NSString stringWithFormat:@"%@/rec_audio.caf",
                                [self documentsDirectory]];
    NSURL *fileUrl = [NSURL fileURLWithPath:filePath];

    [self setRecorder:nil];
    recorder = [[AVAudioRecorder alloc]
                                initWithURL:fileUrl
                                    settings:settings
                                       error:&error];

    [recorder setMeteringEnabled:YES];
    [recorder record];
}
```

Next up is the `playPushed:` action. In this case, all you do is allocate a new audio player, and fire it up to start playing. This is shown in Listing 27.4.

Listing 27.4

Implementation of the playPushed: Action

```
-(IBAction)playPushed:(id)sender;
{
    if([recorder isRecording])
        [recorder stop];
    if([player isPlaying])
        [player stop];

    NSString *filePath =
    [NSString stringWithFormat:@"%@/rec_audio.caf",
                                [self documentsDirectory]];
    NSURL *fileUrl = [NSURL fileURLWithPath:filePath];
    NSError *error = nil;
    [self setPlayer:nil];
    [[AVAudioSession sharedInstance]
        setCategory:AVAudioSessionCategoryPlayback error:&error];
    [[AVAudioSession sharedInstance] setActive:YES error:&error];
    player = [[AVAudioPlayer alloc] initWithContentsOfURL:fileUrl
                                                    error:&error];
    [player setMeteringEnabled:YES];
    [player play];
}
```

Finally, the implementation of your stop action is incredibly simple compared to the previous two. It's shown in Listing 27.5.

Listing 27.5

Implementation of the stopPushed: Action

```
-(IBAction)stopPushed:(id)sender;
{
    if([recorder isRecording])
        [recorder stop];
    if([player isPlaying])
        [player stop];
}
```

If you add this code to your view controller, assuming you have connected everything successfully, you should be able to touch the Record button and record audio. When you tap the Play button, what you've recorded should play back. Most of this code does less error-checking than it probably should. You should be checking the status of the error values on each of the calls where an error object is passed in. In this particular case, though, I'm simply demonstrating how the API works, and so I left it out for simplicity's sake.

Summary

In this chapter, you've seen how to record audio using the AV Foundation Framework. The `AVAudioRecorder` class makes it incredibly easy to capture audio in your applications. You need only to instantiate an instance of this class, and tell it where to store your data. Your sample application demonstrated how to play audio as well. When recording audio, you have access to a very powerful audio encoder that is built into the device. Using this API gives you the ability to create applications that manipulate audio as easily as the built-in applications on the iPhone.

Playing Video in Your Application

You've seen how to play audio in your applications. You've also seen how to play items from a user's iPod library, and how to control the iPod application from your application. The Media Player framework has one more goodie for you to investigate, and this is the `MPMoviePlayerController`, which enables you to play movies within your applications.

The `MPMoviePlayerController` is capable of playing H.264 video at 640x480 resolution at 30 frames per second. It can play media contained within your application bundle, or it can stream video over the Internet from a URL. When the player begins playing, it fades out your user interface (UI) and fades in the media player. When the video has finished, it fades out the media player and brings back your UI. It is only capable of displaying video full screen; it is not able to display it in a portion of the screen.

Let's take a look at how to utilize this class to play video within your application.

Playing Video Files Contained in the App Bundle

In This Chapter

Playing video bundled in your application

Learning how to play cut scenes using MPEG video

Streaming video from the Internet

Playing videos from inside your application bundle using the `MPMoviePlayerController` is a snap. Naturally, you have to be sure that the media is included in your application, so it has to be bundled as a resource. Then you simply use the `NSBundle` method `pathForResource:ofType:` to get the path to the movie within your application's main bundle. You can then use this path to create an NSURL that you can use to initialize the `MPMoviePlayerController`.

NOTE
Remember, `MPMoviePlayerController` is part of the `MediaPlayer` framework. To use it, you must include it as a linked framework and be sure to import the headers needed for it.

Listing 28.1 shows how this is done.

Listing 28.1

Playing a Movie from within Your Application Bundle

```
-(void)playTransitionMovie;
{
    NSBundle *bundle = [NSBundle mainBundle];
    NSString *moviePath =
        [bundle pathForResource:@"transition" ofType:@"mov"];
    NSURL *movieUrl = [NSURL fileURLWithPath:moviePath];
    MPMoviePlayerController *player = [[MPMoviePlayerController alloc]
                                        initWithContentURL:movieUrl];
    [player setMovieControlMode:MPMovieControlModeHidden];
    [player play];
}
```

In this particular case, you're playing a movie to show a transition between two different scenes. Because of this, you don't want the user to be able to control the playback of the movie. To hide the movie controls so that the user is unable to change the playback, you use the property `movieControlMode`. Setting this to `MPMovieControlModeHidden` causes the movie controls to be hidden. Other possible values for this property are `MPMovieControlModeDefault`, which displays the standard controls, and `MPMovieControlModeVolumeOnly`, which displays only the volume control. Figure 28.1 shows the playback controls if you use the default setting.

There are several other properties like this that allow you to control the overall experience as the movie is playing back. Of particular interest are the `backgroundColor` property, which allows you to change the background color of the player, and the `scalingMode` property, which allows you to specify whether the movie should be scaled to fit within the aspect ratio of the iPhone screen.

It may take a while to load the video and start playing it. As a result, the movie player begins loading the video when the initializer is called. When it finishes loading the movie, it sends the `MPMoviePlayerContentPreloadDidFinishNotification` notification using the `NSNotificationCenter`. If you need to know when the video has completed loading, you should register to receive this notification in your application. Once you call the play method, your user interface fades to the background color of the movie player and the video playback begins. If the video has not completed loading, the background color of the video player is shown. It may also display a progress indicator. When playback has completed, your user interface is brought back.

Figure 28.1

The default video playback controls

If you want the video to begin playing at a particular location in the video, you can also set the `initialPlaybackTime` property, which positions the video head at the specified playback time prior to beginning playback.

Playing Video from the Internet

Streaming video from the Internet using the `MPMoviePlayerController` is just as easy as playing video from your application bundle. The only difference is that the URL that you specify should refer to a resource on the Internet. The `MPMoviePlayerController` takes care of all the loading of the video from the Internet for you. You only have to tell it where the video is.

Listing 28.2 shows the same playback that you saw earlier, but instead, loading a video from the Internet.

Listing 28.2

Playing Video from the Internet

```
-(void)playTransitionMovie;
{
    NSBundle *bundle = [NSBundle mainBundle];
    NSURL *movieUrl =
        [NSURL URLWithString:@"http://www.somewhere.com/mymovie.mov"];
    MPMoviePlayerController *player = [[MPMoviePlayerController alloc]
                                      initWithContentURL:movieUrl];
    [player setMovieControlMode:MPMovieControlModeHidden];
    [player play];
}
```

Remember that loading a video file from the Internet can be much slower than loading it from your application bundle. You should therefore take this into account when you consider using this class for streaming video from the Internet.

Summary

In this brief chapter, you've seen how to play video in your application. The built-in video-playing capabilities only allow you to display video full-screen. However, they do provide you with sufficient tools so that you can use that full-screen video capability to do transition videos between scenes in video games, or to display media that the user may want to have control over. If you're looking to display video only in a portion of your application screen, you may want to look at the UIImageView animation capabilities showcased in the chapter on core animation, Chapter 14. However, if you want to display full-screen, 30 fps, full-motion video in your application, MPMoviePlayerController is an excellent choice. It even has the ability to download videos from the Internet and play them as well.

Working with the iPhone Hardware

In This Part

Chapter 29
Discovering Information about the Device

Chapter 30
Getting Your Location Using Core Location

Chapter 31
Working with the Accelerometer

Chapter 32
Interfacing with Peripherals

Discovering Information about the Device

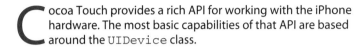

Cocoa Touch provides a rich API for working with the iPhone hardware. The most basic capabilities of that API are based around the UIDevice class.

UIDevice is a singleton that allows basic interaction with some of the core elements of the iPhone hardware. It enables you to find out information about the device orientation, the battery level, and the proximity sensor. It can also tell you things such as what type of device it is, as well as the unique identifier of the device. It's an incredibly powerful little gem in Cocoa Touch, one that can be exploited easily in your applications.

Accessing the Battery State

The first item that I am going to discuss with regard to device hardware knowledge is information about the battery level. Using the UIDevice singleton, you can monitor both the battery charge level and the current battery state. For example, you can find out whether or not the battery is currently being charged from a power adapter.

To utilize this capability, you simply instantiate the UI device singleton, enable battery monitoring, and then retrieve the battery level using the batteryLevel property. The return value from this property ranges from 0.0, indicating a fully discharged battery, to 1.0, indicating a fully charged battery. Additionally, the battery State property can tell you information about whether the battery is charging, full, unplugged, and so on. The possible states for the batteryState property are shown in Table 29.1.

There are applications that might want to behave differently, depending on whether or not the device is plugged into a power adapter. For example, when not plugged into a power adapter, your application may want to reduce its power consumption by turning off certain features, such as GPS.

In This Chapter

How to tell if the device is connected to a power adapter

Working with the proximity sensor

Getting basic information and metadata about the device

Table 29.1 Battery State Values

State	Meaning
`UIDeviceBatteryStateUnknown`	The battery state is unable to be determined.
`UIDeviceBatteryStateUnplugged`	The device is not plugged in, and the battery is currently discharging.
`UIDeviceBatteryStateCharging`	The device is plugged into a power adapter, and the battery is currently charging.
`UIDeviceBatteryStateFull`	The device is plugged into a power adapter, and the battery is currently fully charged.

Accessing the Proximity Sensor

The next hardware element that I'm going to describe is the use of the proximity sensor. This was newly added in iPhone OS 3. The proximity sensor tells the iPhone when you bring the phone close to your face — for example, when you are listening to the headset or speaking into the microphone. The Google application uses this capability very cleverly. It records audio only when the device is held up to your ear. Thus, it reduces the amount of background noise that might confuse its speech recognition system.

Like accessing the battery level information, accessing the proximity sensor state is as simple as retrieving the `UIDevice` singleton, and then enabling proximity sensor monitoring using the `proximityMonitoringEnabled` property. Once you have enabled proximity sensor monitoring, you access the current proximity sensor state utilizing the `proximityState` property. The `proximityState` property is a Boolean value that indicates whether or not the device is being held close to the user's face.

You should only enable proximity monitoring if your application actually needs it. Keeping it disabled when you don't need it helps to conserve battery life.

Working with Device Metadata

The next set of UIDevice instance methods are related to retrieving information about the operating system, the model, and the device. This is more or less metadata about the device, as opposed to hardware information.

As you know, every iPhone has a unique identifier assigned to it. Accessing the unique identifier is as easy as calling the UI device instance method `uniqueIdentifier`. This identifier can be used in cases where you want to be able to uniquely identify a device or user, for example, for storing data on a server. If you give the user access to that data on your server, displaying their unique identifier can be confusing, and so you may want to use the name property as well. The name property on `UIDevice` gives you the user-assigned name of the device. This is the name the user assigned to the device when they initially set it up and activated it.

Table 29.2 shows the "informational" properties on `UIDevice`.

Table 29.2 UIDevice Properties

Method	Purpose
uniqueIdentifier	The device's unique identifier
name	The user-defined name of the device
systemName	The name of the operating system on the device
systemVersion	The version number of the operating system on the device
model	The model name of the device
localizedModel	The model name of the device as a localized string

NOTE
If you want to use the model name to detect certain hardware requirements in your app, you should use the non-localized model name.

Using properties like model and localizedModel, you can even determine what type of device it is, whether it's an iPhone or iPod touch. You can use this information to display dialogs that are tailored specifically for that user and that device. This helps to make your application "feel" more tailor-made to the user's actual environment.

Summary

In this chapter, you've had an introduction to working with the iPhone hardware. The UIDevice API provides a simple interface for accessing some of the information about the device that you're running on. Using it, you can build applications that know more about their environment than they would without it. For example, you can know when the user brings the device up to their face. You can know whether or not the user has connected the device to their power adapter. Additionally, you can find out what the user has named their device so that you can customize your display specifically to that user.

It's details like these that make the difference between a good application and a great application. Use them when possible to enhance your applications.

This has just been a taste of what's to come. In the following chapters, you'll take a closer look at some of the other hardware features, and how to interface with them effectively.

Getting Your Location Using Core Location

One feature that makes developing for the iPhone platform unique and powerful is its support for location awareness. The API used for accessing this feature is called Core Location. With it, you can pinpoint the location of the device anywhere in the world.

Core Location actually uses three methods to determine your location. If you're on a first-generation iPhone, it first uses a combination of cell phone tower triangulation and then Wi-Fi access point lookup. This provides reasonable accuracy both indoors and outdoors but is by no means exact. If you're on a second-generation iPhone, which has a built-in GPS receiver, and if you're outdoors, it instead uses a third method, GPS triangulation. Between these three methods, the iPhone is able to determine your position with reasonable accuracy and is also able to do it both indoors and outdoors, which overcomes a limitation that many devices with only GPS capabilities lack.

Utilizing this ability to determine the user's location enables a new class of applications and provides exciting opportunities for location-aware computing.

The latest iPhone revision, the iPhone 3GS, also adds built-in magnetic compass hardware. The compass can be used in conjunction with the GPS capabilities to determine the device's orientation.

Once you have worked with applications that automatically know the user's location, asking the user for their location in order to display nearby information suddenly begins to feel archaic.

Let's take a look at some of the APIs that enable interaction with this hardware.

In This Chapter

Using GPS and cell tower triangulation to determine your iPhone's location

Using the Core Location API to find where you are

Narrowing the results of Core Location

Using Core Location coordinates to draw your location on a map

Working with the iPhone 3GS compass

Finding Where You Are

As an example, I'm going to have you write an application that monitors the user's location and then opens Google maps to display the user's location on a map. Figure 30.1 shows what the final app looks like when running on the iPhone.

Figure 30.1
The finished Core Location app

Allocating a CLLocationManager

The first thing you need to do to interact with the Core Location API is to allocate a `CLLocation Manager`. The `CLLocationManager` provides the primary entry point to the Core Location system. There's really nothing special about this: simply allocate an instance using `new` or `alloc` and `init`. Once you have an instance of the `CLLocationManager`, you can either get the position for the user at this particular moment or you can register a delegate with the `CLLocationManager` instance to be updated as the user's position changes over time. In both of these situations, however, you must also call the `startUpdatingLocation` method, which causes the `CLLocationManager` to begin gathering location data and updating your delegate. If you don't call this method, your delegate does not receive location notifications, and if you try to access the location directly from the `CLLocationManager` then you receive a nil instead.

In the example application, you're going to allocate the `CLLocationManager` in your `view DidLoad` method and then you're going to have buttons on your UI that allow you to start and stop the updating of the location. Finally, you'll have a button that sends the data to Google Maps to display your location.

TIP
The `viewDidLoad` method gets called automatically after the view is loaded from the nib file. When it's called, all the objects that your `IBOutlets` have been connected to are loaded and initialized, so it's a good place to set their initial values.

Listing 30.1 shows your `viewDidLoad` method.

Listing 30.1

An Example viewDidLoad Implementation

```
- (void)viewDidLoad
{
    locMgr = [[CLLocationManager alloc] init];

    if(![locMgr locationServicesEnabled])
        [status setText:@"Status: Location Services Disabled"];
    else
        [status setText:@"Status: Location Services Enabled"];


    [locMgr setDesiredAccuracy:kCLLocationAccuracyNearestTenMeters];
    [locMgr setDistanceFilter:kCLDistanceFilterNone];
    [locMgr setDelegate:self];

    [super viewDidLoad];
}
```

So, to create the `CLLocationManager`, you simply call `[[CLLocationManager alloc] init]`, assigning it to your member variable.

Once you have an instance of `CLLocationManager`, you should check to determine if the user has location services enabled before actually requesting a location. The user has the option to disable location services in the settings of the iPhone if they prefer not to allow applications to know their location. Calling the method `locationServicesEnabled` returns a Boolean value to tell you whether or not the user has set this setting. If they have disabled location services, then you should probably alert them that if they want to utilize this feature they should change this setting on their device. There is no way for an application to change this setting programmatically.

Next, you set the desired accuracy and the distance filter. I'll talk more about this later, but for now, you can just ignore it.

Setting yourself as the delegate

Finally, you set yourself as the delegate. You've already declared that this object adheres to the `CLLocationManagerDelegate` protocol in the header, as shown in Listing 30.2.

Listing 30.2

The Header File for the CoreLocationDemoViewController

```
@interface CoreLocationDemoViewController :
      UIViewController <CLLocationManagerDelegate>
{
    CLLocationManager *locMgr;
    IBOutlet UILabel *status;

}
@property (retain) CLLocationManager * locMgr;
-(IBAction)startTouched:(id)sender;
-(IBAction)stopTouched:(id)sender;
-(IBAction)sendToGoogleTouched:(id)sender;
- (void)locationManager:(CLLocationManager *)manager
    didUpdateToLocation:(CLLocation *)newLocation
          fromLocation:(CLLocation *)oldLocation;
@end
```

Implementing the CLLocationManagerDelegate protocol

To actually implement this protocol, you implement the methods `locationManager:didUpdateToLocation:fromLocation` and `locationManager:didFailWithError:`, as shown in Listing 30.3.

Listing 30.3

Methods Required for Implementing the Delegate Protocol

```
-(void)locationManager:(CLLocationManager *)manager
   didUpdateToLocation:(CLLocation *)newLocation
          fromLocation:(CLLocation *)oldLocation;
{
    [status setText:[NSString stringWithFormat:@"Status: Got Loc: %f/%f",
                    newLocation.coordinate.latitude,
                    newLocation.coordinate.longitude]];
}
- (void)locationManager:(CLLocationManager *)manager
       didFailWithError:(NSError *)error
{
    [status setText:[NSString stringWithFormat:@"Received error: %@",
                    [error localizedDescription]]];
}
```

The `locationManager:didUpdateToLocation:fromLocation` receives both the new current location and the location from the last update, and so if you need to determine the direction of motion, it's very simple to do so.

Starting and stopping location updates

In this application, you have buttons on the GUI to start and stop the location updates. These buttons are tied to actions in your controller, as shown in Listing 30.4.

Listing 30.4

The Methods That Are Connected to the Start and Stop Buttons in Interface Builder

```
-(IBAction)startTouched:(id)sender;
{
    [locMgr startUpdatingLocation];
    [status setText:@"Status: Waiting for updates."];
}
-(IBAction)stopTouched:(id)sender;
{
    [locMgr stopUpdatingLocation];
    [status setText:@"Status: Stopped receiving."];
}
```

Viewing your location on Google Maps

For your purposes, all you're going to do is update your status messages. Once you've seen the update on receiving data from the `CLLocationManager`, the user can click the button and view the location in Google Maps. Listing 30.5 demonstrates that code.

NOTE
The app displays the coordinates it's received in the status text as it receives data.

Listing 30.5

The Method That Sends the Location to Google Maps

```
-(IBAction)sendToGoogleTouched:(id)sender;
{
    CLLocation *lastLocation = [locMgr location];
    if(!lastLocation) // we haven't received any data yet - so we got a nil!
    {
        UIAlertView *alert;
        alert = [[UIAlertView alloc]
                    initWithTitle:@"No data yet!"
                    message:@"No data has been received by the delegate yet."
                    delegate:nil cancelButtonTitle:nil
                    otherButtonTitles:@"OK", nil];

        [alert show];
        [alert release];
        return;
    }

    NSString *urlString = [NSString stringWithFormat:
                            @"http://maps.google.com/maps?q=Here+I+Am!@%f,%f",
                            lastLocation.coordinate.latitude,
                            lastLocation.coordinate.longitude]
    NSURL *url = [NSURL URLWithString:urlString];

    [[UIApplication sharedApplication] openURL:url];
}
```

As you can see, you access the `lastLocation.coordinate.latitude` and `lastLocation.coordinate.longitude` values to send to Google Maps. With that information, you can then open the URL and you're done. Figure 30.2 shows Safari with Google Maps loaded and your location pinpointed on it.

Figure 30.2

Google Maps indicating your location with a pin

Narrowing the Accuracy of the Coordinates

As I've said before, when working with a mobile device with limited battery power, it's important to try to conserve that power whenever possible. Determining the user's location at high accuracy requires more time and power than determining it with lower accuracy. Because of this, `CLLocationManager` provides the ability to set the accuracy of the data that it returns. This is done using the `desiredAccuracy` property. The values that can be set for this property are listed in Table 30.1. They are relatively self-explanatory for what they represent.

Table 30.1 Possible Values for the DesiredAccuracy Method

Value	Description
`kCLLocationAccuracyBest`	Gives the best possible accuracy
`kCLLocationAccuracyNearestTenMeters`	Accurate to within 10 meters
`kCLLocationAccuracyHundredMeters`	Accurate to within 100 meters
`kCLLocationAccuracyKilometer`	Accurate to within 1 kilometer
`kCLLocationAccuracyThreeKilometers`	Accurate to within 3 kilometers

NOTE
Just because you request a particular level of accuracy does not necessarily mean that you will receive data at that level of accuracy. The device does its best to meet your request, but it may not necessarily be able to accomplish it.

The `CLLocationManager` tries to return at least some kind of data as quickly as possible, so it may return data at a lower level of accuracy to begin with and then return data at a higher level of accuracy later. You should be prepared for this in your application and take appropriate action, such as narrowing your display to the more accurate point after you have received it.

Filtering Location Updates

Finally, again because you're working with a device with limited power, you want to minimize the updates that you get and not necessarily receive an update of the user's location if the user has not moved. The `CLLocationManager` also provides an API for this: the `distance Filter` property.

When you set this property, you only receive updates to the user's location when they pass beyond a certain threshold of distance from the last location received. The value of this property can be set to a number that indicates the distance in meters from the original location, or it can be set to `kCLDistanceFilterNone`, which indicates no filter at all. If you have no filter, then you constantly receive updates.

CAUTION

If you have no filter and therefore receive constant location updates, this drains the device battery very quickly: because of this, you should usually use a distance filter if possible.

Looking at the Final Code

Now that you've looked at all of the individual parts that make up this application, let's take a look at the completed code. Listings 30.6 and 30.7 show your completed Core Location demo application.

NOTE

You can download the full project code from this book's Web site located at:
http://www.wiley.com/go/cocoatouchdevref.

Listing 30.6

CoreLocationDemoViewController.h

```
#import <UIKit/UIKit.h>
#import <CoreLocation/CoreLocation.h>
@interface CoreLocationDemoViewController : UIViewController
    <CLLocationManagerDelegate>
{
    CLLocationManager *locMgr;
    IBOutlet UILabel *status;

}
@property (retain, nonatomic) CLLocationManager * locMgr;
-(IBAction)startTouched:(id)sender;
-(IBAction)stopTouched:(id)sender;
-(IBAction)sendToGoogleTouched:(id)sender;
- (void)locationManager:(CLLocationManager *)manager
    didUpdateToLocation:(CLLocation *)newLocation
           fromLocation:(CLLocation *)oldLocation;
@end
```

Listing 30.7

CoreLocationDemoViewController.m

```objc
@implementation CoreLocationDemoViewController
@synthesize locMgr;
#pragma mark -
#pragma mark IB ACTIONS
-(IBAction)startTouched:(id)sender;
{
    [locMgr startUpdatingLocation];
    [status setText:@"Status: Waiting for updates."];
}
-(IBAction)stopTouched:(id)sender;
{
    [locMgr stopUpdatingLocation];
    [status setText:@"Status: Stopped receiving."];
}
-(IBAction)sendToGoogleTouched:(id)sender;
{
    CLLocation *lastLocation = [locMgr location];
    if(!lastLocation) // we haven't received any data yet - so we got a nil!
    {
        UIAlertView *alert;
        alert = [[UIAlertView alloc]
                  initWithTitle:@"No data yet!"
                  message:@"No data has been received by the delegate yet."
                  delegate:nil cancelButtonTitle:nil
                  otherButtonTitles:@"OK", nil];

        [alert show];
        [alert release];
        return;
    }

    NSString *urlString = [NSString stringWithFormat:
                           @"http://maps.google.com/maps?q=Here+I+Am!@%f,%f",
                           lastLocation.coordinate.latitude,
                           lastLocation.coordinate.longitude]
    NSURL *url = [NSURL URLWithString:urlString];

    [[UIApplication sharedApplication] openURL:url];
}
#pragma mark -
#pragma mark DELEGATE METHODS
- (void)locationManager:(CLLocationManager *)manager
    didUpdateToLocation:(CLLocation *)newLocation
           fromLocation:(CLLocation *)oldLocation;
{
```

```objc
        [status setText:[NSString stringWithFormat:@"Status: Got Loc: %f/%f",
                            newLocation.coordinate.latitude,
                            newLocation.coordinate.longitude]];
}
- (void)locationManager:(CLLocationManager *)manager
        didFailWithError:(NSError *)error
{
    [status setText:[NSString stringWithFormat:@"Received error: %@",
                        [error localizedDescription]]];
}
#pragma mark -
- (void)viewDidLoad
{
    locMgr = [[CLLocationManager alloc] init];

    if(![locMgr locationServicesEnabled])
        [status setText:@"Status: Location Services Disabled"];
    else
        [status setText:@"Status: Location Services Enabled"];


    [locMgr setDesiredAccuracy:kCLLocationAccuracyNearestTenMeters];
    [locMgr setDistanceFilter:kCLDistanceFilterNone];
    [locMgr setDelegate:self];

    [super viewDidLoad];
}

- (void)didReceiveMemoryWarning
{
    [super didReceiveMemoryWarning];
}

- (void)dealloc
{
    [self setLocMgr:nil];
    [super dealloc];
}
@end
```

Working with the iPhone 3GS Compass

Recall that the iPhone 3GS added a new piece of hardware, a magnetic compass. To access the compass, you also use the Core Location API. Just as when working with the location information, the CLLocationManager has two methods, startUpdatingHeading and stopUpdatingHeading, which control whether the CLLocationManager sends heading updates to its delegate. If you enable heading updates using these methods, your delegate will receive calls to location `Manager:didUpdateHeading:`, just as it did for location updates. As with location updates, you can set a filter for heading updates using the property headingFilter.

Because this capability is limited to the iPhone 3GS, you should always make sure heading information is available on the device. You do this by accessing the `CLLocationManager` property `headingAvailable`, which returns YES if it is.

NOTE
Sometimes the compass may need calibration. When it does, your delegate is notified via the method locationManagerShouldDisplayHeadingCalibration:. If your delegate implements this method, it can return NO to suppress display, or YES to allow it.

I won't describe in detail how to update this application with this capability, but if you need to add it to your application, see the Apple documentation for more details.

Summary

In this chapter, you looked at how to use the Core Location API to add location awareness to your applications. The ability of the device to know its location is one of the unique defining features of iPhone development. As a result, leveraging this technology can enable many new and interesting applications.

Working with the Accelerometer

In November 2006, Nintendo introduced the Nintendo Wii. The Nintendo Wii contained a variety of innovative controller mechanisms, but one in particular was especially revolutionary. The controllers that shipped with the Nintendo Wii contained accelerometers, small electronic components inside the controllers that could detect the orientation and motion of the controllers themselves. Using the accelerometers, the Nintendo Wii launched a completely new gaming experience, by unlocking the ability for the players to use their whole bodies as game controllers.

Nearly a year later, when Apple introduced the first iPhone, it also contained an accelerometer, and the accelerometer built into the iPhone was probably as revolutionary for mobile application development as it was for the Nintendo Wii.

Make no mistake; accelerometers are nothing new in the world of electronics. But making accelerometers available to application developers as part of the standard hardware on the device opens up a completely new user experience. It opens up new ways of interacting with applications that were not available before.

In this chapter, you'll learn how to harness the accelerometer so that you can use it in your application.

In This Chapter

Using the accelerometer to determine the orientation of the device

Using a low-pass filter to smooth the input from the accelerometer

Implementing shake motion detection in a custom UIView

Determining Which Way Is Up

The accelerometer built into the iPhone is capable of measuring G-force in any of the X, Y, or Z directions. Figure 31.1 shows the direction of these forces as they relate to the orientation of the device.

As you can see, the X- and Y-axes correspond to the flat-level axis of the iPhone, whereas the Z-axis corresponds to the perpendicular angle to those axes.

Figure 31.1

X, Y, and Z coordinates as related to iPhone orientation

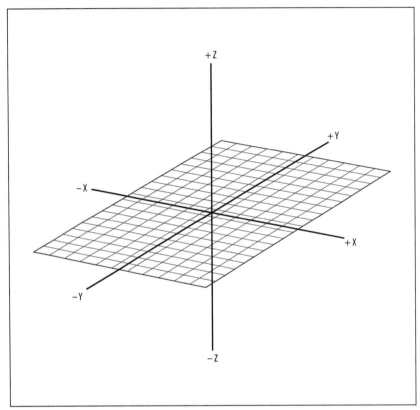

Therefore, when holding your phone flat and level, your X and Y accelerometer values should be zero, and your Z value should be somewhere around −1. This corresponds to no force being exerted in the X and Y directions, but gravity being exerted in the Z direction. The unit of measure for these values is in G-force, with 1 corresponding roughly to 1g.

NOTE
The accelerometer is not exact, and so you'll probably never see exactly 0.0.

Now I'm going to show you how to make an example application using the accelerometer. The example application will be essentially a bubble level. When it's finished, it will look something like Figure 31.2.

Figure 31.2

The completed bubble-level application

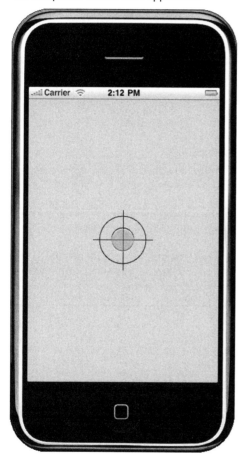

In this application, you will be ignoring the Z values of the acceleration object. Rest assured, however, that if you want to work with the Z values, they are there as well.

Building a bubble level

The first thing to know when working with the accelerometer is that there are actually three core classes involved in the accelerometer API. The first one is the `UIAccelerometer` class. The UI accelerometer class is a singleton and is used to access the accelerometer device. You access the accelerometer so that you can set your delegate object for the accelerometer. You can also set the update interval for the accelerometer. If you do not require constant updates, then you should set the update interval to a longer time period than the default in order to conserve battery life.

You can download the code for the example application on the book's Web site. The example code includes the graphics that are used for the view that will show your level. The application itself is very simple. Essentially, you have a background image that contains a set of crosshairs. The crosshairs represent "level." Over the top of this is a semitransparent, green circle whose position is updated based on input from the accelerometer. You will use the `center` property of the circle image to update its position.

Accessing the UIAccelerometer in viewDidLoad

The important sections of code that you will look at here are the view controller's `viewDidLoad` method, where you are initially accessing the accelerometer, and setting the view controller as your `UIAccelerometerDelegate`. The `UIAccelerometerDelegate` is the object that actually receives the updates from the accelerometer as they occur. By setting it to your view controller, you can easily set the center position of your circle accordingly.

The `viewDidLoad` method is shown in Listing 31.1. The only thing that's really significant about it is your setting of the view controller as the delegate to the `UIAccelerometer` object.

Listing 31.1

Implementation of the viewDidLoad Method

```
- (void)viewDidLoad
{
    origX = [circle center].x;
    origY = [circle center].y;
    [[UIAccelerometer sharedAccelerometer] setDelegate:self];
    [super viewDidLoad];
}
```

You should also make it a point here to store the current center location of the circle. This is simply so that you can return to the center location when the accelerometer is zeroed out.

Implementing the UIAccelerometerDelegate method

The `UIAccelerometerDelegate` protocol defines only one method. That method is `accelerometer:didAccelerate:` and is called based on the update interval that you configure on the `UIAccelerometer`. In your case, the implementation of this delegate method is shown in Listing 31.2.

Listing 31.2

Implementation of the Delegate

```
- (void)accelerometer:(UIAccelerometer *)accelerometer
      didAccelerate:(UIAcceleration *)acceleration
{
    float newX = origX - ([acceleration x] * 100);
    float newY = origY + ([acceleration y] * 100);
    [circle setCenter:CGPointMake(newX, newY)];
}
```

The implementation of this method is extremely simple. All you do is take the `UIAcceleration` object that you are given, normalize the value received from it by multiplying it by 100, and then add or subtract that value to or from your original X and Y values to arrive at your new position values.

You multiply the values by 100 so that you have a new location that is appropriately far away from the center point. If you didn't multiply the values by 100, the circle would barely move at all, as the values that you receive from the accelerometer, just by tilting, range from approximately −1.0 to +1.0.

Finally, you update the center property on your circle object. This causes the circle to move as the accelerometer gives you data.

Go ahead and run the example application with this implementation of the accelerometer delegate method, and observe its behavior.

Now, you may notice that the values that the accelerometer gives you can be slightly spiky. That is to say, if you run this code exactly as I've shown you here, you'll notice that the circle may tend to jump around a little bit and have some jitter. When working with the accelerometer, it can be helpful to filter out this jitter to get a smoother signal.

So let's add a low-pass filter to your code to smooth out that jitter.

Adding a low-pass filter

A low-pass filter simply takes a signal and averages the current sample of that signal with the previous sample of that signal, assigning weights to both the old value and the new value. Using it, you can smooth out a signal because outliers in the signal tend to be averaged back toward the other samples in the signal. In other words, a signal with values of 5, 5, 6, 20, 5, when passed through a low-pass filter, reduces the filtered value of 20 to something more in line with the other values.

In your case, you're going to implement a very simple low-pass filter. The code for that filter is shown in Listing 31.3. Essentially, it adjusts new values such that their value is equal to 60 percent of the new sample, plus 40 percent of the old sample. In this way, new samples can't easily jump to considerably higher or lower values.

Listing 31.3

A Low-Pass Filter

```
-(float)lowPassFilter:(float)inNewValue withLastValue:(float)inLastValue;
{
    float factor = 0.4;
    return (inNewValue * factor + inLastValue * (1 - factor));
}
```

This implementation requires that your delegate method also have some changes. The changes are shown in Listing 31.4.

Listing 31.4

Updated Delegate Method

```
- (void)accelerometer:(UIAccelerometer *)accelerometer
        didAccelerate:(UIAcceleration *)acceleration
{
    float xAccel =
        [self lowPassFilter:[acceleration x] withLastValue:lastX];
    float yAccel =
        [self lowPassFilter:[acceleration y] withLastValue:lastY];
    lastX = xAccel;
    lastY = yAccel;

    float newX = origX - (xAccel * 100);
    float newY = origY + (yAccel * 100);
    [circle setCenter:CGPointMake(newX, newY)];
}
```

Most importantly, you take the accelerometer values and pass them through the low-pass filter, as well as passing the last measured value for each of the vectors you are interested in. You then store your new adjusted values in your last values variables so that you can use them the next time you receive new vectors. The rest of the code stays the same.

If you run the application again, you should see that the signal is smoothed out somewhat. Feel free to tweak the factor value in the low-pass filter to see what kind of effect it has on the application. Another way that you could smooth it out might be to use core animation in updating your position.

CROSS-REF
See Chapter 14 for more information about using core animation.

Now, to make this an application, all you need to do is improve the graphics. You now have in your hand a bubble level that you can use in woodworking projects.

Understanding the UIAcceleration object

You saw in the previous example that the `UIAccelerometerDelegate` receives the notifications of events from the accelerometer via its delegate method. One of the parameters to that delegate method is the `UIAcceleration` object, which contains vital information regarding the current acceleration event that you are receiving. This object contains vectors for the X, Y, and Z forces currently being experienced by the device. Additionally, it contains a timestamp indicating when the measurement took place. This timestamp could be used to do filtering similar to what you did with your low-pass filter.

Capturing Shake Events

You've now seen how to use the accelerometer directly. You can use the capabilities that I've already demonstrated in this chapter when building a game or something where you need to have a fine level of input from the accelerometer. In previous chapters, you've also seen how you can respond to accelerometer events such as tilting the phone from portrait to landscape mode in your view controller.

In iPhone OS 3, however, Apple has introduced a new event implemented on the `UIResponder` class specifically for the purpose of intercepting motion events. As of this writing, the only motion events that are currently available to be captured using this mechanism are shaking events. However, the fact that this capability is on `UIResponder` means that any custom view is able to intercept these events just as if they were intercepting touch events. It's as simple as implementing the appropriate methods on your custom views.

So, in other words, any UI view that you implement can easily detect when the user begins shaking the device and when they stop shaking the device. You do this by implementing the methods `motionBegan:withEvent:` and `motionEnded:withEvent:` on your custom `UIView`. These are similar to the methods `touchesBegan:withEvent:` and `touchesEnded:withEvent:`, except that the event object passed to the method is slightly simpler, as it does not contain a set of touches or motions; it only tells you what type of motion event occurred.

NOTE

The shake event is being implemented as a standard metaphor throughout iPhone OS as "undo," so it's a useful capability to utilize in your own applications.

Let's see how to implement these methods with an example application.

Building a "shake to break" detector

The application that you're going to build here is basically a shake detector. It will have a small, square, custom view in the middle of the screen. Under normal operations, the square will be blue. If the user shakes the phone, the square will turn red. Figure 31.3 shows the completed application.

Implementing drawRect

To build this application, you need to create a custom `UIView`, which forms the square in the middle of the screen. This custom `UIView`'s `drawRect:` method is shown in Listing 31.5.

Listing 31.5

Implementation of the drawRect: Method

```
- (void)drawRect:(CGRect)rect
{
    CGContextRef ctxt = UIGraphicsGetCurrentContext();
    CGContextSetStrokeColorWithColor(ctxt, [[UIColor blackColor] CGColor]);
    CGContextSetFillColorWithColor(ctxt, [currentColor CGColor]);
    CGContextStrokeRect(ctxt, rect);
    CGContextFillRect(ctxt, rect);
}
```

Figure 31.3

The "shake to break" application

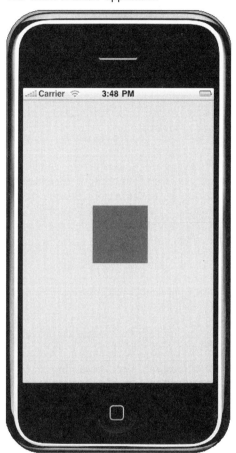

Essentially, you are simply filling the rectangle of the view with whatever color is appropriate. The color is set to blue initially, and then becomes red after some motion events are received.

CAUTION

In order to receive motion events, your view must be the first responder. So be sure to override `canBecomeFirstResponder` on your custom `UIView` so it returns YES and call `becomeFirstResponder` after your view loads.

Implementing motionBegan and motionEnded

The `motionBegan` and `motionEnded` methods are simple to implement. Essentially, you receive the former when the shaking of the device begins, and you receive the latter after it is completed. So, in your `motionBegan` method, you set the color to red, and then when the shaking stops, you set the color back to blue. The code for these two methods is shown in Listing 31.6.

Listing 31.6

Implementing the Motion Detection Events on a Custom UIView

```
- (void)motionBegan:(UIEventSubtype)motion withEvent:(UIEvent *)event
{
    [self setCurrentColor:[UIColor redColor]];
    [self setNeedsDisplay];
}
- (void)motionEnded:(UIEventSubtype)motion withEvent:(UIEvent *)event
{
    [self setCurrentColor:[UIColor blueColor]];
    [self setNeedsDisplay];
}
```

Cancellation of motion events

It is possible to have your motion events cancelled in various ways: by another UI element intercepting the events, by a phone call, or by another action that could make your application no longer active. If this occurs, your view receives a call to `motionCancelled:withEvent:`. Listing 31.7 shows an implementation of this method. In your case, you're going to treat it just like motion ended: you're going to set the color back to blue.

Listing 31.7

Implementation of motionCancelled:withEvent:

```
- (void)motionCancelled:(UIEventSubtype)motion withEvent:(UIEvent *)event
{
    [self setCurrentColor:[UIColor blueColor]];
    [self setNeedsDisplay];
}
```

Once you've implemented this method, you can run the application, shake your phone, and see the square turn from blue to red. When you stop shaking, it returns to blue.

Summary

So now you've seen how to work with the accelerometer. You've seen how to detect changes in the orientation of the device. And you've also seen how you can detect shake events in your custom views. Enabling applications to know more about their surroundings via things such as the accelerometer and the built-in compass leads to applications with better environmental and user awareness.

If you can imagine what it would be like to not be able to see, not be able to touch, not be able to hear, and not be able to sense whether or not you were standing up straight, then you can imagine what it's like being a typical computing device today. Most devices have none of these capabilities. You can then imagine how difficult it would be to work with other people if you were so constrained.

The iPhone, on the other hand, is one of the first handheld devices to open the door to alternative input devices such as the accelerometer, its camera, its microphone, and so on. By leveraging these technologies, you can build applications that are more aware of their surroundings than any applications available on any other platform. This is a powerful concept and one to consider when you're thinking of ideas for iPhone applications.

Interfacing with Peripherals

A new feature in iPhone OS 3 is support for interfacing with third-party accessories through the dock connector or over Bluetooth. The framework responsible for doing this is called the External Accessory framework, and it is this framework that this chapter will discuss.

Creating Accessories That Interface with iPhone

In order for an accessory to interface with iPhone OS, it has to support some form of communication protocol. iPhone OS does not specify the types of protocols that are required to be implemented in order to communicate. When interfacing with an accessory, the API internal to the application is simply a stream API. The data that is streamed (the protocol itself) is entirely up to the developer of the accessory and the developer of the application. The accessory does need to declare what protocols it supports, and the application, when looking for accessories to communicate with, should look at the accessory information to be certain that the accessory supports the protocol it wants to use to communicate with it.

In This Chapter

Listing accessories connected to the device through Bluetooth or the dock connector

Connecting to devices using the External Accessories framework

Talking to accessories using streams

Your application must declare the types of protocols that it supports. To do this, you need to add to your `Info.plist` file a key of `UISupportedExternalAccessoryProtocols`, which should be an array of strings indicating the protocols that your application supports. The protocol identifier strings should be in reverse DNS notation to avoid naming conflicts. If your application declares that it supports a particular protocol, when an accessory requesting that protocol is connected, iPhone OS attempts to launch your application in response. If no application is found supporting the protocol that the accessory has requested, iPhone OS may launch the App Store to recommend applications that do support that protocol.

Finding Accessories That Are Connected Using EAAccessoryManager

Once your application is running, to find connected accessories you use the `EAAccessory Manager`. This class provides a list of connected accessories that your application can communicate with. The expectation is that you will retrieve the list of connected accessories and then check each one's supported protocol list for a protocol that you can use to communicate with it. Once you have found a device that you can communicate with, you can then communicate with it using the `EASession` class.

NOTE
To use the External Accessory framework, remember that you must link it with your project and import its headers.

Listing 32.1 shows how you use the `EAAccessoryManager` to find a device that you can communicate with based on a particular protocol.

Listing 32.1

Finding a Device That Supports Your Protocol

```
{
    NSMutableArray *accessoriesToTalkTo = [NSMutableArray array];
    EAAccessoryManager *manager =
            [EAAccessoryManager sharedAccessoryManager];
    for(EAAccessory *accessory in [manager connectedAccessories])
    {
        if([[accessory protocolStrings] containsObject:myProtocol])
        {
            [accessoriesToTalkTo addObject:accessory];
        }
    }
    // display the list to the user to choose...
}
```

IPhone OS allows you to talk to only one accessory per protocol. If multiple accessories are connected that support the same protocol, your application needs to decide which one to talk to. In this example code, you are simply making a list of the devices that support your protocol. When finished, it displays the list to the user and allows her to select one.

Note that the `EAAccessoryManager` is a singleton and therefore should be accessed using the singleton accessor method.

Understanding the EAAccessory Class

The `EAAccessory` class essentially contains metadata about the connected accessory. It also allows you to find out information about the accessory, such as the accessory name, the manufacturer, model number, serial number, and so on. Additionally, the `EAAccessory` also has a delegate protocol that can tell you if and when the accessory is disconnected. It contains only one method: `accessoryDidDisconnect:`.

Once you have the accessory object, however, it enables you to create an `EASession` object.

Working with EASession

The `EASession` object is the primary means with which you establish communication with the device. To create an `EASession` object, you must previously have acquired an `EAAccessory` object, typically from the `EAAccessoryManager`.

Once you have an `EASession` object, it provides you with a read stream and a write stream, which you can then use to communicate with the device. Again, the External Accessory framework does not provide you with support for the protocol that must be used to communicate with the accessory. Your application has to provide the implementation of that protocol. The External Accessory framework only provides the communication channel.

Listing 32.2 shows how to get the read and write streams from an EASession object.

Listing 32.2

Accessing the Read and Write Streams on an EASession Object

```
{
    EASession *session = [[EASession alloc] initWithAccessory:accessory
                                        forProtocol:myProtocol];

    NSInputStream *inputStream = [session inputStream];
    NSOutputStream *outputStream = [session outputStream];
    [inputStream setDelegate:self];
    [outputStream setDelegate:self];
    [inputStream scheduleInRunLoop:[NSRunLoop currentRunLoop]
                    forMode:NSDefaultRunLoopMode];
    [outputStream scheduleInRunLoop:[NSRunLoop currentRunLoop]
                    forMode:NSDefaultRunLoopMode];

    [inputStream open];
    [outputStream open];

    [session release];
}
```

In this particular case, you are acquiring both an input and an output stream and setting yourself as the delegate for both of them. You schedule the streams to run in the default run loop, and then you open them. Once they're over, you are able to receive data on the input stream and send data on the output stream.

Subsequently, you don't actually have to keep the `EASession` object around. Once the `NSStream` objects have been acquired, it's perfectly acceptable to release the `EASession`.

Talking to Your Device Using NSStreams

Once you have acquired your `NSStream` and set yourself as the delegate, it's a simple matter of implementing the `NSStream` delegate protocol so that you can receive events from the streams.

Listing 32.3 shows the implementation of one of the `NSStream` delegate methods. This method is triggered when an event is received for the `NSStream`. In this particular case, this method is receiving an event indicating that there is data to be read on the stream.

Listing 32.3

Implementing the NSStream Delegate Method

```
- (void)stream:(NSStream *)theStream handleEvent:(NSStreamEvent)streamEvent
{
    if(streamEvent == NSStreamHasBytesAvailable)
    {
        char buf[1024];
        [inputStream read:buf maxLength:1024];
        // do something with the data...
    }
}
```

Obviously, there may be better ways to handle this input. However, this shows you the basic mechanism that is used to read data from the accessory.

It is also simple to write data to the accessory. Listing 32.4 shows an example of how to do this.

Listing 32.4

Writing Data to the Accessory

```
-(void)writeStringToAccessory:(NSString *)inString;
{
    NSData *data = [inString dataUsingEncoding:NSUTF8StringEncoding];

    [outputStream write:[data bytes] maxLength:[data length]];
}
```

If all the data cannot be written, then the `write` method returns the number of bytes that could actually be written. It is up to your application to adjust your buffer accordingly and finish sending the remaining data when you receive an event from the output stream indicating that it can now accept more bytes.

Summary

It's beyond the scope of this book to go into a lot of detail about the various hardware protocols that you might need in order to implement communication with accessories. However, in order to get to the point where you can implement those protocols, you first have to understand how to connect two devices using the iPhone OS API. It is this API that I have introduced to you here.

By using the External Accessory framework, you can communicate with a variety of accessories both through the dock connector and over Bluetooth. For more information, visit the iPhone developer portal.

Handling Distribution

In This Part

Chapter 33
Code Signing Your Apps

Chapter 34
Expanding Your Application Using the In-App Purchase Support

Code Signing Your Apps

So, you've done it. You've learned it all, conquered iPhone OS 3, and you've written your killer app. Now there's just one more thing to learn: how to build an executable that can be uploaded to the App Store.

In the early days of iPhone OS development, this step was perhaps the most frightening to developers. They often felt like it was such a complicated process and so difficult to get right that they would have to resort to voodoo to ensure that the development gods were smiling upon them. They would joke about sacrificing a chicken, or about surrounding their computers with circles of ash, to ward off the evil development spirits. All of this for what should be one of the simpler parts of developing iPhone applications: the process of code signing.

Thankfully, newer versions of the iPhone SDK have improved the code signing process significantly. It's much easier and much simpler to understand today than it was then.

In this chapter, I'm going to guide you through the process of getting a development certificate, installing it in Xcode, and then building your application for distribution.

I'll also go over how to set up an ad hoc build, which you can use to distribute executables to beta testers who do not have Xcode installed.

In This Chapter

Understanding the provisioning process

Understanding the difference between development, distribution, and ad hoc provisioning

Distributing ad hoc builds

Acquiring a Development Certificate

Before you can even think about distributing your application, you need to acquire a development certificate from Apple. To do this, you first have to visit the iPhone developer portal at http://developer.apple.com/iphone. Once there, you can follow their instructions for acquiring a development certificate.

The development certificate consists of an electronic document that you digitally sign using an encryption key on your computer. You then generate a certificate-signing request for that document, and send the certificate-signing request to Apple for them to sign. Once they have signed it, you download the signed version and install it in Xcode.

The signed certificate, once installed in Xcode, is used to sign your applications. Only signed applications can be installed on an iPhone OS device. I will talk shortly about the provisioning process that determines how an application can be distributed, but for now, simply understand that all applications that are installed on iPhone OS devices must be signed.

To generate your original certificate, you use the Keychain Access application in Mac OS X. Using it, you request a certificate from a certificate authority. Doing this generates a private key and a public key. Your private key is stored on your computer and should be backed up. If your private key is lost, you can no longer sign apps.

The iPhone portal has excellent step-by-step instructions for generating this certificate. I suggest that you read through the following few sections so that you gain a better understanding of how the provisioning process works, and then visit the iPhone portal and follow their instructions to generate your development certificate. Once you have your development certificate, you can return to this book and read how to install your development certificate in Xcode.

Understanding the Provisioning Process

A provisioning profile is a collection of digital certificates and device IDs that specify that a particular signing certificate can be used to sign an application and install it on a particular device.

The provisioning process involves entering the unique device ID of devices you want to install your application on into the iPhone development portal. You then use these device IDs, and your certificates, to generate a provisioning profile. The provisioning profile is a file that you download from the iPhone developer portal. After you download the file, you install it in Xcode.

CAUTION
You are limited to 100 device IDs that can be entered into the development portal. You cannot remove or overwrite them, but you can reset your device list once a year.

Understanding development, distribution, and ad hoc

In the realm of iPhone development, there are essentially three types of provisioning profiles.

The first is called the *development provisioning profile*. This provisioning profile is used in conjunction with your development certificate to sign your applications and install them on your development devices. They're meant to be used with only a few devices. When you create the development provisioning profile, you specify the devices that are included in it. The generated profile can only be used in conjunction with those devices. Development provisioning profiles can also only be used with Xcode and only on the machine containing the development certificate that was used to generate the profile. This is the provisioning profile that you use for your day-to-day development.

The second type of provisioning profile is the *distribution provisioning profile*. The distribution provisioning profile is meant to be used for distributing your application on the App Store. Applications signed with a distribution provisioning profile cannot be installed on a test device. They can only be uploaded to the App Store and installed through a purchase.

The third type of provisioning profile is an *ad hoc provisioning profile*. An ad hoc provisioning profile is like a cross between a development provisioning profile and a distribution provisioning profile. That is to say, you can use it to install applications on specific test devices, but the installation process does not require Xcode. Ordinary users can install the application using iTunes. It still requires that devices that you want to install the application on must be included in the provisioning profile. However, you can put as many as 100 devices on an ad hoc provisioning profile. Additionally, when you send the user the application, you also need to send them a copy of the provisioning profile itself. The end user then drags and drops the provisioning profile and the application onto iTunes to install it.

The ad hoc provisioning profile mechanism was originally designed for use in organizations that wanted to do their own distribution of applications and did not want to distribute the applications through the App Store. However, it is commonly used among iPhone developers for doing beta testing. Most beta testers are not developers and do not have the development tools installed. Using an ad hoc provisioning profile enables normal users to install applications using iTunes.

Installing provisioning profiles

In order for an application to run on a particular device, it must contain a provisioning profile for that application. All iPhone OS devices already contain provisioning profiles for applications from the App Store. In order to use your development provisioning profile or ad hoc provisioning profile with a device, you have to install it.

There are two applications that can install provisioning profiles. The first is Xcode. To install a provisioning profile using Xcode, you simply drag-and-drop the provisioning profile file from the Finder into Xcode. Once you do this, the provisioning profile is available in the Xcode organizer. If you configure an application to build with that profile, it automatically installs the profile when you build and run the application with the device connected.

Figure 33.1 shows the Xcode window with the organizer displayed. The provisioning profile group on the left side allows you to manage your provisioning profiles that are installed within Xcode. When a device is connected, you can also see the provisioning profiles that are installed on that particular device. You can also manually add or remove provisioning profiles from that device using this interface.

Figure 33.1

The Xcode organizer window

The second application that can be used to install provisioning profiles is iTunes. iTunes has significantly less functionality when it comes to managing provisioning profiles, but it is much easier to use for nonprogrammers. To install a provisioning profile using iTunes, you simply connect your iPhone OS device to your computer, start iTunes, and drag-and-drop the provisioning profile from the Finder to iTunes. This installs the provisioning profile onto your device as well as into iTunes.

Exploring what happens when an app is signed

All iPhone applications must be signed with a provisioning profile before they can be installed. Code signing is the last step of building an application for iPhone OS. When an application is signed, Xcode uses the certificate associated with the currently active provisioning profile to stamp the executable digitally with the information from the provisioning profile. This makes it so that the application can only be installed on devices that have the provisioning profile installed.

You can configure different Xcode targets for different provisioning profiles. A typical project contains a development target, which is used for installing and testing the application on the developer's devices; a distribution target, which is used for distribution on the App Store; and sometimes, an ad hoc target, which is used for distributing the application to beta testers. In the next sections, I will show you the settings necessary for preparing your build to be signed with a provisioning profile.

Setting up your build to be signed

Once you have your provisioning profile installed, you have to configure your build to use it. Figure 33.2 shows the Xcode build settings window. In the middle of the window are the settings for Code Signing.

Figure 33.2

Xcode build settings

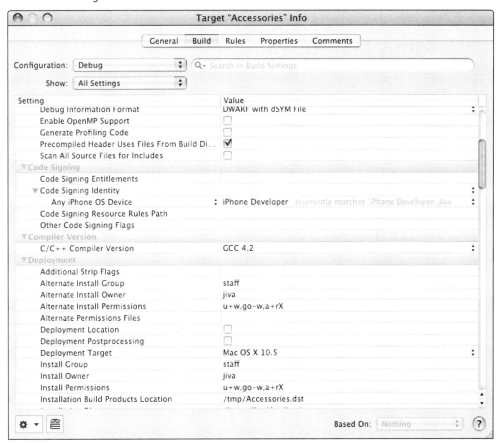

The main setting that influences what provisioning profile will be used is the setting for Code Signing Identity. There are two wildcard settings for this: "iPhone Developer," which matches your development provisioning profile, and "iPhone Distribution," which matches your distribution provisioning profile. If you need to, you can also specify a specific provisioning profile by clicking this drop-down list and selecting a specific provisioning profile instance. In cases where you might have multiple development profiles installed, or when you want to do an ad hoc build, you need to explicitly set the profile. For typical use, however, the wildcard settings are usually adequate.

NOTE
The ad hoc provisioning profile is, in fact, a distribution profile.

The only other setting you may have occasion to change would be the Code Signing Entitlements setting, which I will discuss shortly.

Doing Ad Hoc Builds

Ad hoc builds allow you to distribute a build directly to end users without going through the App Store. You can provision up to 100 devices using an ad hoc provisioning profile.

Doing an ad hoc build requires special configuration of your build, and distributing the provisioning profile to your users along with the application.

Configuring the build in Xcode

Configuring a build for ad hoc distribution involves several steps. First, you should make a copy of whatever build settings you want to use with your ad hoc build. This is easily done in the Configurations tab of your Project Information window, by clicking the Duplicate button for whatever build configuration you want to use. For beta testing, you will probably duplicate the Debug build settings. Once you have your new ad hoc build configuration, you should choose it as your active configuration.

Figure 33.3 shows the build configurations tab with an ad hoc build.

Ad hoc builds require a special property list to be included with the build. This property list is called the `Entitlements` list. Typically, it's called `Entitlements.plist` and should be added to the resources group of your project.

Figure 33.3

The build configurations tab

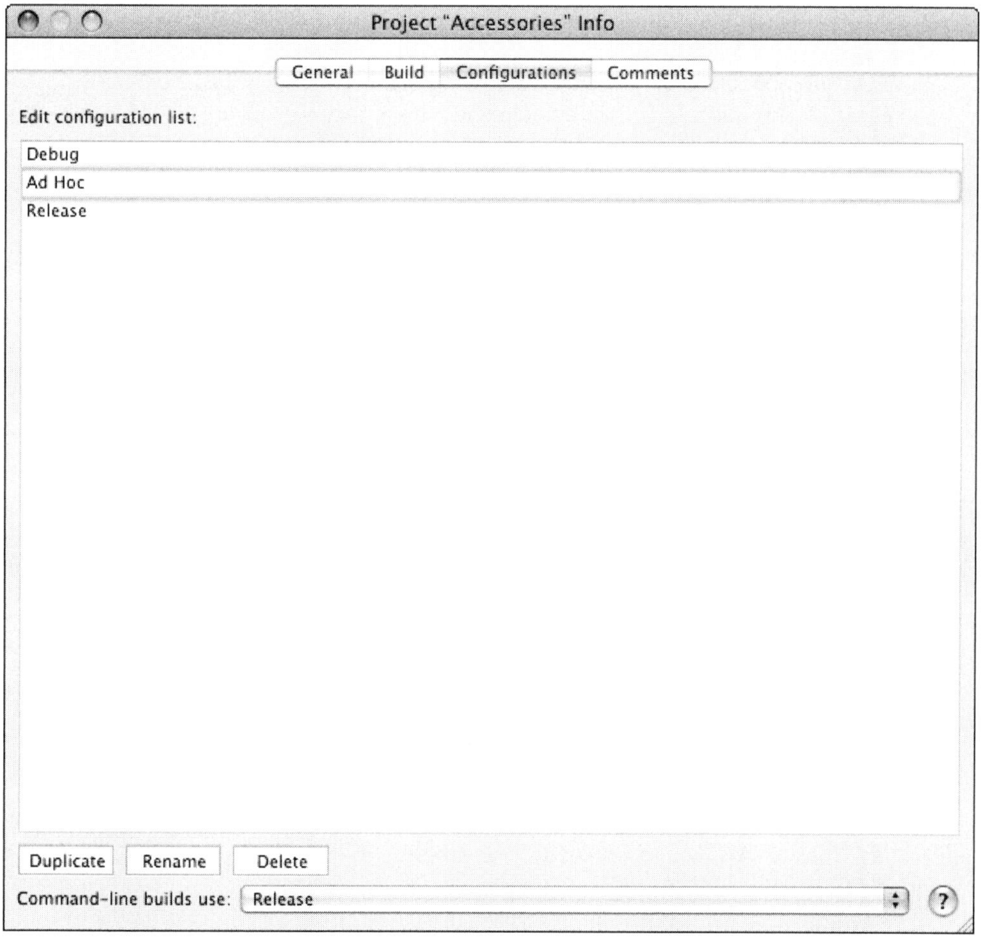

The `Entitlements.plist` file should contain a single property called `get-task-allow`. This property should be a Boolean value and should be set to FALSE. This flag configures whether your application allows the debugger to be attached to it while it is running. Distribution profiles (and remember, ad hoc is considered distribution) require that this flag be set to FALSE.

Once you have the `Entitlements.plist`, you should configure your ad hoc target and set the Code Signing Entitlements property to be the name of the Entitlements property list file, or in the previous example, `Entitlements.plist`. You should also configure your Code Signing Identity to your ad hoc provisioning profile.

If you've configured everything correctly (and sacrificed the requisite chicken), you should be able to build your app. This version of your app is signed with your ad hoc provisioning profile and can be installed on any of the devices listed within it.

Distributing to users

Distributing an ad hoc build to a user involves distributing the application bundle and the provisioning profile to the user. The user then needs to install them both, using iTunes.

When installing the app and provisioning profile, the provisioning profile should be installed first. The user should drag the provisioning profile to iTunes and then drag the application bundle to iTunes. If it's installed correctly, the user then sees the application in the Applications group in iTunes.

NOTE
iTunes does a version check and alerts the user if the version they are installing is older than or the same version as the one that is currently installed. Because of this, you may want to increase the version number when distributing new ad hoc builds.

Summary

This chapter has enlightened you a bit about how the code signing process and distribution work. Code signing is one of the most complicated parts of iPhone development. Fully understanding the terms used, and the process involved, can help make it easier to work with. After reading this chapter, you should have that additional understanding, which will make your builds smoother and easier.

Expanding Your Application Using the In-App Purchase Support

One of the most exciting new features in iPhone OS 3 for developers is the addition of in-app purchase support. Using it, you can leverage the power of the built-in App Store purchase system to allow the user to purchase additional content for your application. This content can be in the form of additional levels, additional features, and so forth. This capability enables new kinds of business models to be developed around iPhone OS and the App Store.

In this chapter, you're going to take a look at how to work with the in-app purchase system. You're going to see what kinds of products you can sell through your application, and how you build that capability into your application. You're also going to see how you work with the Store Kit API for making purchases.

When you are finished, you should be able to deploy an application that includes in-application purchases.

Knowing the Types of Things You Can Sell

The first thing that you need to know when working with the in-app purchase system is what kind of content you can sell. There are three categories of content that you can sell as in-app purchasable content: non-consumables, consumables, and subscriptions.

Non-consumable content is content that, once purchased, is available permanently to the user. Examples of this kind of content might be new levels for games, or new features in an application. In addition to being available to the user forever, that non-consumable content must also be available to all of the user's other devices. For example, if the user owns both an iPhone and an iPod touch, then non-consumable, purchased content must be available to the user on both devices at no additional charge.

In This Chapter

Learning the business models you can use with In-App purchasing

Understanding the types of things you can sell

Integrating a store with your app

Making purchases using Store Kit

Displaying prices in the localized currency

The second category, consumables, represents items that can be used within the application. Consumable items need not be available on all of the user's devices. Consumable items are something that the user might purchase and use within the application, and then when he has used them up, he would purchase more. Examples of consumable items would be ammunition or poker chips.

The third and final category is the subscription category. Subscriptions work just like consumables, except that they need to be available on all of the user's devices. So, for example, a user might purchase a subscription to a magazine for a period of one month. That subscription should be available to the user in the application on all of the user's devices.

CAUTION
The term *subscription* may suggest to you that a repeating billable cycle is part of the capability built into the App Store. This is not the case; the App Store provides no support for repeated billing.

So, the first thing to realize when considering adding support for in-app purchases to your application is that the content that you're selling must fall into one of those three categories.

The second thing to realize when working with in-app purchases is that, as usual, Apple enforces certain rules with regard to the content that you can sell.

I could list off different kinds of things that Apple allows and disallows. But most importantly, I think it's important to keep one key rule in mind. The main goal that Apple has with regard to selling content through your application is that the content that you're selling must be delivered through the application to the device. This means that anything that is not delivered through the device is not appropriate to be sold using the in-app purchase system.

For example, Apple has specifically said that they will not allow applications to sell real-world goods. For example, you cannot make an application that allows you to sell a book that is then delivered via UPS to someone's house. On the other hand, you can make an application that sells a book that is then delivered electronically, directly to the user's device. Again, the goal is that anything that you sell should be deliverable to, and directly usable on, the device.

Working with Unlockable Content

Once the purchase is made on the device, it is up to you to unlock the content that the user has purchased. Where that content comes from is up to you. It's important to remember, however, that the old rules still apply; you cannot modify your existing executable. Therefore, whatever new content you are delivering must be either non-executable or unlockable through configuration.

The unlocked content can already be compiled into your application or bundled in your application bundle. Alternatively, it can be content that you download from the Internet. If you download it, however, realize that you cannot download executable content to the device.

Let's take a look at a few examples so that you get a feel for how it works.

Let's say that you have a racing game that ships with five tracks. You can sell additional tracks using the in-app purchase system. The additional tracks might already be bundled as part of your application. In this model, the application need not even contact your server to complete the purchase. The Store Kit can perform the entire purchase on the device and unlock the levels that are already there.

As a second example, let's say that you have an e-book reader. Perhaps you sell the e-book reader with one book already unlocked, but you want to give the user the ability to purchase additional books inside the app. In this case, your catalog of books might actually exist on your server. When the user purchases a book, Store Kit performs the purchase and then sends the receipt to your server, and your server gives the application the ability to download the content that the user has purchased.

As a third and final example, consider the case where the user purchases the ability to save a document to your server for a period of time. In this model, the user makes the purchase using the Store Kit on the device and then sends the receipt to your server; your server then enables that functionality within the application.

Although the Store Kit does maintain a record of the user's purchases that can be retrieved at any time, it is up to you to maintain whatever persistence mechanism is necessary to show that the user has unlocked a particular purchase. In order to download the list of the user's existing purchases, the application has to prompt the user for her App Store username and password. Therefore, you can't use this as a mechanism for determining what has and has not been purchased on that particular device. Instead, it is better to use something like `NSUserDefaults` to store this information. In the event that the user causes her `NSUserDefaults` to be reset, either through restoring her iPhone to factory settings or deleting and reinstalling your application, the Store Kit provides a mechanism for retrieving all of the user's prior purchases, and you should use this mechanism to update your settings and re-unlock whatever content the user has purchased.

Setting up Purchasable Content in iTunes Connect

Now that you understand the types of things that you can sell, the final stage before you actually start implementing code is to set up your products in iTunes Connect. However, before you do that, you need to gather some specific information about the products that you're going to sell.

Firstly, you obviously need a name for what you're going to sell. The name is simply for your reference. The user does not necessarily see it, but you need it for setting up your product and iTunes Connect.

Secondly, you need a description for your product. Again, this is something that the user may not necessarily see it, but it is needed to set up the product in iTunes Connect.

The third item that you need is a price tier for your application content. The price tiers used in iTunes Connect are the same as those used for applications. You should already be familiar with these tiers and how they work.

Finally, the last item that you need is a unique item ID. In order to ensure that these IDs are unique, Apple recommends that you use reverse DNS notation, for example, "com.example-company.product-name.content-name." This unique ID appears on your sales reports and allows Apple to track purchases of this particular item.

Once you have all of this information, you log into iTunes Connect and visit the "Manage Your Applications" link. Within this area, each application has a button for Managing in App Purchases. Using this, you configure each of your products that you want to have available for in-app purchase. Once you have configured your products, it's time to begin building support for your purchases into your application. To do this, you work with the Store Kit API, and it is this framework that I will focus on for the remainder of this chapter.

Working with Store Kit

Now that you've figured out what you're going to sell, and set it up in iTunes Connect, you can begin integrating the Store Kit framework with your application. Remember that, in order to use the Store Kit, you have to first add it as a linked framework to your application build.

The Store Kit framework consists of a group of classes that enable you to interact with the App Store. The classes primarily center on interacting with payments, products, and the record of transactions the user has already completed. I will discuss each of these in the following sections.

The process for actually accepting payments consists of five distinct parts: verification of the ability to purchase items; displaying the items that can be purchased; requesting payment; verifying the transaction; and finally, unlocking the content. Additionally, you should provide a mechanism within the application to restore purchases, as I have previously mentioned.

Let's step through each one of these parts now.

Verifying app purchase availability

Before you even think about allowing the user to purchase within your application, you need to verify that they have not disabled in-app purchases. iPhone OS 3 has the ability to enable or disable purchases from within applications. If the user has disabled this capability, either you should inform them that in order to purchase additional content they need to enable it or you should simply not display the ability to purchase additional content at all.

To do this, you use the `SKPaymentQueue` class method `canMakePayments`. If in-app purchases are disabled, this method returns NO.

Listing 34.1 shows how to use this method.

Listing 34.1

Determining if in-app Purchases Are Enabled

```
-(void)presentStore;
{
    if([SKPaymentQueue canMakePayments])
    {
        // display the store
    }
    else
    {
        // inform the user they can't make purchases.
    }
}
```

Once you have determined that the user can utilize the in-app purchase system, you should present the user with the products you have for sale.

Presenting your store

The implementation of how you present your products to the end user for sale is very specific to your application and your user interface. There is no right or wrong way to present the products to the user. A game might have a simple dialog to prompt the user to purchase additional levels. An e-book reader might have a complicated catalog with previews of books and so on. How you implement your store is entirely up to you.

That said, however, you should still use the Store Kit API to retrieve the details of your products for displaying to the end user. This is because you have to remember that the iPhone is sold in dozens of countries, with a variety of localized currencies. You want to present to the user the appropriate localized price for the product they are purchasing. Therefore, you should request the list of products from the Store Kit API and use the localized information from that list for displaying your catalog.

Listing 34.2 shows how to use the Store Kit API to retrieve localized information about your products prior to displaying your store.

Listing 34.2

Getting Localized Pricing Information

```
-(void)retrieveProductInfo;
{
    NSSet *productIds = [NSSet setWithObjects:
                            @"com.company.product1",
                            @"com.company.product2",
                            @"com.company.product3",
                            nil];
    SKProductsRequest *catalogRequest =
        [[SKProductsRequest alloc] initWithProductIdentifiers:productIds];
    [catalogRequest setDelegate:self];
    [catalogRequest start];
    [catalogRequest release];
}
- (void)productsRequest:(SKProductsRequest *)request
    didReceiveResponse:(SKProductsResponse *)response
{
    // formatting the price can be done with a number formatter to get
    // a properly localized price string
    NSNumberFormatter *numberFormatter = [[NSNumberFormatter alloc] init];
    [numberFormatter setFormatterBehavior:NSNumberFormatterBehavior10_4];
    [numberFormatter setNumberStyle:NSNumberFormatterCurrencyStyle];
    NSArray *products = [response products];
    // display to the user...
    for(SKProduct *product in products)
    {
        NSString *desc = [product localizedDescription];
        NSString *name = [product localizedTitle];

        [numberFormatter setLocale:[product priceLocale]];
        NSString *price =
                [numberFormatter stringFromNumber:[product price]];
        // populate the UI.
    }
    [numberFormatter release];
}
```

Like most other things, accessing the App Store is done asynchronously. Therefore, you initialize your product request with the product identifiers that you want to retrieve information about, and then you start the request. Once the request is started, your delegate method is called when it has completed.

When the delegate method is called, the parameter to the delegate method is of type `SKProductsResponse`, which contains a list of your products. That list contains objects of type `SKProduct`. The `SKProduct` class contains methods you can use to get localized information about your products, including the correct localized price. As shown in this code, you should use a number formatter to format the actual price value into the appropriate format for the user's locale.

Making the purchase

Actually making the purchase in code is simple. You instantiate an instance of `SKPayment` and then add it to the default `SKPaymentQueue`. This is shown in Listing 34.3.

Listing 34.3

Sending a Payment to the Store

```
SKMutablePayment *payment = [SKMutablePayment paymentWithProductIdentifier:
                             @"com.example.product1"];
// optionally, set the quantity if needed - default is 1.
payment.quantity = 2;
[[SKPaymentQueue defaultQueue] addPayment:payment];
```

Processing the payment

In order to actually receive a payment, your application needs to have an instance of a transaction observer added to the default `SKPaymentQueue` object. This transaction observer must implement the `SKPaymentTransactionObserver` protocol and is responsible for actually unlocking the content purchased by the user.

Probably the best place to implement this observer is in your application's initialization. This is a good place to do it because the Store Kit remembers transactions that are in process, even if the application should crash or be interrupted before the transaction has completed. Therefore, when your application starts up and reinitializes, it should be prepared to handle whatever completed transactions may still be in the queue.

Listing 34.4 shows an implementation of `applicationDidFinishLaunching:` where the application delegate is being set as the transaction observer.

Listing 34.4
Setting up the Transaction Observer

```
- (void)applicationDidFinishLaunching:(UIApplication *)application
{
    [[SKPaymentQueue defaultQueue] addTransactionObserver:self];

    [window addSubview:viewController.view];
    [window makeKeyAndVisible];
}
```

Once the transaction observer has been set, that object must implement the appropriate delegate methods for handling the transactions as they come in.

Listing 34.5 shows how you might implement the most important of the delegate methods in the observer, `paymentQueue:updatedTransactions:`.

Listing 34.5
Receiving the Updated Payments

```
- (void)paymentQueue:(SKPaymentQueue *)queue
 updatedTransactions:(NSArray *)transactions
{
    SKPayment *pmt = nil;
    NSString *productId = nil;
    NSError *error = nil;
    for (SKPaymentTransaction *transaction in transactions)
    {
        switch ([transaction transactionState])
        {
            case SKPaymentTransactionStatePurchased:
                pmt = [transaction payment];
                productId = [pmt productIdentifier];
                // unlock the content purchased
                break;
            case SKPaymentTransactionStateFailed:
                error = [transaction error];
                // inform the user of the failed transaction
                break;
            case SKPaymentTransactionStateRestored:
                pmt = [[transaction originalTransaction] payment];
                productId = [pmt productIdentifier];
                // unlock the content
```

```
                break;
        default:
            break;
    }
    [queue finishTransaction: transaction];
  }
}
```

NOTE
You must mark the transaction as finished in order to not receive notification of that transaction again the next time your app launches. This removes the transaction from the queue until and unless the user later performs a restore.

Remember that this method is called both on completed transactions as they occur and to restore transactions. Therefore, you should differentiate between the two, as I have shown here. The primary difference in handling the two different types of transactions is that for restored transactions, you want to get the product identifier by looking at the original transaction, which is included in the new restored transaction instance.

Verifying the transaction

In cases where the product that the user is purchasing is then downloaded from your server, you need a mechanism for your server to verify that the transaction is valid. In other words, the user has purchased the product on your application. They then send the receipt to your server to retrieve the unlocked data. You need a way to verify that the receipt is valid. Fortunately, Apple provides you with a mechanism to do exactly this.

To verify a receipt from your application, you can send a request to the Apple App Store. This request should be a JSON object containing the receipt data retrieved from the `SKPayment Transaction` object `transactionReceipt` property. This data should be encoded in the JSON object and associated with a key of "receipt-data." Using this JSON object, you post an HTTP POST request to the URL `https://buy.itunes.apple.com/verifyReceipt`. The response from the server is also a JSON object containing a key called "status." If the status is 0, then the receipt is valid. If it is anything else, then the receipt is invalid. The resulting JSON object also contains details of the actual transaction receipt encoded in the "receipt" key.

By using this mechanism, you can ensure that a transaction is valid before providing content from your server to the user.

Unlocking the content

Finally, once you have processed the transaction, and verified that the receipt is valid at your server (if needed), it's time to unlock the content.

Just as when dealing with the catalog user interface, this area of implementation is completely up to you. The important thing is that it must be persisted. The Store Kit framework does not provide any mechanism for persistence of purchases. You should store, either in your `NSUserDefaults` or some other mechanism, a record of exactly what the user has unlocked and purchased. You should check the settings when your application starts up and ensure that whatever content the user is entitled to they have access to.

An important item to note here, however, is that when the transaction is completed, you should unlock the content right away. Users expect their content to be immediately available in the application after purchase. You should not make them wait to receive their new functionality.

Restoring purchased content

The last component of working with in-app purchases is that of restoring purchases when the user wipes his phone or deletes and reinstalls your application. If you choose to provide in-app purchase support, you should also be sure to support restoring the user's purchases. To do this, you should include a Restore Purchases button somewhere in your user interface. When the user pushes the button, you should use the `SKPaymentQueue` instance method, `restoreCompletedTransactions`. This causes your `SKPaymentQueue` delegate to receive new transactions for all purchases the user has ever made with your application. The transactions state is `SKPaymentTransactionStateRestored`, and you are expected to process the transactions as I showed you in the previous section — that is, unlocking the content.

NOTE
Restoring purchases is also how a user can transfer purchases from one device to another without repurchasing the item.

Listing 34.6 shows how this is done.

Listing 34.6

Restoring the User's Purchases

```
[[SKPaymentQueue defaultQueue] restoreCompletedTransactions];
```

NOTE
You might ask the question, "why can't I simply restore completed transactions every time my application starts up?" The answer is that when you request to restore completed transactions, the user is prompted for his App Store username and password. This would be very annoying to have to do every time your application launches. Therefore, only do it in response to the user specifically requesting it.

Understanding In-App Purchasing testing

Finally, there are a few things to understand about testing applications with in-app purchasing.

Obviously, during development, you don't want to actually be accessing the real App Store. It would be better if you could access a sandbox, where you could test everything, including your transactions, without fear of accidentally charging a credit card. Thankfully, Apple has already provided for this as part of the development process related to the App Store. Your application, when signed with your developer certificate, automatically connects only to the sandbox store. When you sign your application with your distribution certificate, it actually connects to the real store. Therefore, you can test without fear when signing with your development certificate.

You do need to configure a test Apple ID for working in the sandbox. Details for how to do this are located in iTunes Connect.

CAUTION
When you're testing on your device using the sandbox, you need to make sure that you log out of your normal account before testing your application. Otherwise, it attempts to use your account to verify your purchases, and it doesn't work.

Additionally, to test your server-side implementation, the URL for verifying receipts should be changed to `https://sandbox.itunes.apple.com/verifyReceipt`.

Summary

In this chapter, I've shown you how to expand your applications with the in-app purchase system. This enables you to explore new business models and customize your applications to your users' needs. Applications can be designed such that they are à la carte, with the user picking and choosing the features that she needs. Alternatively, you can release an application with minimal functionality to start, and then release updates with unlockable additional content. Finally, you can even make applications that work with online services to deliver paid digital content directly to the device.

Using this capability opens a whole new world of business opportunity, one that I hope you will be successful in for years to come.

Index

Special Characters

`@property(nonatomic, readonly)NSTimeInterval timestamp` method, 139
`@property(nonatomic, readonly)NSUInteger tapCount` method, 139
`@property(nonatomic,getter=isSecureTextEntry)BOOLsecureTextEntry` property, 165
`@property(nonatomic)BOOLenablesReturnKey Automatically` property, 165
`@property(nonatomic) UIKeyboardAppearancekeyboard Appearance` property, 165
`@property(nonatomic) UIKeyboardTypekeyboardType` property, 165
`@property(nonatomic) UIReturnKeyTypereturnKeyType` property, 165
`@property(nonatomic) UITextAutocapitalizationType autocapitalizationType` property, 165
`@property(nonatomic)UITextAutocorrectionType autocorrectionType` property, 165
`@synthesize` declaration, 48
- symbol, 202
+ symbol, 202

A

accelerometer, 20, 23
`accelerometer:didAccelerate:` method, 365
`acceptCallID:error:` method, 285
`acceptConnectionFromPeer:error:` method, 282
access, asynchronous, 395
accessory view, 86
`accessoryDidDisconnect:` method, 375
`action-loc-key`, dictionary alerts, 262
ad hoc network, Bluetooth
　`GKPeerPickerController` class, 278
　`GKSession` class, 280
ad hoc provisioning profile, 383
`addAnnotations:` method, 303
`addCity` method, 115
`addFilterPredicate:` method, 326
`addItems:` method, 292
`addressDictionary` property, 305
`addresses` property, 254
`administrativeArea` property, 305
advertising services, Bonjour, 255–256
`albumsQuery` method, 326
`alcCreateContext` method, 318
`alcMakeCurrentContext` method, 318
`alcOpenDevice` method, 318
alert messages, 163
`alGenSources` function, 318
`alListenerfv` function, 318
`alSourcefv` function, 318
`alSourcePlay` function, 318
`alSourceStop` function, 318
animation
　Core Animation
　　`CALayer`, 178–181
　　`UIImageViews`, animating, 181–182
　　`UIViews` animation methods, 171–177
　keyframe, 179
　OpenGL ES, 187–188
　for rotation transition, 43–44
　`UINavigationController`, 104
animation blocks, 171–172, 177
animation curves, 174–175
animation paths, 178–179
`animationImages` method, 181
annotating maps, 301–303
APIs (application programming interfaces)
　`CFHost`, 244, 247–249
　`CFSocket`, 244–247
　`CFStream`, 244, 249–252
　Foundation, 25
　Game Kit
　　Bluetooth network, 284
　　in-game voice, 284–287
　　peer-to-peer connectivity, 277–284
　Maps
　　annotating maps, 301–303
　　coordinates, converting, 303–305
　　`MKMapView`, 297–299
　　regions, specifying, 299–301
　overview, 16–17, 21–22
APNS SSL certificate generation, 268
.app folder, 24
App Store
　accessing asynchronously, 395
　distribution provisioning profile, 383
"appearance" setting, 163
AppKit framework, 25
Apple iPhone
　applications
　　GUI, building, 11–13
　　launching, 16–17
　　templates, creating projects from, 7–10
　　`UIViewController`, writing, 14–16

Apple iPhone *(continued)*
    Settings application, 203, 206
    UI
        built-in applications, 34–36
        cinematic user experience, 33
        navigation, 36–37
        overview, 31
        screen, 32–33
    Web browser, opening URLs with, 189–190
Apple iPhone Developer Center, 266
Apple iPhone Developer Program Portal, 266–268, 271
Apple iPhone simulator
    multi-touch events in, 140
    need to test applications on iPhone, 16–17
    versus real device
        classes, 25
        performance, 26–27
    UI
        Core Location, 21–22
        filesystem, 23–24
        Hardware menu, 23
        multi-touch, 20–21
        overview, 19–20
Apple iPod, 34, 309, 327
Apple iPod Library
    media artwork, accessing, 328
    media library
        accessing, 320–323
        searching, 323–326
    Media Player framework, 319–320
    player controllers, 327
Apple iTunes
    ad hoc provisioning profile, 383
    installing provisioning profiles, 384, 388
Apple Push Notification Service (APNS)
    architecture
        communications, 259–264
        security, 264–266
    certificates, acquiring, 266–268
    integrating with iPhone client
        receiving notifications in applications, 274–275
        registering for notifications, 273–274
    overview, 257–258
    server-side services, developing
        delivery feedback, 272
        pushing notifications, 272
        Ruby supplier, implementing, 269–272
Apple servers, 259
Apple Web site, 266
application bundle, 24, 201–202, 207
application delegate
    modifying, 212–213
    notifications
        overview, 120–121
        phone calls, handling, 122
        remote notifications, handling, 123
        resource alerts, handling, 121–122
        sleep, handling, 122
        status, changing, 122–123
    overview, 8
    purpose of, 117
    startup and shutdown
        applicationWillTerminate method, 119–120
        launch methods, 118–119
    `UINavigationController`, 108
    `UITableView`, 73
    `UITableViewDelegate`, 75–76
    `UIViewController`, 56
application IDs, iPhone Developer Program Portal, 266–268
application music player mode, `MPMusicPlayerController` class, 327
application programming interfaces (APIs)
    `CFHost`, 244, 247–249
    `CFSocket`, 244–247
    `CFStream`, 244, 249–252
    Foundation, 25
    Game Kit
        Bluetooth network, 284
        in-game voice, 284–287
        peer-to-peer connectivity, 277–284
    Maps
        annotating maps, 301–303
        coordinates, converting, 303–305
        `MKMapView`, 297–299
        regions, specifying, 299–301
    overview, 16–17, 21–22
`applicationDidBecomeActive:` message, 122
`application:didChangeStatusBarFrame:` method, 122
`application:didChangeStatusBarOrientation:` method, 122
`application:didFailToRegisterForRemoteNotificationsWithError:` method, 123
`applicationDidFinishLaunching:` method, 54, 104, 118, 212–213, 222–223, 395–396
`application:didFinishLaunchingWithOptions:` method, 118–119, 274
`applicationDidFinishLoading` method, 53, 108
`application:didReceiveRemoteNotification:` method, 123, 275
`application:didRegisterForRemoteNotificationsWithDeviceToken:` method, 123
`application:handleOpenURL:` method, 119
Applications folder, 24
`applicationSignificantTimeChange:` method, 121–122
`application:willChangeStatusBarFrame:` method, 122
`application:willChangeStatusBarOrientation:duration:` method, 122

Index

applicationWillResignActive: message, 122
applicationWillTerminate: method, 119–120, 122, 212, 215
apView:regionWillChangeAnimated: method, 301
arrays of dictionaries, 53
artistsQuery method, 326
artwork, media, 328
asynchronous accessing, 395
asynchronous hostname resolution, 248
asynchronous mechanisms for networking, 235
asynchronous streams, 264
attachments, email, 240–241
audio
　decoding errors, 317
　formats
　　compressed, 309
　　uncompressed, 310
　input devices, 330
　playing
　　AV Foundation Framework, 311–317
　　OpenAL, 317–318
　recording
　　available input methods, 313
　　AVAudioRecorder, 330–332
　　AVAudioSession, 329–330
　　voice recorder, 332–337
　sample rates, 313–314
　sessions, initializing, 285
audio metering, 316, 332, 334
audiobooksQuery method, 326
audioPlayerBeginInterruption: method, 317
audioPlayerDecodeErrorDidOccur:error: method, 317
audioPlayerDidFinishPlaying:successfully: method, 317
audioPlayerEndInterruption: method, 317
auto correction, 162
AV Foundation Framework
　AVAudioPlayer, 315–317
　AVAudioPlayerDelegate, 317
　AVAudioSession, 311–314
　AVAudioSessionDelegate, 314
AVAudioPlayer class, 315–317
AVAudioPlayerDelegate class, 317
AVAudioRecorder class, 330–332
AVAudioSession class, 285, 311–314, 329–330
AVAudioSessionCategoryAmbient category, 312
AVAudioSessionCategoryPlayAndRecord category, 312, 329
AVAudioSessionCategoryPlayback category, 312
AVAudioSessionCategoryRecord category, 312, 329
AVAudioSessionCategorySoloAmbient category, 312
AVAudioSessionDelegate class, 314
AVEncoderAudioQualityKey key, 331
AVEncoderBitDepthHintKey key, 331
AVEncoderBitRateKey key, 331
averagePowerForChannel: method, 316, 332
AVFormatIDKey key, 331
AVLinearPCMBitDepthKey key, 331
AVLinearPCMIsBigEndianKey key, 331
AVLinearPCMIsFloatKey key, 331
AVNumberOfChannelsKey key, 331
AVSampleRateConverterAudioQualityKey key, 331
AVSampleRateKey key, 331
awakeFromNib method, 134, 142

B

backgroundColor property, 340
BasicIPhoneAppViewController.h file, 14
BasicIPhoneAppViewController.m file, 14
BasicIPhoneAppViewController.xib file, 11, 15
battery state values
　overview, 345
　UIDeviceBatteryStateCharging, 346
　UIDeviceBatteryStateFull, 346
　UIDeviceBatteryStateUnknown, 346
　UIDeviceBatteryStateUnplugged, 346
batteryLevel property, 345
batteryState property, 345
becomeFirstResponder method, 164
beginAnimations:context: method, 177
beta testing
　ad hoc provisioning profile, 383
　Debug build settings, 386
billing, repeated, 390
binary format, push message, 262–263
blocks, animation, 171–172, 177
Bluetooth ad hoc network
　GKPeerPickerController class, 278
　GKSession class, 280
body key, dictionary alerts, 262
Bonjour
　NSNetServices, 255–256
　overview, 252
　services, browsing for, 253–255
-(BOOL)application:(UIApplication *) application didFinishLaunching WithOptions:(NSDictionary *) launchOptions method, 118
+(BOOL)areAnimationsEnabled method, 174
-(BOOL)boolForKey:(NSString *)defaultName method, 199
-(BOOL)shouldAutorotateToInterface Orientation:(UIInterfaceOrientation) interfaceOrientation method, 42
-(BOOL)textFieldShouldBeginEditing: (UITextField *)textField method, 166
-(BOOL)textFieldShouldClear:(UITextField *) textField method, 166
-(BOOL)textFieldShouldEndEditing:(UITextFi eld *)textField method, 166

-(BOOL)textFieldShouldReturn:(UITextField *) textField method, 166
-(BOOL)textField:(UITextField *)textField shouldChangeCharactersInRange: (NSRange)range replacementString: (NSString *)string method, 166
browsers, Web
 opening URLs, 189–190
 UIWebView
  adding to applications, 191
  HTML content, loading from application bundle, 193
  loading URL, 191
  UIWebViewDelegate, implementing, 192–193
BSD sockets, 233, 243
bubble-level application, 363
Build and Go button, 9, 16–17
built-in applications, 34–36
bundle
 application, 24, 201–202, 207
 settings, 203–207
buttons
 iPhone UI, 36
 moving with animation, 172
 removal with animation, 175–176

C

C structures, 129
CALayer class, 178–181
calibrating compass, 360
callback types, 245–247
callouts, map, 302
canMakePayments method, 393
canSendMail class method, 240
capitalization, configuring, 161
cell reuse, 94–98
cell styles
 overview, 86
 UITableViewCellStyleDefault, 86
 UITableViewCellStyleSubtitle, 86
 UITableViewCellStyleValue1, 86
 UITableViewCellStyleValue2, 86
certificates
 development, 17
 digital, 382
 iPhone Developer Program Portal, 271
certificate-signing request, 382
CFAllocatorRef allocator parameter, 244
CFHost API, 244, 247–249
CFNetwork function, 243
CFOptionFlags callBackTypes parameter, 245
CFRelease function, 129–130
CFRetain function, 129–130
CFRunLoopAddSource function, 246
CFRunLoopSourceRef object, 246

CFSocket API, 244–247
CFSocketCallBack callout parameter, 245
CFSocketConnectToAddress function, 246
CFSocketCreate method, 244–245
CFSocketCreateRunLoopSource function, 246
CFStream API, 244, 249–252
CGAffineTransform type, 128
CGAffineTransformMake method, 132–133
CGContextRef method, 128, 130
CGContextRestoreGState( ) method, 130
CGContextSaveGState( ) method, 130
CGContextSetFillColorWithColor( ) function, 131
CGContextSetStrokeColorWithColor( ) function, 131
CGFloat type, 128
CGImageRef type, 128
CGPathRef type, 128
CGPoint type, 128
-(CGPoint)locationInView:(UIView *)view method, 139
-(CGPoint)previousLocationInView:(UIView *) view method, 139
CGRect type, 128
CGRectZero type, 91
CGSize type, 128
Child Pane value type, 204, 207
cinematic user interface
 CALayer, 178–181
 overview, 33
 UIImageViews, animating, 181–182
 UIViews animation methods, 171–177
Class Identity property, 127
Class property, 107
Code Signing Entitlements property, 388
Code Signing Identity, 386, 388
codecs
 hardware, 309
 software, 309–310
codes, mail result, 242–243
color property, 291
colors, drawing, 131
colors property, 291
compilationsQuery method, 326
completed transactions, 397
composersQuery method, 326
Configurations tab, Project Information window, 386–387
Configure App ID screen, iPhone Developer Program Portal, 267–268
Connection Trust mechanism, 264–265
connection:didFailWithError: method, 238–239
connectionDidFinishLoading: method, 238–239
connection:didReceiveData: method, 238–239
const CFSocketContext *context parameter, 245
constant strings, static, 200
consumables, 390
content view, 86
Continuous property, 148

controllers, in MVC design pattern, 41
coordinates
    converting, 303–305
    iPhone screen, 125–126
copy: method, 293–294
copying
    copy: method, 293–294
    with standard controls, 289–290
Core Animation
    CALayer, 178–181
    UIImageViews, animating, 181–182
    UIViews animation methods, 171–177
Core Data
    adding to Favorite Cities application
        Managed Object Model, 216–219
        modifying app delegate, 212–215
    components of, 210–211
    CRUD
        creating, 220
        deleting, 222
        overview, 222–230
        reading, 220–221
        updating, 221–222
    overview, 209
    uses for
        impractical, 230–231
        practical, 230
Core Foundation
    object ownership, 129
    sockets
        CFHost, 247–249
        CFSocket, 244–247
        CFStreams, 249–252
core frameworks memory management model
    CFRelease function, 129–130
    CFRetain function, 129–130
    Core Foundation object ownership, 129
Core Graphics
    context, 130–131
    core frameworks memory management model
        CFRelease function, 129–130
        CFRetain function, 129–130
        object ownership, 129
    drawing operations, 131–132
    mixing with OpenGL, 184
    transforms, 132–133
    types, basic
        CGAffineTransform, 128
        CGContextRef, 128
        CGFloat, 128
        CGImageRef, 128
        CGPathRef, 128
        CGPoint, 128
        CGRect, 128
        CGSize, 128
        overview, 128–129

Core Location, 21–22
country property, 305
countryCode property, 305
Create App ID screen, iPhone Developer Program Portal, 266–267
createFramebuffer method, 187
Creating, Reading, Updating, Deleting (CRUD)
    creating, 220
    deleting, 222
    overview, 222–230
    reading, 220–221
    updating, 221–222
currentHardwareSampleRate method, 313
currentTime property, 316
curves, animation, 174–175
custom buttons, iPhone UI, 36
custom GKSession instances, 280
custom views, 60, 126–127
customizableViewControllers property, 153
CustomUITableViewCell header, 91–93

D

data models, 217–219
data source
    implementation file, 68–70
    interface, 67–68
DataToAllPeers:withDataMode:error: method, 284
dealloc method, 219
declarations, property, 47–48
decoding errors, audio, 317
Default Value parameter, Settings application, 204
default video playback controls, 341
defining paths in Core Graphics, 132
delegate, application
    modifying, 212–213
    notifications
        overview, 120–121
        phone calls, handling, 122
        remote notifications, handling, 123
        resource alerts, handling, 121–122
        sleep, handling, 122
        status, changing, 122–123
    overview, 8
    purpose of, 117
    startup and shutdown
        applicationWillTerminate method, 119–120
        launch methods, 118–119
    UINavigationController, 108
    UITableView, 73
    UITableViewDelegate, 75–76
    UIViewController, 56
delegate method, EditViewController, 112–114
deleteObject instance method, 222
deleteRecording method, 332

deleteRowsAtIndexPaths:withRowAnimation:
        method, 64
deleting
    objects, 222
    rows, 66, 116
    table views, 64
delivery feedback, APNS, 272
denyCallID: method, 285
denyConnectionFromPeer: method, 282
dequeueReusableAnnotationViewWithIdentifier:
        method, 301–302
dequeueReusableCellWithIdentifier: method, 98
deselecting selected rows, 77
deselectRowAtIndexPath:animated: method, 65
desiredAccuracy property
    kCLLocationAccuracyBest value, 356
    kCLLocationAccuracyHundredMeters value, 356
    kCLLocationAccuracyKilometer value, 356
    kCLLocationAccuracyNearestTenMeters value, 356
    kCLLocationAccuracyThreeKilometers value, 356
    overview, 356
development certificate, 17
development provisioning profile, 382–383
development server, Apple, 259, 267
device IDs, 382
device tokens, 261–263, 272–274
dictionary alerts, 262
DidFinishLaunching: method, 73
didReceiveMemoryWarning: message, 44, 121
didRegisterForRemoteNotifications
        WithDeviceToken: method, 261
didRotateFromInterfaceOrientation method, 43
didSelectRowAtIndexPath: method, 77, 81
digital certificate, 382
dismissing keyboard, 164
distanceFilter property, 356
distribution provisioning profile, 383
DNS notation, reverse, 290, 292, 392
Dock, iPhone screen, 33
Document window, Interface Builder, 11
Documents folder, 24, 212
doneTouched: method, 111
-(double)doubleForKey:(NSString *)
        defaultName method, 199
downloading
    images and displaying in UIImageView, 237
    NSURLConnection delegate methods, 238–239
    URL into NSString, 234–235
    using NSURLConnection, 235
drawing
    basic operations, 131–132
    efficiency, 95
drawRect: method, 127, 133–135, 141
drawView method, 187
duration property, 316

E

EAGLContext object, 186–187
EAGLView interface, 186–187
EAGLView.h file, 185
EAGLView.m file, 185
e-book reader, 391
editedItem property, 114
editing
    data, iPhone UI, 37
    rows
        adding rows, 115
        deleting rows, 116
        EditViewController, 109–111
        MyView, moving into, 107–108
        overview, 111–114
    table views, 64
editing window, Xcode, 4–5
EditViewController object
    editing rows, 112–115
    making, 109–111
editViewFinishedEditing: method, 111
efficiency, drawing, 95
e-mail
    attachments, 240–241
    keyboard for, 162
    sending from within apps with Message UI framework, 239–243
embedded maps, 297
Entitlements.plist file, 386–387
environment mapping system, OpenAL, 317
errors
    audio decoding, 317
    handling, 192–193
    managed object, 220
event handling
    keyboard
        moving view in response to appearance of, 167–168
        UITextFieldDelegate, 166–167
    Next button, 48–49
executeFetchRequest:error instance method, 220–221
External Accessory framework, 375

F

"favorite applications" dock, iPhone screen, 33
Favorite Cities application
    Managed Object Model, 216–219
    modifying app delegate, 212–215
feedback messages, APNS, 263–264, 272
feedback servers, 259–260, 272
fetched properties, Core Data, 219
FHostSetClient function, 248–249
file organization view, Xcode, 4
File Owner object, 15–16, 51
filesystem, iPhone simulator, 23–24

fill color, 131
filtering, query object, 326
first responder, 163–164
flip transitions, 177
-(float)floatForKey:(NSString *)defaultName method, 199
Foundation API, 25
frame, status bar, 122

G

Game Kit API
    Bluetooth network, 284
    in-game voice, 284–287
    peer-to-peer connectivity
        overview, 277
        peers, finding, 278–280
        sessions, 281–284
generalPasteboard method, 290
genresQuery method, 326
get-task-allow property, 387
getters, 198–200
G-force, measuring, 361
GKPeerPickerController class, 278, 283
GKSession class, 279–281
GKVoiceChatClient protocol, 285
GKVoiceChatService singleton, 285–287
globally unique identifier (GUID), 24
goBack: action, 192
goForward: action, 192
Google application, proximity sensor, 346
GPS, 349
graphical user interface (GUI), 11–13
Group value type, Settings application, 204
grouped table view, 78–82
GUI (graphical user interface), 11–13
GUID (globally unique identifier), 24

H

hardware, iPhone versus desktop, 26–27
hardware codecs, 309
Hardware menu, iPhone simulator, 23
header
    UI, 14–16
    UITableView, 62–63
    UIViewController, 46–47
Heads-Up-Display (HUD)-style window, 16
home screen, iPhone, 32–33
hostName property, 254
hostname resolution, asynchronous, 248
HTML content, loading from application bundle, 193
HUD (Heads-Up-Display)-style window, 16

I

IBActions (Interface Builder actions), 14, 16, 51, 151
IBOutlets (Interface Builder outlets)
    MyViewController, 51
    overview, 14, 16
    UINavigationController, 108
    UIViewController, 46–47, 53, 56
identifier, accessing, 346
identifier parameter, Settings application, 204
-(id)objectForKey:(NSString *)defaultName method, 199
image property, UIPasteboard class, 291
image views, animating, 180–182
images
    adding to UITableViewCells, 87–90
    as email attachments, 240–241
    UITabBar, 153
images property, 291
immutable objects, 200
implementation file, UI, 15
info icon, Weather application, 36
Info.plist file, 373
in-game voice, 284–287
inheritance, 90–91
init initializer, 320
init method, 253
Initial value property, 151
initialize method, 201–202
initializer
    awakeFromNib method, 142
    custom view, 127
    data source, 70
    init, 320
    initWithContentsOfURL:encoding:error:, 234
    initWithMediaTypes:, 320
    initWithSessionID:displayName:
        sessionMode:, 281
    initWithURL:settings:error:, 331
    UITableViewCell, 91
initialPlaybackTime property, 341
initWithAPI: method, 187
initWithContentsOfURL:encoding:error: initializer, 234
initWithContentsOfURL:error: constructor, 315
initWithData:error: constructor, 315
initWithFavoriteCities: method, 70
initWithFrame: method, 127, 134
initWithMediaTypes: initializer, 320
initWithSessionID:displayName:sessionMode: initializer, 281
initWithURL:settings:error: initializer, 331
input devices, audio, 330
input stream, 376

`inputIsAvailable` method, 313, 330
`inputIsAvailableChanged` method, 330
`insertRowsAtIndexPaths:withRowAnimation:` method, 64
installing provisioning profiles, 383, 384, 388
interactions, iPhone UI, 36
Interface Builder
    custom views, adding to project, 126–127
    `EditViewController`, 109
    grouped table view, 78
    GUI, building simple, 11–13
    keyboard, configuring
        auto correction, enabling and disabling, 162
        capitalization, configuring, 161
        overview, 159–161
        Return key, setting behavior of, 163
        type, setting, 162–163
    `MKMapView`, 297–298
    overview, 5–6
    `UIPickerView`, 155
    `UISlider`, 148–151
    `UITabBar`, 153
    `UIViewController`
        adding, 51–53
        writing, 15–16
    voice recorder UI, 332–333
Interface Builder actions (`IBActions`), 14, 16, 51, 151
Interface Builder outlets (`IBOutlets`)
    `MyViewController`, 51
    overview, 14, 16
    `UINavigationController`, 108
    `UIViewController`, 46–47, 53, 56
Internet
    connections, 280
    `NSURLConnection`, 235–239
    `NSURLRequest`, 235–239
    Safari
        opening URLs, 189–190
        `UIWebView`, 191–193
    URLs
        opening with iPhone browser, 189–190
        using with foundational classes, 234–235
        video, playing, 341–342
interruptions, `AVAudioPlayer`, 317
iPhone
    Settings application, 203, 206
    UI
        built-in applications, 34–36
        cinematic user experience, 33
        navigation, 36–37
        overview, 31
        screen, 32–33
    Web browser, opening URLs with, 189–190
iPhone applications
    GUI, building, 11–13
    launching, 16–17

    templates, creating projects from, 7–10
    `UIViewController`, writing, 14–16
iPhone client, integrating with APNS
    receiving notifications in applications, 274–275
    registering for notifications, 273–274
iPhone Developer Center, 266
iPhone Developer Program Portal, 266–268, 271
iPhone Developer setting, 386
iPhone Distribution setting, 386
iPhone simulator
    multi-touch events in, 140
    need to test applications on iPhone, 16–17
    versus real device
        classes, 25
        performance, 26–27
    UI
        Core Location, 21–22
        filesystem, 23–24
        Hardware menu, 23
        multi-touch, 20–21
        overview, 19–20
iPod, 34, 309, 327
iPod Library
    media artwork, accessing, 328
    media library
        accessing, 320–323
        searching, 323–326
    Media Player framework, 319–320
    player controllers, 327
`isAvailable` property, 281
item IDs, 392
iTunes
    ad hoc provisioning profile, 383
    installing provisioning profiles, 384, 388

J

JSON object, 261, 397

K

`kCFSocketAcceptCallBack` callback value, 245
`kCFSocketConnectCallBack` callback value, 245
`kCFSocketDataCallBack` callback value, 245
`kCFSocketNoCallBack` callback value, 245
`kCFSocketReadCallBack` callback value, 245
`kCFSocketWriteCallBack` callback value, 245
`kCLDistanceFilterNone` value, 357
`kCLLocationAccuracyBest` value, 356
`kCLLocationAccuracyHundredMeters` value, 356
`kCLLocationAccuracyKilometer` value, 356
`kCLLocationAccuracyNearestTenMeters` value, 356
`kCLLocationAccuracyThreeKilometers` value, 356

keyboard
    configuring through Interface Builder
        auto correction, enabling and disabling, 162
        capitalization, configuring, 161
        overview, 159–161
        Return key, setting behavior of, 163
        type, setting, 162–163
    dismissing, 164
    event handling
        moving view in response to keyboard appearance, 167–168
        `UITextFieldDelegate`, 166–167
    making appear, 163–164
    types of, 162
    `UITextInputTraits`, 164–165
Keychain Access application, 268, 382
keyframe animation, creating, 179

L

landscape mode, simulator, 23
`lastLocation.coordinate.latitude` value, 354
`lastLocation.coordinate.longitude` value, 354
launch methods, 118–119
launching iPhone applications, 16–17
`launchOptions` parameter, 119
`layoutSubviews` method, 91, 127
`leftBarButtonItem` property, 106
level metering, 316, 332, 334
Library, iPod
    accessing, 320–323
    searching
        media property predicates, 323–326
        queries, building, 326
Library folder, 24
Library window, Interface Builder, 11
`loadData:MIMEType:textEncodingName:baseURL:` method, 193
`loadHTMLString:baseURL:` method, 193
`loadImage:` method, 237–238
`locality` property, 305
localized pricing, 394
`localizedModel` property, 347
`loc-args` key, dictionary alerts, 262
`locationInView:` method, 139
`locationManager:didFailWithError:` method, 352–353
`locationManager:didUpdateToLocation:fromLocation` method, 352–353
`locationServicesEnabled` method, 352
`loc-key`, dictionary alerts, 262
low-memory warnings, 44, 120–121

M

magnetic compass hardware, 349, 359–360
mail result codes, 242–243
`mailComposeController:didFinishWithResult:error:` method, 242
`mailto:` URL, 239–240, 242
main interaction area, iPhone screen, 33
`MainWindow.nib` file, 55
`MainWindow.xib` file, 185
`makeModel` method, 53–54, 78–79
Manage Your Applications link, iTunes Connect, 392
Managed Object Model
    entities
        defining, 217
        generating classes for, 218–219
        properties, 219
        relationships among, 217–218
    overview, 210–211, 216
    schema, designing in Xcode, 217
`ManagerShouldDisplayHeadingCalibration:` method, 360
map callouts, 302
`MapKit` framework, 298
Maps API
    annotating maps
        overview, 301–303
        performance, 303
    coordinates, converting, 303–305
    `MKMapView`
        creating, 297–299
        showing embedded map with, 297
    regions, specifying, 299–301
`mapView:annotationView:calloutAccessoryControlTapped:` method, 301
`mapView:didAddAnnotationViews:` method, 299
`mapView:regionDidChangeAnimated:` method, 301, 303
`mapView:viewForAnnotation:` method, 301
Maximum value, `UISlider`, 151
media
    audio
        decoding errors, 317
        input devices, 330
        playing, 311–318
        recording, 329–337
        sample rates, 313–314
        sessions, initializing, 285
    iPod, 34, 309, 327
    iPod Library
        media artwork, accessing, 328
        media library, 320–326
        Media Player framework, 319–320
        player controllers, 327
    streaming video, 341

media *(continued)*
   supported audio formats
      compressed, 309
      uncompressed, 310
   supported video formats, 310
Media Player framework, iPod, 319–320
`mediaPicker:didPickMediaItems:` method, 322–323
memory
   alerts regarding, 120–121
   core frameworks memory management model
      `CFRelease` function, 129–130
      `CFRetain` function, 129–130
      Core Foundation object ownership, 129
   handling memory warnings, 44, 120–121
   iPhone versus desktop machine, 26–27
   shared model, 184
Message UI framework, 239–243
metadata style approach, Interface Builder, 5
metering information, audio, 316, 332, 334
`meteringEnabled` property, 316
method signatures
   `UIApplicationDelegate`
      `-(BOOL)application:(UIApplication *) application didFinishLaunchingWithOptions:(NSDictionary *)launchOptions` method, 118
      `-(void)applicationDidFinishLaunching (UIApplication *)application` method, 118
      `-(void)applicationWillTerminate: (UIApplication *)application` method, 118
   `UITextFieldDelegate`
      `-(BOOL)textFieldShouldBeginEditing: (UITextField *)textField` method, 166
      `-(BOOL)textFieldShouldClear: (UITextField *)textField` method, 166
      `-(BOOL)textFieldShouldEndEditing: (UITextField *)textField` method, 166
      `-(BOOL)textFieldShouldReturn: (UITextField *)textField` method, 166
      `-(BOOL)textField:(UITextField *) textField shouldChangeCharactersInRange:(NSRange)range replacementString:(NSString *)string` method, 166
      `-(void)textFieldDidBeginEditing: (UITextField *)textField` method, 166
      `-(void)textFieldDidEndEditing: (UITextField *)textField` method, 166
   `UIViewController`
      `-(BOOL)shouldAutorotateToInterfaceOrientation: (UIInterfaceOrientation) interfaceOrientation` method, 42
      `-(void)didAnimateFirstHalfOfRotationToInterface Orientation: (UIInterfaceOrientation)toInterface Orientation` method, 42
      `-(void)didReceiveMemoryWarning` method, 42
      `-(void)didRotateFromInterface Orientation:(UIInterfaceOrientation) fromInterfaceOrientation` method, 42
      `-(void)loadView` method, 41
      `-(void)viewDidAppear:(BOOL)animated` method, 42
      `-(void)viewDidDisappear:(BOOL)animated` method, 42
      `-(void)viewDidLoad` method, 41
      `-(void)viewDidUnload` method, 41
      `-(void)viewWillAppear:(BOOL)animated` method, 42
      `-(void)viewWillDisappear:(BOOL)animated` method, 42
      `-(void)willAnimateFirstHalfOfRotationTo InterfaceOrientation:(UIInterfaceOri entation)toInterfaceOrientationdurat ion:(NSTimeInterval)duration` method, 42
      `-(void)willAnimateRotationToInterface Orientation:(UIInterfaceOrientation) interfaceOrientationduration: (NSTimeInterval)duration` method, 42
      `-(void)willAnimateSecondHalfOfRotation FromInterfaceOrientation: (UIInterfaceOrientation)from InterfaceOrientationduration: (NSTimeInterval)duration` method, 42
      `-(void)willRotateToInterface Orientation:(UIInterfaceOrientation) toInterfaceOrientationduration: (NSTimeInterval)duration` method, 42
`MFMailComposeResultCancelled` code, 243
`MFMailComposeResultFailed` code, 243
`MFMailComposeResultSaved` code, 243
`MFMailComposeResultSent` code, 243
`MFMailComposeViewController` class, 240–241
Minimum value, `UISlider`, 151
`MKAnnotation` protocol, 297, 301
`MKAnnotationView` class, 297, 301–302
`MKCoordinateRegionMakeWithDistance` function, 299
`MKMapView` class
   creating, 297–299
   showing embedded map with, 297
`MKMapViewDelegate` protocol, 301–303
`MKPinAnnotationViews`, 302
`MKPlacemark` class, 304–305
`MKReverseGeocoder` class, 303–304
model, creating in app delegate, 53–55
Model, View, Controller design pattern, 40–41
`model` property, 347
model-editing window, Managed Object Model, 216–218
`motionBegan:withEvent:` method, 368
`motionCancelled:withEvent:` method, 370–371
`motionEnded:withEvent:` method, 368
`movieControlMode` property, 340

moving
    buttons with animation, 172
    shapes, 141–144
`MPMediaItem` class, 319–320, 325, 328
`MPMediaItemArtwork` object, 328
`MPMediaItemCollection` object, 325
`MPMediaItemPropertyAlbumArtist` property, 324
`MPMediaItemPropertyAlbumTitle` property, 324
`MPMediaItemPropertyAlbumTrackCount` property, 324
`MPMediaItemPropertyAlbumTrackNumber` property, 324
`MPMediaItemPropertyArtist` property, 324
`MPMediaItemPropertyArtwork` property, 324
`MPMediaItemPropertyComposer` property, 324
`MPMediaItemPropertyDiscCount` property, 324
`MPMediaItemPropertyDiscNumber` property, 324
`MPMediaItemPropertyGenre` property, 324
`MPMediaItemPropertyIsCompilation` property, 324
`MPMediaItemPropertyLyrics` property, 324
`MPMediaItemPropertyMediaType` property, 324
`MPMediaItemPropertyPersistentID` property, 324
`MPMediaItemPropertyPlaybackDuration` property, 324
`MPMediaItemPropertyTitle` property, 324
`MPMediaPickerController` class, 319, 320–323
`MPMediaPredicate` class, 319
`MPMediaPropertyPredicate` class, 319, 323–325
`MPMediaQuery` class, 319, 323, 325–326
`MPMediaTypeAny` media type, 320
`MPMediaTypeAnyAudio` media type, 320
`MPMediaTypeAudioBook` media type, 320
`MPMediaTypeMusic` media type, 320
`MPMediaTypePodcast` media type, 320
`MPMovieControlModeDefault` value, 340
`MPMovieControlModeHidden` value, 340
`MPMovieControlModeVolumeOnly` value, 340
`MPMoviePlayerContentPreloadDidFinish Notification` notification, 340
`MPMoviePlayerController` class, 339, 341–342
`MPMusicPlayerController` class, 320, 327
`MPMusicPlayerControllerNowPlayingItemDid ChangeNotification` notification, 327
`MPMusicPlayerControllerPlaybackStateDid ChangeNotification` notification, 327
`MPMusicPlayerControllerVolumeDidChange Notification` notification, 327
multicast DNS resolution service, 252
multi-touch events
    handling in custom `UIView`, 140
    `touchesBegan:withEvent:` method, 139
multi-touch screen, simulating, 20–21
`multiTouchEnabled` property, 140
Multi-Value value type, Settings application, 204
mutable objects, 200
`myCallback` callback method, 247
`MyView` class, 107–108
`MyViewController` class, 51, 111, 116, 223–230

N

`name` property, 254, 347
named pasteboard, 290
naming collisions, 290
navigation, iPhone UI, 36–37
navigation bar, 106
navigation controller
    architecture, 101–102
    configuring, 105–106
    creating, 102–104
    editing rows
        adding rows, 115
        deleting rows, 116
        `EditViewController`, 109–111
        `MyView`, moving into, 107–108
        overview, 111–114
    overview, 33, 39
    `UIToolbar`, adding, 107
    `UIViewControllers`, pushing and popping, 106–107
`netServiceBrowser:didFindService:moreComing:` method, 253–254
`netServiceBrowser:didRemoveService:moreCom ing:` method, 255
`netService:didNotPublish:` method, 255–256
`netService:didNotResolve:` method, 254
`netServiceDidPublish:` method, 255–256
`netServiceDidResolveAddress:` method, 254
networking
    Bonjour
        `NSNetServices`, 255–256
        overview, 252
        services, browsing for, 253–255
    Core Foundation sockets
        `CFHost`, 247–249
        `CFSocket`, 244–247
        `CFStreams`, 249–252
    e-mail, sending from within apps with Message UI framework, 239–243
    overview, 233
    Web, accessing
        `NSURLConnection`, 235–239
        `NSURLRequest`, 235–239
        URLs, using with foundational classes, 234–235
New Project dialog, 7–8
nib files, 5, 51
Nintendo Wii, 361
non-consumables, 389
non-executable content, 390
notifications
    Apple Push Notification Service
        architecture, 259–266
        certificates, acquiring, 266–268
        integrating with iPhone client, 273–275
        overview, 257–258
        server-side services, developing, 269–272

notifications *(continued)*
   `MPMusicPlayerController` class, 327
   pushing, 272
   registering for, 260
   remote, 118, 121, 123, 273–275
   in response to keyboard appearance, 167
   types of, 261
   `UIApplicationDelegate`
      overview, 120–121
      phone calls, handling, 122
      remote notifications, handling, 123
      resource alerts, handling, 121–122
      sleep, handling, 122
      status, changing, 122–123
`-(NSArray *)arrayForKey:(NSString *)defaultName` method, 199
`NSBinaryStoreType` persistent store, 215
`-(NSData *)dataForKey:(NSString *)defaultName` method, 200
`-(NSDictionary *)dictionaryForKey:(NSString *)defaultName` method, 200
`NSDictionary` parameter, 201–202
`NSEntityDescription` method, 220
`NSFetchRequest` object, 220–221
`NSIndexPath` object, 60–61
`NSInMemoryStoreType` persistent store, 215
`-(NSInteger)integerForKey:(NSString*)defaultName` method, 199
`-(NSInteger)numberOfSectionsInTableView:(UITableView *)tableView` method, 62
`-(NSInteger)tableView:(UITableView*)tableView numberOfRowsInSection:(NSInteger)section` method, 62
`NSManagedContext` object, 222
`NSManagedObjectContext` object, 211–212
`NSManagedObjects` object, 211, 219
`NSMutableArray` object, 45
`NSNetService` object, 254
`NSNetServiceBrowser` class, 253
`NSNetServices` object, 255–256
`NSNotificationCenter` object, 340
`NSObject` child class, 75
`NSPersistentStore` class, 211
`NSPredicate` class, 221
`NSSet` object, 139
`NSSQLiteStoreType` persistent store, 215
`-(NSString *)stringForKey:(NSString *)defaultName` method, 200
`NSString` class, 234–235
`NSTimer` object, 332, 334
`NSURLConnection` object, 235–239
`-(NSURLRequest *)connection:(NSURLConnection *)connectionwillSendRequest:(NSURLRequest *)request`

`redirectResponse:(NSURLResponse *)redirectResponse` method, 236
`NSURLRequest` object, 191, 235–239
`NSUserDefaults` object
   acquiring, 197–198
   reading values to, 198–201
   storing purchase information, 391
   writing values to, 198–201
`numberOfComponentsInPickerView:` method, 156
`numberOfLoops` property, 315
`numberOfSectionsInTableView:` method, 62, 68, 79–80
numeric keyboard type, 162
numeric keypad keyboard type, 162

O

object modeling tool, Xcode, 210
Objective-C class, 67
objects
   allocation of, 96
   creating, 220
   deleting, 222
   File Owner, 15–16, 51
   immutable, 200
   JSON, 261, 397
   Managed Object Model
      entities, 217–219
      overview, 210–211, 216
      schema, designing in Xcode, 217
   Managed Objects, 211
   mutable, 200
   reading, 220–221
   serialized, 198
   updating, 221–222
   vertex buffer, 184
opacity, 95–96
OpenAL, playing audio with, 315, 317–318
OpenGL ES
   capabilities of, 183–184
   OpenGL view, 184–188
output stream, 376
overlay view, during table view search, 84
overriding methods, on `UIViewController`
   memory warnings, handling, 44
   overview, 41–43
   view rotations, handling, 43–44

P

.p12 format, 271
packets for APNS servers, 270–272
`paste:` method, 294
pasteboards
   overview, 289–290
   UIPasteboard, 291–292

`pasteboardWithName:create:` method, 290
pasting
    `paste:` method, 294
    with standard controls, 289–290
`pathForResource:ofType:` method, 339
paths
    animation, 178–179
    defining in Core Graphics, 132
payload, APNS, 262–263
`paymentQueue:updatedTransactions:` method, 396
payments, accepting, 392
`peakPowerForChannel:` method, 316, 332
`peerPickerControllerDidCancel:` method, 280
`peerPickerController:didConnectPeer:toSession:` method, 279
`peerPickerController:didSelectConnectionType:` method, 280
`peerPickerController:sessionForConnectionType:` method, 280
peer-to-peer connectivity
    overview, 277
    peers, finding, 278–280
    sessions, 281–284
.pem format, 271
performance
    of iPhone simulator versus real devices, 26–27
    UITableViewCells, 94–96
persistence of purchase, 398
Persistent Store Coordinator, 211
persistent stores, Core Data, 211–212, 214–215
phone calls, handling, 122
`pickerView:didSelectRow:inComponent:` method, 157
`pickerView:numberOfRowsInComponent:` method, 156
`pickerView:titleForRow:forComponent:` method, 157
`pickerView:viewForRow:forComponent:reusingView:` method, 157
`pickerView:widthForComponent:` method, 157
pinch gesture, 144–146
player controllers, 327
playing audio
    AV Foundation Framework, 311–317
    OpenAL, 317–318
`playlistsQuery` method, 326
`playPushed:` action, 332, 336
`podcastsQuery` method, 326
polygons, 184, 187
popping `UIViewControllers`, 106–107
`popToRootViewControllerAnimated:` method, 107
`popToViewController:animated:` method, 107
`popViewControllerAnimated:` method, 107
`port` property, 254
portrait mode, simulator, 23
`postalCode` property, 305

power consumption, reducing, 345
`predicateWithValue:forProperty:` method, 324
`predicateWithValue:forProperty:comparisonType:` method, 324
`prepareToPlay` method, 315
`prepareToRecord` method, 331–332
`presentModalViewController:animated:` method, 240
price, localized, 394
price tier, 392
private key, 382
product presentation, 393
production server, Apple, 259, 267
profiles, provisioning
    ad hoc, 383
    creating, 268
    development, 382–383
    distribution, 383
Project Information window, Configurations tab, 386–387
`Prompt` property, 106
Properties window, Interface Builder, 11
property declarations, 47–48
property list editor, Xcode, 203
property lists in application bundles, 201–202, 207
`@property(nonatomic, readonly)NSTimeInterval timestamp` method, 139
`@property(nonatomic, readonly)NSUInteger tapCount` method, 139
`@property(nonatomic, getter=isSecureTextEntry)BOOLsecureTextEntry` property, 165
`@property(nonatomic)BOOLenablesReturnKeyAutomatically` property, 165
`@property(nonatomic)UIKeyboardAppearancekeyboardAppearance` property, 165
`@property(nonatomic)UIKeyboardTypekeyboardType` property, 165
`@property(nonatomic)UIReturnKeyTypereturnKeyType` property, 165
`@property(nonatomic)UITextAutocapitalizationTypeautocapitalizationType` property, 165
`@property(nonatomic)UITextAutocorrectionTypeautocorrectionType` property, 165
protocol identifier strings, 373
provisioning profile
    ad hoc, 383
    creating, 268
    development, 382–383
    distribution, 383
`proximityMonitoringEnabled` property, 346
`proximityState` property, 346
proxy, view controller, 54–55
public key, 382
`publish` method, 255

publishing services, Bonjour, 255–256
Push Me button, Xcode, 16
push notification
    architecture
        communications, 259–264
        security, 264–266
    certificates, acquiring, 266–268
    integrating with iPhone client
        receiving notifications in applications, 274–275
        registering for notifications, 273–274
    overview, 257–258
    server-side services, developing
        delivery feedback, 272
        pushing notifications, 272
        Ruby supplier, implementing, 269–272
pushing `UIViewControllers`, 106–107
`pushViewController:animated:` method, 106–107

R

racing game, 391
read streams, 251–252
reading objects, 220–221
receipt key, 397
`receiveData:fromPeer:inSession:context:` method, 284
`receivedData:fromParticipantID:` method, 286, 287
receiving data, Game Kit API, 284, 286
`recordForDuration` method, 332
recording audio
    available input methods, 313
    `AVAudioRecorder`, 330–332
    `AVAudioSession`, 329–330
    voice recorder, 332–337
`recordPushed:` method, 332, 334–335
`registerDefaults:` method, 201
`registerForRemoteNotificationTypes:` method, 260, 273
registering for notifications, 260
reliable transmission modes, 284
remote notifications, 118, 121, 123, 273–275
`removeAnnotations:` method, 303
render buffers, 187
`renderbufferStorage:fromDrawable:`, 187
rendering
    core graphics, 184
    tile-based deferred, 184
reordering table view rows, 64
repeated billing, App Store, 390
`resignFirstResponder` method, 164
`resolutionCallback` callback, 248–249
`resolveWithTimeout:` method, 254
resource alerts, handling, 121–122
responder chain, 164
Restore Purchases button, 398

`restoreCompletedTransactions` method, 398
restoring
    graphics context, 130–131
    transactions, 397
Return key, setting behavior of, 163
`returnKeyType` property, 163
reuse concept, 94–95
`reuseIdentifier` object, 97–98
reverse DNS notation, 290, 292, 392
reverse geocoder, Map Kit, 303–304
`rightBarButtonItem` property, 106
root view controller, 103, 107
`Root.plist` file, 203
"Rotate Left" option, simulator, 23
"Rotate Right" option, simulator, 23
row method, 60
rows
    adding, 115
    deleting, 66
    deselecting, 77
    editing, 112–115
    reordering, 64
    table view, 62, 64–66, 68–69
    `UIPickerView`, 156–157
Ruby supplier, implementing, 269–272
runtime notification messages
    overview, 120–121
    `-(void)applicationDidReceiveMemoryWarning:(UIApplication*)application` method, 120
    `-(void)applicationSignificantTimeChange:(UIApplication*)application` method, 120
    `-(void)application:(UIApplication*)applicationdidChangeStatusBarFrame:(CGRect)oldStatusBarFrame` method, 120
    `-(void)application:(UIApplication*)applicationdidChangeStatusBarOrientation:(UIInterfaceOrientation)oldStatusBarOrientation` method, 120
    `-(void)application:(UIApplication *)applicationwillChangeStatusBarFrame:(CGRect)newStatusBarFrame` method, 120
    `-(void)application:(UIApplication*)applicationwillChangeStatusBarOrientation:(UIInterfaceOrientation)newStatusBarOrientationduration:(NSTimeInterval)duration` method, 120

S

Safari
    opening URLs, 189–190
    `UIWebView`

adding to applications, 191
HTML content, loading from application bundle, 193
loading URL, 191
`UIWebViewDelegate`, implementing, 192–193
sample rates, audio, 313–314
sandbox, application, 24
sandbox store, 399
saving graphics context, 130–131
scalar values, 198–200
scaling shapes, 144–146
`scalingMode` property, 340
schema, Managed Object Model, 210–211
screen, iPhone UI, 32–33
scrolling, 94
`searchBarTextDidBeginEditing:` method, 84
`searchBar:textDidChange:` method, 84
`searchForServicesOfType:inDomain:` method, 253
section method, 60
`sectionIndexTitlesForTableView:` method, 83
sections, table view, 62, 68, 79–80
secure text fields, 163
security, APNS
  connection, 264–265
  token, 265–266
`sendData:toPeers:withDataMode:error:` method, 284
sending data, Game Kit API, 284
serialized objects, 198
servers
  APNS, 259–260, 264–265, 270–272
  Apple, 259
  development, 259, 267
  feedback, 259–260, 272
  production, 259, 267
`session:didReceiveConnectionRequestFromPeer:` method, 282
sessionIDs, 281
`session:peer:didChangeState:` method, 282–283
sessions, peer-to-peer connectivity, 281–284
`setActive:error:` method, 329
`setAnimationCurve:` method, 174
`setAnimationDelegate:` method, 175
`setAnimationDidStopSelector:` method, 175–176
`setAnimationTransition:forView:` method, 177
`setAnimationTransition:forView:cache:` method, 177
`setCategory:error:` method, 329
`setCircleCenter:` method, 142
`setCircleSizeFromTouches:` method, 145
`setData:forPasteboardType:` method, 291
`setDataReceiveHandler:withContext:` method, 284
`setEditing:animated:` method, 64, 111
`setLabel:` method, 14–15
`setLabelPushed:` method, 15–16
`setPosition` method, 180
`setPreferredHardwareSampleRate:error:` method, 313

`setSelected:animated:` method, 93–94
setters, 198–200
Settings application
  Default Value parameter, 204
  Group value type, 204
  identifier parameter, 204
  iPhone home screen, 203
  settings bundle
    adding, 203
    adding settings to, 204–207
  Slider value type, 204
  Text Field value type, 204
  title parameter, 204
  Title value type, 204
  Toggle Switch value type, 204
  value types, 204
settings dictionary, 331
`setToolbarHidden:animated:` method, 107
`setToolbarItems:animated:` method, 107
`setupPersistentStore` method, 214
`setValue:forPasteboardType:` method, 291
`setViewControllers:animated:` method, 153
shake event, 368
shapes
  moving, 141–144
  scaling, 144–146
shared memory model, 184
sharedInstance method, 312
`shouldAutoRotateToInterfaceOrientation` method, 43
simulator
  multi-touch events in, 140
  need to test applications on iPhone, 16–17
  versus real device
    classes, 25
    performance, 26–27
  UI
    Core Location, 21–22
    filesystem, 23–24
    Hardware menu, 23
    multi-touch, 20–21
    overview, 19–20
singletons
  AVAudioSession, 285, 312–313
  GKVoiceChatService, 285–287
  NSUserDefaults, 197–198
`SInt32 protocol` parameter, 245
`SInt32 protocolFamily` parameter, 244
`SInt32 socketType` parameter, 244
`SKPaymentQueue` class, 393, 395
`SKPaymentTransaction` object, 397
`SKPaymentTransactionObserver` protocol, 395
`SKPaymentTransactionStateRestored` state, 398
`SKProduct` class, 395
sleep mode, handling, 122
Slider value type, Settings application, 204

sliding control
    configuring through Interface Builder, 148–151
    overview, 147
    status, updating, 151
Smart groups, 4
sockets
    BSD, 233, 243
    Core Foundation
        CFHost, 247–249
        CFSocket, 244–247
        CFStreams, 249–252
        overview, 233
    creating and connecting to port, 246
    TCP, 245, 262
    UDP, 245
    UNIX domain, 245
software codecs, 309–310
song selection, Media Player framework, 319
songsQuery method, 326
SQL database, Core Data, 211
startAnimating method, 181
startAnimation method, 187
starTouched: method, 111
startUpdatingHeading method, 359–360
startUpdatingLocation method, 351
startVoiceChatWithParticipantID:error: method, 287
state objects, creating, 220
static constant strings, 200
status bar, iPhone screen, 32, 120, 122–123
status key, 397
status message, updating, 354
stopAnimating method, 181
stopPushed: action, 332, 336
stopUpdatingHeading method, 359–360
stopVoiceChatWithParticipantID: method, 287
Store Kit API, 389, 391
streamCallback object, 249–251
streaming video, 341–342
streams, asynchronous, 264
string property, 291
strings
    alerts, 262
    NSNetServiceBrowser class, 253
    protocol identifier, 373
    static constant, 200
strings property, 291
stroke color, 131
style parameter, 86
subAdministrativeArea property, 305
subLocality property, 305
subscriptions, 390
subThoroughfare property, 305
subtitle method, 301

subviews
    custom view, 127
    UITableViewCell, 86–87, 91
[super didReceiveMemoryWarning] method, 44
"swipe to delete" feature, 64, 66
synchronize method, 198
@synthesize declaration, 48
systemName property, 347
systemVersion property, 347

T

Tab bar, iPhone screen, 32–34
table views
    adding to applications
        overview, 66–67
        running application, 73–74
        taking action when row is touched, 75–77
        UITableViewDataSource, 67–73
    configuring
        alphabetical list, adding to side, 83
        search function, adding, 84
        state, adding, 78–79
        UITableViewDataSource, 79–82
    deleting, 64
    editing, 64
    grouped, 78–82
    NSIndexPath, 60
    overview, 59–60
    rows, 62, 64–66, 68–69
    sections, 62, 68, 79–80
    UITableViewDataSource
        behavioral methods in, 63–64
        core methods in, 61–63
        overview, 60–61
    UITableViewDelegate
        editing behavior, customizing, 65–66
        overview, 64–65
        reordering of rows, handling, 66
        selection behavior, customizing, 65
tableHeaderView property, 84
tableView:canEditRowAtIndexPath: method, 64, 116
tableView:canMoveRowAtIndexPath: method, 64
tableView:cellForRowAtIndexPath: method, 62, 69, 80–81, 87–89, 93
tableView:commitEditingStyle:forRowAtIndexPath: method, 64
tableView:commitEditingStyle:forRowAtIndexPath: method, 64, 116
tableView:didEndEditingRowAtIndexPath: method, 66
tableView:didSelectRowAtIndexPath: method, 65, 77, 81, 111–112

`tableView:didUpdateTextFieldForRowAtIndexPath:withValue:` method, 64
`tableView:editingStyleForRowAtIndexPath:` method, 65
`tableView:moveRowAtIndexPath:toIndexPath:` method, 64
`tableView:numberOfRowsInSection:` method, 62, 68
`tableView:targetIndexPathForMoveFromRowAtIndexPath:toProposedIndexPath:` method, 64, 66
`tableView:titleForDeleteConfirmationButtonForRowAtIndexPath:` method, 66
`tableView:titleForFooterInSection:` method, 62–63
`tableView:titleForHeaderInSection:` method, 62–63, 80
`tableView:willBeginEditingRowAtIndexPath:` method, 66
`tableView:willSelectRowAtIndexPath:` method, 65, 76
TCP socket, 245, 262
templates
    creating projects from, 7–10
    OpenGL ES, 185
text editor, Xcode, 203
Text Field value type, Settings application, 204
text input, through keyboard
    configuring through Interface Builder
        auto correction, enabling and disabling, 162
        capitalization, configuring, 161
        overview, 159–161
        Return key, setting behavior of, 163
        type, setting, 162–163
    dismissing, 164
    event handling
        moving view in response to keyboard appearance, 167–168
        `UITextFieldDelegate`, 166–167
    making appear, 163–164
    types of, 162
    `UITextInputTraits`, 164–165
Text Input Traits values, 161
text-editing window, Xcode, 4–5
`textFieldDidEndEditing` method, 166–167
`textFieldShouldEndEditing` method, 166
`thoroughfare` property, 305
thumb, 151
tile-based deferred rendering, 184
time event notifications, 120–122
`time_t` component, feedback message, 263
timestamp, `UIAcceleration` object, 367
`title` method, 301
title parameter, Settings application, 204
`Title` property, 106

Title value type, Settings application, 204
TLS (Transport Layer Security), 264
Toggle Switch value type, Settings application, 204
Token Trust mechanism, 264–266
tokens, device, 261–263, 272–274
toolbar, UINavigationController, 107
`toolbarItems` property, 107
touch events
    handling in custom `UIView`
        multi-touch events, 140
        NSSet, 139
        overview, 137–138
        `touchesBegan:withEvent:`, 138–139
        `touchesEnded:withEvent:`, 140
        `touchesMoved:withEvent:`, 140
    updating custom `UIView` with
        moving circle, 141–144
        scaling, 144–146
`touchesBegan:withEvent:` method, 137–140, 143, 368
`touchesEnded:withEvent:` method, 137–140, 144, 146, 292–293, 368
`touchesMoved:withEvent:` method, 137–138, 140, 143, 145
`trackImageForState` property, 151
`transactionReceipt` property, 397
transactions, 397
transforms, Core Graphics, 132–133
transitions
    between data views, 36
    flip, 177
    rotation, 43–44
    `UIView`, 177
transparency, 95–96
Transport Layer Security (TLS), 264
triangular polygons, 184
triangulation, 349

U

UDP socket, 245
UI (user interface)
    cinematic
        CALayer, 178–181
        overview, 33
        `UIImageViews`, animating, 181–182
        `UIViews` animation methods, 171–177
    creating project from template, 7–8
    fading out, 339
    graphical, 11–13
    Interface Builder
        custom views, adding to project, 126–127
        `EditViewController`, 109
        grouped table view, 78
        GUI, building simple, 11–13

UI (user interface) *(continued)*
    keyboard, configuring, 159–163
    `MKMapView`, 297–298
    overview, 5–6
    `UIPickerView`, 155
    `UISlider`, 148–151
    `UITabBar`, 153
    `UIViewController`, 15–16, 51–53
    voice recorder UI, 332–333
  iPhone
    built-in applications, 34–36
    cinematic user experience, 33
    navigation, 36–37
    overview, 31
    screen, 32–33
  simulator
    Core Location, 21–22
    filesystem, 23–24
    Hardware menu, 23
    multi-touch, 20–21
    overview, 19–20
`UIAccelerometer` class, 364
`UIAccelerometerDelegate` protocol, 364
`UIApplication sharedApplication` method, 190
`UIApplicationDelegate` class
  method signatures
    `-(BOOL)application:(UIApplication *) application didFinishLaunchingWithOptions:(NSDictionary *)launchOptions` method, 118
    `-(void)applicationDidFinishLaunching:(UIApplication *)application` method, 118
    `-(void)applicationWillTerminate:(UIApplication *)application` method, 118
  notifications
    overview, 120–121
    phone calls, handling, 122
    remote, handling, 123
    resource alerts, handling, 121–122
    sleep, handling, 122
    status, changing, 122–123
  purpose of, 117
  startup and shutdown
    `applicationWillTerminate` method, 119–120
    launch methods, 118–119
`UIApplicationLaunchOptionsRemoteNotificationKey` value, 119
`UIApplicationLaunchOptionsSourceApplicationKey` value, 119
`UIApplicationLaunchOptionsURLKey` value, 119
`UIApplicationStatusBarFrameUserInfoKey` value, 119

`UIApplicationStatusBarOrientationUserInfoKey` value, 119
`UIDatePicker` class, 155
`UIDevice` class
  `localizedModel` property, 347
  `model` property, 347
  `name` property, 347
  overview, 345–347
  `systemName` property, 347
  `systemVersion` property, 347
  `uniqueIdentifier` property, 347
`UIDeviceBatteryStateCharging` state, 346
`UIDeviceBatteryStateFull` state, 346
`UIDeviceBatteryStateUnknown` state, 346
`UIDeviceBatteryStateUnplugged` state, 346
`UIGraphicsGetCurrentContext` function, 130
`+[UIImage imageNamed:]` method, 88
`UIImageView` class
  animating, 181–182
  enabling, 87
`UIKeyboardDidHideNotification` notification, 167
`UIKeyboardDidShowNotification` notification, 167–168
`UIKeyboardWillHideNotification` notification, 167
`UIKeyboardWillShowNotification` notification, 167
`UIKit` framework, 25, 184
`UILabel` object, 49
`UIMenuController` class, 292–294
`UIMenuController setTargetRect:inView:` method, 293
`UINavigationBar` class, 106
`UINavigationController` class
  architecture, 101–102
  configuring, 105–106
  creating, 102–104
  editing rows
    adding rows, 115
    deleting rows, 116
    `EditViewController`, 109–111
    `MyView`, moving into, 107–108
    overview, 111–114
  `UIToolbar`, adding, 107
  and `UIViewController`, overview, 39
  `UIViewControllers`, pushing and popping, 106–107
`UINavigationItem` class
  `leftBarButtonItem` property, 106
  overview, 106, 111
  `Prompt` property, 106
  `rightBarButtonItem` property, 106
  `Title` property, 106
`UIPasteboard` class, 290, 291–292
`UIPickerView` class
  configuring through Interface Builder, 155
  delegate, creating, 157
  overview, 148, 155
  `UIPickerViewDataSource`, creating, 156

Index

UIPickerViewDataSource class, 156
UIRemoteNotificationTypeAlert notification, 261
UIRemoteNotificationTypeBadge notification, 261
UIRemoteNotificationTypeNone notification, 261
UIRemoteNotificationTypeSound notification, 261
UIResponder class, 367
UIResponderStandardEditActions protocol, 293–294
UISearchBar method, 84
UISlider class
    configuring through Interface Builder, 148–151
    overview, 147
    status, updating, 151
UISupportedExternalAccessoryProtocols object, 373
UITabBar class
    adding UITabBarItems to, 153–155
    advanced configuration of, 155
    configuring through Interface Builder, 153
    overview, 148, 151–152
    UIPickerView
        configuring through Interface Builder, 155
        delegate, creating, 157
        overview, 155
        UIPickerViewDataSource, creating, 156
UITabBarController class, 153
UITabBarControllerDelegate object, 155
UITabBarItems class, 153–155
UITableView class
    adding to applications
        overview, 66–67
        running application, 73–74
        taking action when row is touched, 75–77
        UITableViewDataSource, 67–73
    configuring
        alphabetical list, adding to side, 83
        search function, adding, 84
        state, adding, 78–79
        UITableViewDataSource, 79–82
    NSIndexPath, 60
    overview, 59–60
    UITableViewDataSource
        behavioral methods in, 63–64
        core methods in, 61–63
        overview, 60–61
    UITableViewDelegate
        editing behavior, customizing, 65–66
        overview, 64–65
        reordering of rows, handling, 66
        selection behavior, customizing, 65
-(UITableViewCell *)tableView:(UITableView *)tableView cellForRowAtIndexPath: (NSIndexPath *)indexPath method, 62
UITableViewCell class, 69
UITableViewCellEditingStyle values, 65

UITableViewCells class
    customizing, 90–94
    images, adding to, 87–90
    parts of, 85–87
    performance, 94–96
    reusing, 96–98
UITableViewCellStyleDefault class, 86
UITableViewCellStyleSubtitle class, 86
UITableViewCellStyleValue1 class, 86
UITableViewCellStyleValue2 class, 86
UITableViewController class, 64, 67, 71–73, 77
UITableViewDataSource class
    behavioral methods in, 63–64
    configuring in Interface Builder, 70–73
    core methods in, 61–63
    -(NSInteger)numberOfSectionsInTable View:(UITableView *)tableView method, 62
    -(NSInteger)tableView:(UITableView *)table ViewnumberOfRowsInSection:(NSInte ger)section method, 62
    overview, 59, 60–61, 67–68
    -(UITableViewCell *)tableView:(UITableView *)tableView cellForRowAtIndexPath: (NSIndexPath *)indexPath method, 62
    writing implementation, 68–70
UITableViewDelegate class
    allowing user to select row, 76
    creating, 75–76
    editing behavior, customizing, 65–66
    overview, 59–60, 64–65
    reordering of rows, handling, 66
    selection behavior, customizing, 65
    updating model when row is touched, 77
UITextField class, 159–161
UITextFieldDelegate object
    method signatures
        -(BOOL)textFieldShouldBeginEditing: (UITextField *)textField method, 166
        -(BOOL)textFieldShouldClear: (UITextField *)textField method, 166
        -(BOOL)textFieldShouldEndEditing: (UITextField *)textField method, 166
        -(BOOL)textFieldShouldReturn: (UITextField *)textField method, 166
        -(BOOL)textField:(UITextField *) textField shouldChangeCharactersInRa nge:(NSRange)range replacement String:(NSString *)string method, 166
        -(void)textFieldDidBeginEditing: (UITextField *)textField method, 166
        -(void)textFieldDidEndEditing: (UITextField *)textField method, 166
    overview, 166–167

`UITextInputTraits` class, 159, 164–165
`UITextView` class, 159–160, 164
`UIToolbar` class, 107
`UITouch` class, 139
`UIView` class
    adding to project, 126–127
    animation methods, 171–177
    `CALayer` property, 178
    copying and pasting, 292
    Core Graphics
        context, 130–131
        core frameworks memory management model, 129–130
        drawing operations, 131–132
        transforms, 132–133
        types, 128–129
    `drawRect:` method, 133–135
    Tag feature, 95
    touch events
        handling in, 137–140
        updating with, 140–146
    view geometry, 125–126
    Xcode OpenGL ES template, 185
`UIViewAnimationCurveEaseIn` curve type, 174
`UIViewAnimationCurveEaseInOut` curve type, 174
`UIViewAnimationCurveEaseOut` curve type, 175
`UIViewAnimationCurveLinear` curve type, 175
`UIViewController` class
    code, creating, 45–46
    header file, creating, 46–47
    implementation, creating, 47–51
    Interface Builder, adding to, 51–53
    method signatures
        `-(BOOL)shouldAutorotateToInterface Orientation:(UIInterfaceOrientation) interfaceOrientation` method, 42
        `-(void)didAnimateFirstHalfOfRotationTo InterfaceOrientation:(UIInterfaceOri entation)toInterfaceOrientation` method, 42
        `-(void)didReceiveMemoryWarning` method, 42
        `-(void)didRotateFromInterface Orientation:(UIInterfaceOrientation) fromInterfaceOrientation` method, 42
        `-(void)loadView` method, 41
        `-(void)viewDidAppear:(BOOL)animated` method, 42
        `-(void)viewDidDisappear:(BOOL)animated` method, 42
        `-(void)viewDidLoad` method, 41
        `-(void)viewDidUnload` method, 41
        `-(void)viewWillAppear:(BOOL)animated` method, 42
        `-(void)viewWillDisappear:(BOOL)animated` method, 42
        `-(void)willAnimateFirstHalfOfRotationTo InterfaceOrientation:(UIInterfaceOri entation)toInterfaceOrientationdurat ion:(NSTimeInterval) duration` method, 42
        `-(void)willAnimateRotationToInterface Orientation:(UIInterfaceOrientation) interfaceOrientationduration:(NSTime Interval)duration` method, 42
        `-(void) willAnimateSecondHalfOfRotationFrom InterfaceOrientation:(UIInterfaceOri entation)fromInterfaceOrientationdur ation:(NSTimeInterval) duration` method, 42
        `-(void)willRotateToInterfaceOrientation :(UIInterfaceOrientation)toInterface Orientationduration:(NSTimeInterval) duration` method, 42
    model, creating in app delegate, 53–55
    Model, View, Controller design pattern, 40–41
    overriding methods on
        memory warnings, handling, 44
        overview, 41–43
        view rotations, handling, 43–44
    overview, 39, 41
    pushing and popping, 106–107
    `UIApplicationDelegate`, 53
    `UINavigationController`
        configuring, 106
        pushing, 106–107
        replacing with, 107–108
        using custom in, 103
    view, adding to window, 55–57
    writing for iPhone applications, 14–16
`UIWebView` class
    adding to applications, 191
    HTML content, loading from application bundle, 193
    loading URLs, 191
    `UITextInputTraits`, 160
    `UIWebViewDelegate`, 192–193
`UIWebViewDelegate` object, 192–193
uniqueIdentifier method, 346
uniqueIdentifier property, 347
UNIX
    domain socket, 245
    overview, 233
unreliable transmission modes, 284
`updateMeters` method, 316, 332
`updateStar` method, 110
updating
    accelerometer, 364
    objects, 221–222
URL keyboard type, 162

URL property, 291
URLs
    loading with `UIWebView`, 191
    opening with iPhone Web browser, 189–190
    previously registered, 118–119
    using with foundational classes, 234–235
URLs property, 291
user defaults
    `NSUserDefaults` object
        acquiring, 197–198
        reading and writing values to, 198–201
    setting, 201–202
    Settings application, 203–207
user interface (UI)
    cinematic
        `CALayer`, 178–181
        overview, 33
        `UIImageViews`, animating, 181–182
        `UIViews` animation methods, 171–177
    creating project from template, 7–8
    fading out, 339
    graphical, 11–13
    Interface Builder
        custom views, adding to project, 126–127
        `EditViewController`, 109
        grouped table view, 78
        GUI, building simple, 11–13
        keyboard, configuring, 159–163
        `MKMapView`, 297–298
        overview, 5–6
        `UIPickerView`, 155
        `UISlider`, 148–151
        `UITabBar`, 153
        `UIViewController`, 15–16, 51–53
        voice recorder UI, 332–333
    iPhone
        built-in applications, 34–36
        cinematic user experience, 33
        navigation, 36–37
        overview, 31
        screen, 32–33
    simulator
        Core Location, 21–22
        filesystem, 23–24
        Hardware menu, 23
        multi-touch, 20–21
        overview, 19–20
`userLocation` method, 299
utility method, 212

V

value types, Settings application, 204
vertex buffer objects (VBOs), 184
video
    Internet, 341–342
    streaming, 310
    supported formats, 310
view appearance methods, 42–43
view controller
    code, creating, 45–46
    header file, creating, 46–47
    implementation, creating, 47–51
    Interface Builder, adding to, 51–53
    model, creating in app delegate, 53–55
    Model, View, Controller design pattern, 40–41
    overriding methods on
        memory warnings, handling, 44
        overview, 41–43
        view rotations, handling, 43–44
    overview, 7–8, 39, 41
    pushing and popping, 106–107
    `UIApplicationDelegate`, 53
    `UINavigationController`
        configuring, 106
        pushing, 106–107
        replacing with, 107–108
        using custom in, 103
    view, adding to window, 55–57
    writing for iPhone applications, 14–16
`viewDidLoad` method, 351, 364
view geometry, 125–126
view hierarchy, removal of buttons from, 176
view property, 51
view rotations, handling, 43–44
`ViewControllerDemoAppDelegate.h` file, 52
`ViewControllerDemoAppDelegate.m` file, 53
`viewDidLoad` method
    adding rows, 115
    `EditViewController`, 111
    loading URLs, 191
    `MKMapView`, 298–299
    `NSTimer`, 332, 334
    overview, 47–48
    `UITableView`, 72, 75–76
`viewDidUnload` method, 44
views
    adding to window, 55–57
    custom, 126–127
    MVC design pattern, 40
    transitions between, 36

voice, in-game, 284–287
voice chats, Game Kit, 284–287
voice recorder, 332–337
voiceChatService:didReceiveInvitationFrom
    ParticipantID:callID: method, 285
voiceChatService:didStopWithParticipantID:
    error: method, 287
voiceChatService:sendData:toParticipantID:
    method, 285–286, 287
void CGContextAddArcToPoint(CGContextRef c,
    CGFloat x1, CGFloat y1, CGFloat x2,
    CGFloat y2, CGFloat radius); method,
    132
void CGContextAddLineToPoint(CGContextRef c,
    CGFloat x, CGFloat y); method, 132
void CGContextClearRect (CGContextRef
    c,CGRect rect); function, 131
void CGContextClosePath (CGContextRef c);
    method, 132
void CGContextFillPath (CGContextRef c);
    function, 131
void CGContextFillRect (CGContextRef
    c,CGRect rect); function, 131
void CGContextMoveToPoint(CGContextRef c,
    CGFloat x, CGFloat y); method, 132
void CGContextStrokeEllipseInRect(CGContext
    Ref context, CGRect rect); function, 132
void CGContextStrokePath (CGContextRef c);
    function, 132
void CGContextStrokeRect (CGContextRef
    c,CGRect rect); function, 132
-(void)addAttachmentData:(NSData*)attachment
    mimeType:(NSString*)mimeType
    fileName:(NSString*)filename
    method, 240
(void)applicationDidBecomeActive:(UI
    Application*)application method, 121
-(void)applicationDidFinishLaunching:
    (UIApplication *)application
    method, 118
-(void)applicationDidReceiveMemoryWarning:
    (UIApplication *)application
    method, 120
-(void)applicationSignificantTimeChange:(UI
    Application *)application method, 120
-(void)application:(UIApplication *)
    application
    didChangeStatusBarFrame:(CGRect)
    oldStatusBarFrame method, 120
-(void)application:(UIApplication *)
    applicationdidChangeStatus
    BarOrientation:(UIInterface
    Orientation)oldStatusBarOrientation
    method, 120

-(void)application:(UIApplication *)
    applicationdidFailTo
    RegisterForRemoteNotifications
    WithError:(NSError *)error method, 121
(void)application:(UIApplication *)
    applicationdidReceiveRemote
    Notification:(NSDictionary *)
    userInfo method, 121
-(void)application:(UIApplication *)
    applicationdidRegisterForRemote
    NotificationsWithDevice
    Token:(NSData *)deviceToken method, 121
-(void)application:(UIApplication *)
    applicationwillChangeStatusBarFrame:
    (CGRect)newStatusBarFrame method, 120
-(void)application:(UIApplication *)
    applicationwillChangeStatusBar
    Orientation:(UIInterfaceOrientation)
    newStatusBarOrientationduration:
    (NSTimeInterval)duration method, 120
(void)applicationWillResignActive:
    (UIApplication*)application method, 121
-(void)applicationWillTerminate:
    (UIApplication *)application
    method, 118
+(void)beginAnimations:(NSString *)
    animationID context:(void *)context
    method, 173
-(void)beginInterruption method, 314
-(void)categoryChanged:(NSString*)category
    method, 314
+(void)commitAnimations method, 173
-(void)connectionDidFinishLoading:
    (NSURLConnection *)connection
    method, 236
-(void)connection:(NSURLConnection *)
    connection didCancelAuthentication
    Challenge:(NSURLAuthentication
    Challenge *)challenge method, 236
-(void)connection:(NSURLConnection *)
    connection didReceiveAuthentication
    Challenge:(NSURLAuthentication
    Challenge *)challenge method, 236
-(void)connection:(NSURLConnection *)connect
    iondidFailWithError:(NSError *)error
    method, 236
-(void)connection:(NSURLConnection *)connect
    iondidReceiveData:(NSData *)data
    method, 236
-(void)connection:(NSURLConnection *)connect
    iondidReceiveResponse:(NSURL
    Response *)response method, 236

-(void)currentHardwareInputNumberOfChannels
    Changed:(NSInteger)numberOfChannels
    method, 314
-(void)currentHardwareOutputNumberOfChannels
    Changed:(NSInteger)numberOfChannels
    method, 314
-(void)currentHardwareSampleRateChanged:
    (double)sampleRate method, 314
-(void)didAnimateFirstHalfOfRotationTo
    InterfaceOrientation:(UIInterface
    Orientation)toInterfaceOrientation
    method, 42
-(void)didReceiveMemoryWarning method, 42
-(void)didRotateFromInterfaceOrientation:
    (UIInterfaceOrientation)
    fromInterfaceOrientation method, 42
-(void)endInterruption method, 314
-(void)inputIsAvailableChanged:(BOOL)
    isInputAvailable method, 314
-(void)loadView method, 41
+(void)setAnimationBeginsFromCurrentState:
    (BOOL)fromCurrentState method, 173
+(void)setAnimationCurve:(UIViewAnimation
    Curve)curve method, 173
+(void)setAnimationDelegate:(id)delegate
    method, 173
+(void)setAnimationDidStopSelector:(SEL)
    selector method, 173
+(void)setAnimationDuration:(NSTimeInterval)
    duration method, 173
+(void)setAnimationRepeatAutoreverses:(BOOL)
    repeatAutoreverses method, 173
+(void)setAnimationRepeatCount:(float)
    repeatCount method, 173
+(void)setAnimationsEnabled:(BOOL)enabled
    method, 173
+(void)setAnimationStartDate:(NSDate *)
    startTime method, 173
+(void)setAnimationTransition:
    (UIViewAnimationTransition)
    transition forView:(UIView *)
    viewcache:(BOOL)cache method, 174
+(void)setAnimationWillStartSelector:(SEL)
    selector method, 173
-(void)setBccRecipients:(NSArray*)
    bccRecipients method, 240
-(void)setBool:(BOOL)valueforKey:(NSString
    *)defaultName method, 199
-(void)setCcRecipients:(NSArray*)
    ccRecipients method, 240
-(void)setDouble:(double)valueforKey:
    (NSString *)defaultName method, 199
-(void)setFloat:(float)valueforKey:(NSString
    *)defaultName method, 199
-(void)setInteger:(NSInteger)value
    forKey:(NSString *)defaultName
    method, 199
-(void)setMessageBody:(NSString*)body
    isHTML:(BOOL)isHTML method, 240
-(void)setObject:(id)valueforKey:(NSString
    *)defaultName method, 199
-(void)setSubject:(NSString*)subject
    method, 240
-(void)setToRecipients:(NSArray*)
    toRecipients method, 240
-(void)textFieldDidBeginEditing:(UITextField
    *)textField method, 166
-(void)textFieldDidEndEditing:(UITextField
    *)textField method, 166
-(void)viewDidAppear:(BOOL)animated method, 42
-(void)viewDidDisappear:(BOOL)animated
    method, 42
-(void)viewDidLoad method, 41
-(void)viewDidUnload method, 41
-(void)viewWillAppear:(BOOL)animated method, 42
-(void)viewWillDisappear:(BOOL)animated
    method, 42
-(void)willAnimateFirstHalfOfRotationTo
    InterfaceOrientation:(UIInterface
    Orientation)toInterfaceOrientation
    duration:(NSTimeInterval)duration
    method, 42
-(void)willAnimateRotationToInterface
    Orientation:(UIInterfaceOrientation)
    interfaceOrientationduration:
    (NSTimeInterval)duration method, 42
-(void)willAnimateSecondHalfOfRotationFrom
    InterfaceOrientation:(UIInterface
    Orientation)fromInterfaceOrientation
    duration:(NSTimeInterval)
    duration method, 42
-(void)willRotateToInterfaceOrientation:
    (UIInterfaceOrientation)to
    InterfaceOrientationduration:
    (NSTimeInterval)duration method, 42

W

Weather application, 34–36
Web
    NSURLConnection, 235–239
    NSURLRequest, 235–239
    Safari
        opening URLs, 189–190
        UIWebView, 190–193

Web *(continued)*
    URLs
        opening with iPhone browser, 189–190
        using with foundational classes, 234–235
`webView:didFailLoadWithError:` delegate method, 192–193
`webViewDidFinishLoad` method, 192
Wii, 361
wildcard settings, Code Signing Identity, 386
`willAnimateRotationToInterfaceOrientation:duration:` method, 43–44
`willRotateToInterfaceOrientation` method, 43
`willSelectRowAtIndexPath:` method, 76
Window-Based Application template, 44–45
`write` method, 377
write streams, 249–251

X

X-axis, 361–362
Xcode
    build settings, 385
    Core Data templates, 211
    creating project from template, 7–10
    designing schema in, 217
    development certificate, 381–382
    development provisioning profile, 383
    installing provisioning profiles, 383
    launching on iPhone, 16–17
    object modeling tool, 210
    organizer window, 384
    overview, 3–5
    property list editor, 203
    template-generated code for OpenGL ES-based application, 184–185
    text editor, 203
    `UINavigationController`, creating, 102–103
    `UIViewController`, 14–16, 44–45, 55–57
.xib files, 5

Y

Y-axis, 361–362

Z

Z-axis, 361–362
zooming in/out, Maps API, 303

Everything You Need to Craft Killer Code for Apple Applications

Whether you are a seasoned developer or just getting into the Apple platform, Wiley's Developer Reference series is perfect for you. Focusing on topics that Apple developers love best, these well-designed books guide you through the most advanced and very latest Apple tools, technologies, and programming techniques. With in-depth coverage and expert guidance from skilled authors who are proven authorities in their fields, the Developer Reference series will quickly become your indispensable Apple development resource.

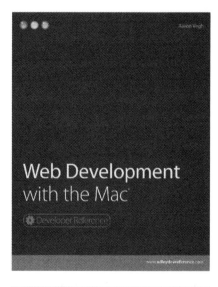

The Developer Reference series is available wherever books are sold.

Now you know.

Take the Book with You, Everywhere

How to purchase

Go to www.wileydevreference.com and follow the link to purchase the app in iTunes.

Wiley's Developer Reference app is just 99¢ and includes Chapter 21, "Using the Game Kit API" from *Cocoa Touch for iPhone OS 3*. When you're ready for a full Developer Reference book, you can purchase any title from the series directly in the app for $19.99.

Want tips for developing and working on Apple platforms on your iPhone? Wiley's Developer Reference app puts you in touch with the new Wiley Developer Reference series. Through the app you can purchase any title in the series and then read, highlight, search, and bookmark the text and code you need. To get you started, Wiley's Developer Reference app includes Chapter 21 from *Cocoa Touch for iPhone OS 3*, which offers fantastic tips for developing for the iPhone and iPod touch platforms. If you buy a Wiley Developer Reference book through the app, you'll get all the text of that book including a searchable index and live table of contents linked to each chapter and section of the book.

Here's what you can do

- Jump to the section or chapter you need by tapping a link in the Table of Contents
- Click on a keyword in the Index to go directly to a particular section in the book
- Highlight text as you read so that you can mark what's most important to you
- Copy and paste, or email code samples, out of the book so you can use them where and when needed
- Keep track of additional ideas or end results by selecting passages of text and then creating notes and annotations for them
- Save your place effortlessly with automatic bookmarking, which holds your place if you exit or receive a phone call
- Zoom into paragraphs with a "pinch" gesture

Cocoa Touch and iPhone are trademarks or registered trademarks of Apple